ENERGY MANAGEMENT AND EFFICIENCY FOR THE PROCESS INDUSTRIES

ENERGY MANAGEMENT AND EFFICIENCY FOR THE PROCESS INDUSTRIES

Edited by
Alan P. Rossiter
Beth P. Jones

WILEY

Cover Image: iStockphoto©chictype

A Joint Publication of the American Institute of Chemical Engineers and John Wiley & Sons, Inc.

Published by John Wiley & Sons, Inc., Hoboken, New Jersey.
Published simultaneously in Canada.

Library of Congress Cataloging-in-Publication Data is available.

ISBN: 978-1-118-83825-9

CONTENTS

FOREWORD

Energy sustainability is the cornerstone to the health and competitiveness of industries that produce and manufacture in our global economy. There are many other considerations such as access to markets and availability of labor, but the effective utilization of secure and affordable energy is what positions companies and societies for competitive and sustainable leadership.

Energy sustainability is much more than the process of being environmentally responsible and earning the right to operate as a business. It is the ability to utilize and optimize multiple sources of secure and affordable energy for the enterprise, and then continuously improve utilization through systems analysis and through an organizational drive for continuous improvement as a core principle. It is achieved through management discipline and technical excellence and it never stops.

Today, the choices and external conditions are changing more rapidly than at any other time in our history. Management commitment to ensure the best energy efficiency management in existing process operations, as well as a dedicated pursuit of new system technologies and processes, is the only recipe for excellence. I commend the authors in this endeavor and their work to provide insights and best practices, and to challenge the reader to aspire to even greater energy performance and sustainability.

THE HONORABLE CHARLES D. MCCONNELL

Former Assistant Secretary of Energy, U.S. Department of Energy

Executive Director, Energy and Environment Initiative, Rice University

PREFACE

This is an exciting time to be involved in industrial energy management. While the oil crisis of the 1970s precipitated a knee-jerk reaction from industry to cut dependence on foreign oil, the trend over the past 20 years has been toward a more considered and systematic approach to energy efficiency. Many companies, especially the larger ones, have developed comprehensive programs that include corporate energy policies, reporting systems, benchmarking, various types of energy audits, and integration of energy efficiency elements into engineering procedures and purchasing protocols. ExxonMobil, for example, started its Global Energy Management System (GEMS) in 2000. By 2009 the company reported that the program had identified savings opportunities of between 15 and 20% at its manufacturing sites, and had captured over 60% of these savings [1]. Many other companies are working on energy efficiency initiatives, and numerous software and consulting organizations provide support for the various aspects of energy efficiency and energy management.

It is not only industry that has taken an interest in this area. In the United States, both the Department of Energy and the Environmental Protection Agency have been active in developing and promoting energy efficiency practices. Meanwhile, the International Organization for Standardization launched ISO 50001:2011 in 2011 to "support organizations in all sectors to use energy more efficiently, through the development of an energy management system (EnMS)" [2].

THE LAWS OF INDUSTRIAL ENERGY EFFICIENCY

For 30 years I have been active in industrial energy management in various forms. Over that time a couple of general principles have become apparent to me, and they are so powerful that I venture to call them "the laws of industrial energy management."

Law 1: There is no silver bullet—no single method or technology that ensures that energy use will be optimized. Those of us with a background in engineering tend to think in terms of technological options to improve energy performance—better heat recovery, more selective catalysts, higher efficiency motors, and fundamentally better process design—and indeed these are very powerful ways of lowering energy usage. However, real-world energy efficiency is also a function of the behavior of both individuals and organizations. It follows that management and motivation are also critical factors, and they require very different types of expertise than solutions that are purely technological.

Law 2: I don't know it all—and neither do you. I am constantly amazed at the breadth and depth of my own ignorance. For many years, I have specialized in heat integration using pinch analysis, and I would like to think I have become quite good at it—yet still every project brings surprises. Energy management is a multidisciplinary activity, and over the past 30 years I have been privileged to work alongside people with expertise in a great variety of energy-related fields. I have learned a lot from them, and I have greatly expanded my own expertise. However, there are still many areas where I know virtually nothing. It takes a long time to become truly proficient in any given field, and life is a constant learning process.

These two laws are at the root of the structure of this book and the approach we have taken in writing it. The book consists of two main sections—one focusing on management and organizational issues, and the other on technology. I have spent most of my career as a freewheeling consultant specializing in the technical aspects of energy efficiency. It was very clear that the book also needed the perspective of a corporate insider with experience in managing an energy program in a large company. I was delighted that Beth Jones, who until recently filled that role at LyondellBasell, agreed to be my coauthor. Both of us then set about writing and also recruiting other leading program managers and technical specialists—mostly people who have worked with us in the past in various energy management activities—with the expertise needed to cover the major facets of contemporary energy management in the process industries.

ENERGY AND THE PROCESS INDUSTRIES

To many people the term "process industries" is synonymous with continuous, large-scale, petroleum and petrochemical processing—and indeed these types of operations are well represented in this book. However, the Institute of Industrial Engineers defines the process industries much more broadly as "those industries where the primary production processes are either continuous, or occur on a batch of materials that is indistinguishable" [3]. This includes not just oil refining and petrochemicals, but also a wide range of other sectors such as food and beverages, inorganic chemicals, pharmaceuticals, base metals, plastics, rubber, wood and wood products, paper and paper products, textiles, and many others.

Within the wide range of sectors in the process industries, there is a great diversity of operations. World-scale refineries and petrochemical facilities have annual energy bills in the hundreds of millions or even billions of dollars, and energy is a major component of variable cost. In contrast, raw materials and labor tend to be much more dominant in many of the smaller, more specialized facilities, where annual energy bills can be on the order of a million dollars or less.

In many of the smaller facilities, lighting and space heating and cooling are often the dominant energy users. These are just a tiny percentage of the cost at large petrochemical sites, where the dominant energy users are associated with moving, transforming, and separating feed and product materials.

Energy management is also very different for batch processes than it is for continuous ones. In particular, where many batch operators can apply "downtime shutdown strategies" to eliminate unneeded energy use for much of the time and thus gain "free" energy savings, this approach is not generally applicable in continuous processing.

There are also both similarities and differences in the equipment and systems across the process sector. Steam systems are virtually ubiquitous, and so are electric motors, heat exchangers, pumps, and compressors. However, beyond these few common elements there is a great diversity, from distillation columns, catalytic reactors, and centrifuges to cookers, toasters, and belt dryers. Each system and each type of equipment has its own issues that need to be considered within a comprehensive energy management program.

OUR SCOPE

Our goal in this book is to provide a concise overview of energy management principles and techniques for the process industries. This necessarily means that we cannot cover every possible process type or piece of equipment, but we have tried to provide the basics that are needed by most energy managers and technical specialists.

Irrespective of the size of the energy bill, the continuous or batch nature of the processes, or the types of equipment employed, energy efficiency is a must. Understandably, though, management gives the greatest amount of attention to the largest costs. The basic principles of energy management and energy efficiency are universal, but different types of facilities require different types of energy management programs.

Within the pages of this book, we have tried to capture both the common threads and the diversity. The book consists of two main parts:

Section 1 focuses on energy management principles and systems. This includes an exploration of the role of the energy manager, examples of successful corporate energy management programs from diverse companies, and also benchmarking and management systems.

Section 2 looks at the technologies of energy efficiency. It covers some of the most widely used types of equipment, utility systems, and process-wide approaches for improving energy efficiency.

There is one very important area that we have not covered in any great depth. Breakthrough technologies—new equipment or processes that radically improve efficiency—do appear periodically, and they can lead to drastic reductions in energy use. For example, the development of the low-pressure polyethylene process in the 1950s was a major technological advance over the older high-pressure process, and it uses much less energy per unit of production. A more familiar example for most people outside of the polymer industry is the rise in recent years of compact fluorescent lights and light-emitting diodes, which provide dramatic energy savings compared with the familiar incandescent bulbs.

Breakthrough technologies are an important piece of the energy management puzzle, but they tend to be very specific to individual processes or equipment types. Furthermore, they tend to arise through lengthy research and development programs—or occasionally through serendipity—and this makes it very difficult to anticipate them. An alert energy manager should always be on the lookout for breakthroughs that are relevant to his or her field, but we do not devote much space to this topic.

We invite you now to join us as we examine the how and the what of energy management in the process industries. We trust that you will be able to reinforce your current knowledge, and also learn some new things that will help you rise to greater heights of energy efficiency.

February 2015 ALAN ROSSITER

REFERENCES

1. ExxonMobil Annual Report, 2009.
2. International Organization for Standardization, *ISO 50001: Energy management*, http://www.iso.org/iso/home/standards/management-standards/iso50001.htm (accessed May 14, 2014).
3. Institute of Industrial Engineers, Process Industries Division, https://www.iienet2.org/details.aspx?id=887 (accessed April 26, 2014).

ACKNOWLEDGMENTS

Beth would like to thank her friends on the past and present LyondellBasell energy team, especially Brian Goedke, for imagining and implementing an excellent energy management program and for his coaching and encouragement, and also Brian Finnegan, Mike Carlson, Matt Michnovicz, and Cheryl Carouth for their enthusiasm, commitment, and outstanding individual gifts and graces. She'd also like to thank her dear husband James for providing all the kinds of support needed for her to retire, move, and work on the book, Alan for "interrupting her quiet life in England" with this project, and all of the gracious and expert contributors—especially those who have worked with her before and knew what they might be getting into.

Alan would like to thank:

- his wife Belinda, not only for tolerating his distraction while working on this manuscript, but also for her help with proofreading and graphics;
- Beth, for joining him in this project, and who *definitely* had no idea what she was getting into;
- all of our contributing authors, whose depth of knowledge and experience has enriched these pages; and
- Bob Esposito and Michael Leventhal of John Wiley & Sons, who have answered innumerable questions and kept us on track throughout the preparation of this book.

Disclaimer

CONTRIBUTORS

Celestina (Tina) Akinradewo, KBC Process Technology, Northwich, UK

Joe A. Almaguer, The Dow Chemical Co., League City, TX, USA

Bruce Bremer, Bremer Energy Consulting Services, Inc., Union, KY, USA

Mike Carlson, LyondellBasell Industries, Houston, TX, USA

Glenn T. Cunningham, Mechanical Engineering Department, Tennessee Tech University, Cookeville, TN, USA

Joe L. Davis, PSC Industrial Outsourcing, LP, Houston, TX, USA

Bala S. Devakottai, Chevron Phillips Chemical Company, Houston, TX, USA

Elizabeth Dutrow, ENERGY STAR Industrial Sector Partnerships, U.S. Environmental Protection Agency, Washington, DC, USA

Mark Eggleston, Phillip Townsend Associates, Houston, TX, USA

Kathey Ferland, Texas Industries of the Future, The University of Texas, Austin, TX, USA

Joe Ghislain, Ford Motor Company, Dearborn, MI, USA; Ghislain Operational Efficiency, LLC, Milford, MI, USA

Beth P. Jones, LyondellBasell (retired), Guildford, Surrey, UK

Thomas Lestina, Heat Transfer Research Inc., College Station, TX, USA

Sharon L. Nolen, Eastman Chemical Company, Kingsport, TN, USA

Bruce L. Pretty, KBC Advanced Technologies, Inc., Houston, TX, USA

R. Tyler Reitmeier, Soteica Visual MESA LLC, Houston, TX, USA

James R. Risko, TLV Corporation, Charlotte, NC, USA

Alan P. Rossiter, Rossiter & Associates, Bellaire, TX, USA

Paul E. Scheihing, Advanced Manufacturing Office, U.S. Department of Energy, Washington, DC, USA

Graziella F. Siciliano, Office of International Affairs, U.S. Department of Energy, Washington, DC, USA

Graham Thorsteinson, General Mills, Atlanta, GA, USA

Jon S. Towslee, EFT Energy, New York, NY, USA

William (Bill) Turpish, W.J. Turpish and Associates, PC, Consulting Engineers, Shelby, NC, USA

Ven V. Venkatesan, VGA Engineering Consultants Inc., Orlando, FL, USA

Jonathan P. Walter, TLV Corporation, Charlotte, NC, USA

UNITS OF MEASURE

Except where otherwise noted, conventional U.S. units of measure are used throughout this book. However, there is no universally accepted standard across U.S. industry and engineering disciplines for reporting energy consumption. In this book, the standard unit for energy consumption is the British thermal unit (Btu). The multiples of the Btu used in this book are as follows:

10^3 Btu: k(kilo)Btu
10^6 Btu: M(mega)Btu
10^9 Btu: G(giga)Btu
10^{12} Btu: T(tera)Btu

For extremely large quantities of energy (e.g., national consumption), the "quad" or "quadrillion Btu" (10^{15} Btu) is used.

Prices, costs, savings, and other amounts of money are quoted in U.S. dollars ($), unless otherwise stated.

SECTION 1

ENERGY MANAGEMENT PROGRAMS

1

ENERGY MANAGEMENT IN PRACTICE

Beth P. Jones

LyondellBasell (Retired), Guildford, Surrey, UK

Attention to energy efficiency seems to rise and fall in corporate priority with both the rise and fall of energy prices and the rise and fall in manufacturing margins. The life of the corporate energy manager is much the same. If energy costs are high and margins are slim, you are the man or woman of the hour, and if times are good and natural gas supplies are booming, no one will return your calls. Resources are allocated similarly. Bad times and poor prospects for capital spending free up lots of people to look for opportunities for more efficient operations. When times are better, those human resources are reallocated to opportunities with higher expected returns. A well thought-out energy management program identifies opportunities, sets the bar for ongoing performance, and maintains improvements with a minimum of resources. Otherwise, over time and with reduced scrutiny, efficiencies decline and relative costs go up. A few years later, the cycle repeats itself and a new team relearns all the lessons from the last cycle.

Several years ago, a technician received the company's highest "attaboy" award from its then president. His achievement included tuning up all the company's huge olefin furnaces and saving millions of dollars per year in energy costs. To quote the president's comment in this regard, "Great job rediscovering what we already knew! Next time maybe we'll just punish the people who stopped doing it." The technician took

Energy Management and Efficiency for the Process Industries, First Edition. Edited by Alan P. Rossiter and Beth P. Jones.

the comment to heart and wrote a "Furnace Manifesto" to preserve and institutionalize the knowledge.

This book is intended to provide an overview of industrial energy management, particularly for the process industries. Section 1 is focused on management issues—how to start and maintain an effective program, identify the components of a successful program, benchmark, and create management systems—and case studies that provide practical insights from successful and experienced energy managers. The rest of the book provides expert technical help in what to do to save energy, with particular focus on the energy users most significant to the process industries.

This chapter describes a practical overall approach to starting an energy management program and identifying and implementing energy efficiency improvements in a sustainable way. The chapter is particularly focused on helping new energy managers get started: What do they need to know to be effective in their role? Since most of these issues are also relevant to experienced managers and engineers with energy management responsibilities, we hope it will also be useful to all readers.

ASSESSING THE VALUE OF AN ENERGY MANAGEMENT PROGRAM

In most process industries, energy costs are second only to raw material costs. Entire departments are devoted to optimizing raw material choices and product slates, using planning models, supply strategies, and online optimization. Apart from buying energy at the lowest possible cost, most companies also consider energy to be an inevitable cost of doing business. However, energy use is not just a concern for the utilities department, and you, as an energy manager, must separate the cost of doing business from the cost of doing business well. Other benefits flow from focusing on energy, such as a reduced environmental impact and a cultural change toward reducing waste, but the scope of any energy management program is determined by its economic value.

Determine your company's energy use and energy cost, beginning with large sections of the organization and drilling down as far as it is practical. Use the data that are available now and are already understood by the organization. Energy information can come from site utility bills, internal cost accounting, and/or the utilities procurement group. Often the search for the data is as enlightening as the actual data. Was the information easy or difficult to collect? What does it tell you about each site? Are total company or division costs combined and analyzed already?

Draw a material balance box around each unit, site, or division, based on the details that are available (Figure 1.1). Include all external sources of energy that cross the boundary, such as purchased electric power, fuel gas, solid or liquid fuel, purchased steam, or whatever else is consumed within the box. If feed or product streams are burned or consumed to produce energy, include those as well. Collect both the nameplate capacity of each unit and its average operating rate.

Now that you have an initial fix on the company's energy profile, you can proceed toward estimating the value of potential energy savings with a combination of data and engineering judgment. From industry publications or data searches, determine the benchmark-specific energy use for the technologies in each box, and scale the specific

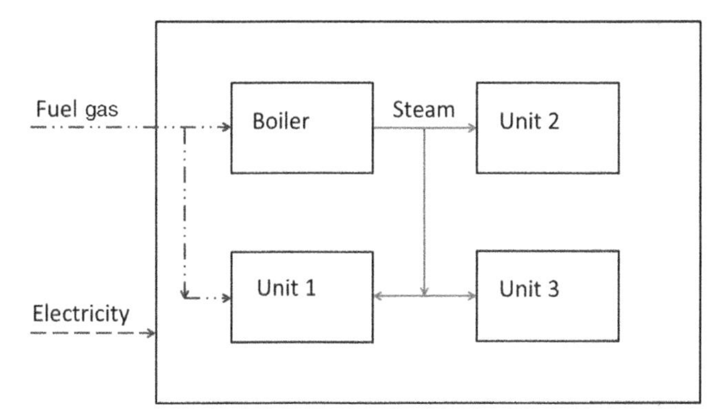

Figure 1.1. A simple material balance provides great insights into overall energy usage.

Site energy use = fuel gas rate × heating value + electricity rate
Site energy cost = fuel gas rate × price + electricity rate × electricity price
Site-specific energy use = site energy use/average feed(or major product)rate

energy to the nameplate capacity of each process. Chapter 5 gives a more detailed look at benchmarking.

Subtract the actual energy use from the benchmark use for each box and add up the potential savings. If no benchmark is available or the available data are not detailed enough to apply a benchmark, choose a reasonable percentage of current use. If no energy improvement program was active in the last 5–10 years, a 10% saving is feasible. Even relatively new and efficient units generally have a bit of refinement available (1–2%) through active energy management.

Compare your estimate of potential savings with the current total energy bill.

Can you afford to focus some resources on energy? Can you afford not to?

LAUNCHING THE PROGRAM

Like most corporate objectives, energy efficiency goals need support from above and heavy lifting from below. Managers will not generally support a new initiative unless they think that the issue is important enough to displace other priorities, and benefits cannot be achieved unless people are motivated to work toward them. An energy management program is a circular process of identifying opportunities, scoping solutions, and implementing improvements, and by doing so, refining the view for another pass (Figure 1.2). Getting started is the important thing. Significant energy improvements can be achieved without the final polishing of the process, so do not lose the good enough in pursuit of the perfect. Early successes prove that focusing on energy is worthwhile and inspire people to go out and look for opportunities in their areas.

Starting a program can be easier if you begin at the top. Presumably, upper management has both the vision to see the benefits of energy efficiency and command

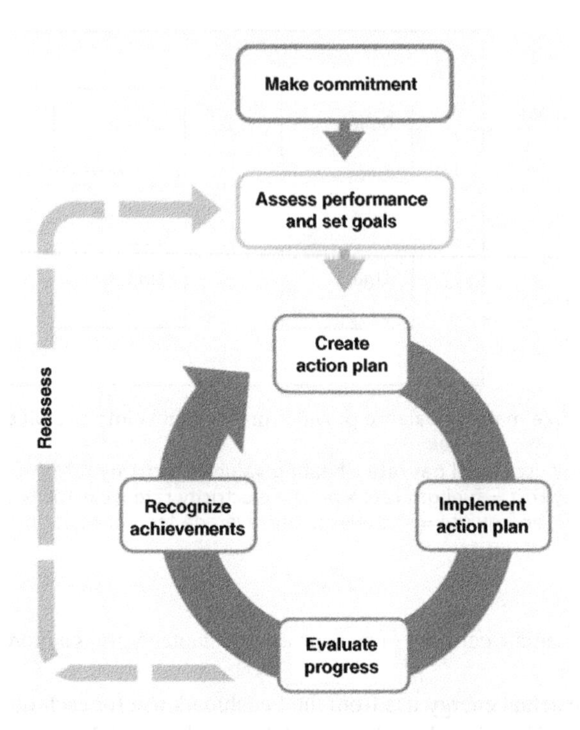

Figure 1.2. The "circular process" of energy management. (*Source:* www.energystar.gov/guidelines)

of the resources needed to do the work. Management commitment is key to the success of the effort. An energy policy statement from the executives demonstrates that commitment and provides the justification to get every site involved in the energy management process.

Some groundwork is necessary before you present your request for support. It is not enough just to estimate potential savings. You will need a rough action plan and an idea of the resources needed to achieve the savings in a given time. Data, plans, and resource requirements will improve with time and study, but first you must get the program off the ground.

Your credibility will also be enhanced by some judicious energy information gathering. You may encounter people within your company who do not believe that significant improvements in energy use are possible. Show them what your competitors are doing. Most large corporations discuss their energy improvements, and the resulting environmental improvements, on their websites. Case studies of successful programs from three companies in very different businesses and with very different energy usage profiles are discussed in Chapters 2–4. Energy policy statements are also displayed on corporate websites. Public organizations, such as ENERGY STAR from the U.S. Environmental Protection Agency (Chapter 7), also showcase companies that have made significant improvements. It is important to point out to management that spending resources on energy always provides some return. Energy prices fluctuate, but the risk associated with that spending is low compared to many other types of spending.

Involving the people you need to actually work on the energy efficiency issue will help ensure that your plans are feasible. Each company is different, but process or production engineers are usually the best points of contact for an energy-saving initiative. Each functional group has a role to play in the program, but the leaders of the energy-saving initiative need to understand the process flow and how energy consumption is affected by process variables. A senior process engineer makes a terrific site leader for the energy program. Site leaders must also have access to energy use data and the skills to develop and justify projects to improve energy efficiency. They also need the support of their managers. Choose a site with potential and spend some time talking to both the engineers and the operations managers. Share the site energy costs and any benchmarking data and brainstorm ways to identify saving opportunities and implement improvements. The exercise should both test your assumptions about achievable benefits and help you to determine the resources needed to realize the savings. Extrapolate these discussions to the entire company, and you will have a good start on an action plan.

Test the plan against the opportunities the group identified during the brainstorming at your "pilot" site. Will the plan be successful in implementing the opportunitics? Is the site willing to follow the plan and try to capture those opportunities? Encourage them to start right away. Early successes increase buy-in at all levels and help to sell, and fund, the program.

Another potential selling point is the environmental benefit of an energy-saving program. The U.S. EPA provides some conversion factors for the greenhouse gas impact of different types of energy use [1]. You can use these factors to estimate the environmental impact associated with the energy-saving path you expect. Table 1.1 shows a few handy factors.

The next stage of planning adds timing and targeting, which leads to initial goals. Take a first pass at prioritizing your company's energy opportunities and determining what could be achieved in the first year or two by applying the resources you have just identified. Start with the sites with both significant energy use and a significant gap between benchmark and actual energy use, and work outward from there.

Initial goals should be achievable but not necessarily easy. The organization should have to act, even to stretch itself, to achieve the goal. As the energy program ramps up, the resource–return relationship will become clearer, but setting a target will get people moving in the right direction. Your management will not hesitate to weigh in on how much can be accomplished in a given time.

TABLE 1.1. Greenhouse Gas Impacts for Various Common Fuels [1]

Natural gas	53.02 kg CO_2/MBtu
Propane	61.46 kg CO_2/MBtu
No. 2 fuel oil	73.96 kg CO_2/MBtu
Bituminous coal	93.4 kg CO_2/MBtu

1 kg/MBtu = 1 kilogram per million Btu.

Now set up that management meeting and state your case:

> We believe that as a company we could save A% of our total energy use over the next B years. At today's prices, the savings are worth $C million per year by the end of that period. Other companies in our industry have achieved similar results—and have advertised the savings and the environmental benefits on their websites. The energy savings bring with them about D tons per year in CO_2 reductions.
>
> Your commitment to a policy of tracking and saving energy, and setting an energy-saving goal, is critical. Setting both a short-term reduction goal and a corporate energy policy that explains how we view energy will kick-start the program. Your continuing attention will ensure that the program is sustained. Here are some suggestions for a goal and a policy statement.
>
> (For example: WeCo pledges itself to monitoring our energy consumption, instituting an energy reduction program, and reporting our progress to employees and shareholders. Our initial goal is to reduce energy consumption by 10% over the next 6 years, based on last year's energy consumption and throughput.)
>
> And here are some ideas on how we could structure a program here at WeCo.
>
> (For example: We could begin with a small central energy resource group to identify initial opportunities through site visits and surveys. To help to implement those opportunities and find others, we could name an energy leader at each site. At most sites, that role would take about 25% of an engineer's time, and at larger sites, 50% of an engineering resource.)

And now, armed with management approval and input, and start-up resources, you can begin an energy program for real. Let us start with resources.

ENERGY RESOURCES: ROLES AND RESPONSIBILITIES

A well thought-out energy management program identifies opportunities, sets the bar for ongoing performance, and maintains improvements with a minimum of resources. Resources are necessary, however. First, we will discuss the people resources. It takes the right resources to understand energy data, identify opportunities, act, and track the consequences of acting on the opportunities. It also takes people to design and implement policies and procedures to maintain efficiency improvements.

Energy Manager and Energy Staff

Your job as an energy manager is to help identify, facilitate, and reward energy improvements, using whatever means you have at your disposal. The energy management role generally provides an opportunity to influence others to take action. A good energy management program provides tools and requirements that identify opportunities. Rarely does an energy manager have the authority and budget to actually realize those opportunities. Instead, the energy manager gets to persuade and assist other people in completing the actions or projects needed.

The energy manager has access to all the company's energy information and is responsible for communicating progress toward goals. Formal communications keep management informed about the value of their investment in energy resources and also keep the participants in the energy-saving process informed and, sometimes, rewarded. Informal communications are more important in getting people engaged in the process and producing useful activity. How you do your job is as important as what you do. Here are a few bits of personal advice, from both the "this worked for me" and the "do what I say, not what I did" columns:

- Remember that you are there to help, not to judge.
- Reward successes before you chide failures.
- Be respectful of competing priorities, but be persistent. Give people every opportunity to do the right thing.
- Give credit generously and enthusiastically. Having the idea is not as important as completing the work to deliver the savings. Be a cheerleader.
- Ensure that savings are "owned" by the site where they are generated. You are a helper, not a pirate who swoops in to claim thc trcasurc.
- Look for links between energy and other functions and resources. You can recruit other groups to consider energy in their normal activities, and you can also help out other groups with what you know and the data you keep.

The next section goes into energy accounting more deeply, but a few more bits of advice are useful while discussing the energy management function:

- Create metrics that fairly measure performance and ensure that the metrics are publicized.
- Account for savings clearly and fairly. Energy calculations should be easy to understand and should be based on the same prices or forecasts as are used for any other project or opportunity. There should be no easy opportunity for hand waving or fogging the results. It is better to underestimate (and over deliver) than to inflate expectations and lose credibility.
- Weave energy efficiency into existing processes and systems—both for tracking savings and for maintaining the savings. It is better to make a pretty good addition to a system people understand than to create a perfect but new and unfamiliar system—especially if it takes time away from looking for savings.
- Likewise, consider related process and maintenance benefits as you consider energy projects. For example, improving the efficiency of a major and limiting piece of equipment reduces energy use at the current throughput, but the change could also be used to increase throughput—and still achieve a lower specific energy.
- Bring the energy message to the people and do your best to set their minds on conservation. Remind employees of simple things such as reporting steam leaks,

turning off lights when they leave their offices, and turning off outside lights in the daytime. Posters, publications, and "toys" are useful tools.

- Automate what you can and think ahead for sustainability.

It is very helpful to amass a small group of energy specialists with specialized skills to get the process off the ground. Experience in energy assessment, database construction and management, accounting, process engineering, and best practice gathering is useful. Depending on the types of energy opportunities you find, specific technical skills could be a tremendous asset. In any of the process industries, solid furnace and boiler skills are helpful—both a good understanding of process- and firebox-side efficiency and good furnace maintenance knowledge. Enthusiasm for the tasks ahead, persuasive capabilities, and a desire to help are important personal qualities.

The energy manager and the energy group also represent the company in external associations and forums. Associating with peers brings in new ideas and proven methods for starting and maintaining energy management programs. Public energy meetings also bring out a wide range of service and equipment providers, and create opportunities to discuss the usefulness of the providers and products with their customers. The energy group can also take the lead in seeking public recognition for the company's energy achievements: for example, by certification through the ENERGY STAR program (Chapter 7) or ISO50001/SEP (Chapter 6) or by applying for energy-saving awards. Public recognition also goes a long way toward improving the internal focus on energy efficiency.

Site Energy Leads

It is important for each site to name a local leader for the energy reduction effort. Ideally, the site energy lead will report directly to the site manager, and energy efficiency will be a topic of discussion at each management staff meeting. The fraction of a person's time required at each site will depend on the size and complexity of the site and on the time line set in the energy goal.

An energy lead's duties include collecting and reporting energy data; identifying, justifying, and supporting energy project implementation; and acting as the advocate for energy efficiency at the site. The energy advocate role requires belief in the value of the activity and active engagement with other functions at the site.

If your company has multiple units or sites, bringing the energy leaders together into a team (a best practice team, a center of excellence, or a business improvement team—whatever your corporate culture supports) is an effective way to leverage improvements between sites. Team meetings are helpful, but organizers should be respectful of time and ensure that each meeting provides value to the participants. Statistics can be handled in reports. Regular virtual meetings keep both the team focused and engaged and the team members up to date on issues and plans. For example, the background behind changes in energy statistics or the reasons for a new initiative can be discussed. If Plant A has a highly successful compressed air audit, the Plant A representative can share justification, process, and results with the rest of the team.

Occasional face-to-face meetings provide more space for team members to share ideas and results, ask questions, and help each other with issues. Face-to-face meetings also provide an opportunity to bring in subject matter experts to build technical proficiency, upper management to demonstrate support and value for the program, or colleagues who can discuss the impact of the energy program on their area of business.

The energy manager should provide each team member or site energy lead with a "job description" and a set of ground rules for energy accounting and energy goals. The description and rules should cover reporting requirements (energy use, actions taken to reduce energy, savings, etc.) and frequency of reporting, guidelines accounting for usage and savings, and team input requirements. Providing a resource guide with the available internal and external energy tools and services is important. It is also useful, over time, to check to see if each energy lead has taken advantage of all the energy resources (audits, tools, etc.) available. And most importantly, you as the energy manager should check in with each team member on a regular basis and see what you and your team can do to help.

Operations Management

Operations managers "own" both the energy-using equipment and the people and systems that can control a large part of energy use. In a perfect world, the operations staff provides enthusiastic support and interweaves energy awareness into processes and procedures. Operations managers have approval authority for operating changes and oversight of the operating technicians. Technicians often have excellent energy-saving ideas, and their vigilance maintains energy improvements. Managers' interest ensures the technicians remain interested, and profitability from improved energy performance accrues visibly to their site's economics. It is important to have the operations staff's active buy-in, and in order to achieve that, they must have a say in how the energy program is implemented in their area. This does not mean allowing a handcrafted "I will do this, but not that" approach; rather, it is a collaboration on "how can we best achieve these corporate objectives in your area." Work together, and share your vision and agreement with the entire operations staff.

Maintenance

The maintenance department has the power to control a surprisingly large amount of site energy waste. Active, enthusiastic management of antiwaste programs like steam trap testing, repair, and replacement, steam leak repair, and insulation maintenance have a huge impact. Chapters 9–16 contain a number of energy-critical maintenance suggestions. Maintenance groups are often given a fixed annual budget, and unexpected repairs can whittle away at the discretionary portion of the budget. Waste minimization programs are often the first choice for "saving" maintenance funds. It is important to ensure both that the maintenance group understands the value these programs provide and that the group gets credit for their savings. Often the maintenance group sees no consequence for cutting these programs and no credit for saving them. The energy manager can provide both. Chapters 13 and 14 discuss steam traps and leaks in more detail.

In the longer term, maintenance standards can also have a large impact on ongoing energy costs. Ensure that standards reflect the current economics of equipment replacement and refitting. For example, it is frequently more cost-effective to replace old electric motors with newer, more efficient motors than to continue to rewind older motors. Ensure that insulation standards include appropriate services and thicknesses based on the marginal costs of the heat—or refrigeration—that could be lost. Insulation is discussed in more detail in Chapter 16.

Measurement is also important in energy management. Many plants have meters only at the utility system level, rather than providing meters and measurements for all energy sources at the unit or subunit level. Even when sufficient meters are available to close a unit energy balance, maintaining utility meters is generally given a lower priority than maintaining feed or product meters. Regular attention and good preventive maintenance standards can improve the situation.

Process Control

It is worth the time and effort to ensure that reasonable marginal costs of energy are included in process-side advanced process control (APC) and optimization calculations. Advanced control of boilers, furnaces, and utility systems can also bring substantial benefits. Real-time optimization of complex utility systems can also be worthwhile, as discussed in Chapter 19.

Collaboration between the energy manager, site energy leads, unit operators, and process control engineers can highlight opportunities to use APC methods to balance energy needs and energy generators. Better control of utility systems can even allow the shutdown of individual utility generators or plants for increased savings.

COMPONENTS AND SYSTEMS

We have identified the key people; now we will discuss the systems necessary to success for the energy management program. The following are key components of an energy management program:

- Data collection and analysis
- Goal setting
- Identification of opportunities
- Implementation of opportunities
- Reporting and recognizing achievements
- Re-evaluation and renewal

While each component is important, some components will evolve faster than others in a new energy management program. It is more important to get started than to have each step fully planned out and methodically executed. Set a basic plan and get going, and refine the plan as you go along. Early savings provide motivation and secure resources, as well as provide learning that can be applied to the evolving program. Even mature and successful programs continue to evolve to adapt to changing situations.

Data Collection and Analysis

Data collection and analysis depends on measurement, accounting, and benchmarking. Handling the data also requires an understanding of the program's ground rules for accounting.

Ground Rules: Basis and Calculation Methods. Savings in pursuit of a goal must be clear and calculable, from an easily understood baseline. The full year before the program starts makes a useful baseline. If that year's operating rate was unusual for a plant, such as the effect of a major turnaround, an average of 3 years' data may be more appropriate.

Progress against goals should be measured in terms controllable and understood by the participants. For example, a change in energy consumption (Btu) is a better goal than a change in spending on energy ($) because of fluctuations in market prices. Sites can control processes, however, and can be accountable for the way the processes run. For some processes, specific energy use per pound or ton of product is an easily understood goal. For other processes, optimum feedstock choices or reaction severities are more valuable than specific energy, and the optimum energy use will change as situations change. Is there a way to easily measure the gap between the optimum and actual energy use? Discuss this with the process control or production planning group and incorporate their input into the energy goal.

If the savings goal is expressed as a percentage of base year energy use, the annual goal becomes harder to achieve as energy use declines. Should the basis be reset at some point? Should each year's goal be relative to the year before?

Savings must also be calculated in a defensible fashion. While reduced usage is a good thing, a site or unit should be able to show that it took action in order to achieve the savings. Reducing waste by changing an operating condition is as valid an action as installing a new heat exchanger. In most cases, the savings are best represented at average conditions—average operations and average reactor conditions—rather than at design or actual operating conditions. To be completely accurate would require the database savings to be a calculated function of operating rate and reactor severity, but to track those fluctuations would generally take more time and effort than the value the accuracy could provide. If a project is so material to the company's energy use that it is worth the effort, by all means you should be more accurate—preferably by performing the calculations in the plant data historian.

Justifications for energy-saving projects should reflect the company's official economic outlook, meaning that energy projects should be justified using the same pricing assumptions as every other project. It is necessary, however, to represent relative risk appropriately. Energy projects often have a much lower risk than new products or expansions, where product price or demand changes can eliminate returns. If the energy price drops, project returns can drop, but rarely do the energy savings drop to zero.

This lower risk can be expressed through sensitivity calculations or in the manner your company normally treats risk.

Measurement. The measurement of energy use and energy savings must be balanced between data accuracy and the effort and expense required to provide the data. Survey the energy metering and the details that are available to you now at each site. Do the data from various sections of the plant add up to the site total? Are the data detailed enough to provide a clear picture of the energy use of significant portions of the plant, and of the significant users? Can you create a history of energy use that is meaningful to the site? Accurate-enough measurement of energy use provides a clear-enough picture of the performance of the unit or portion. It can also identify opportunities for improvement, and it allows savings to be measured. In an ideal world, the energy consumption of individual pieces of high-energy-use equipment, individual units, and meaningful sections of units and plants would be measured. In reality, there may be one electricity meter for the whole plant, and one steam flow meter on the main boiler. It can be difficult to justify new utility metering, but you should be able to make a case for clarity when the usage is material and variable. Utility meters are generally low on plant preventive maintenance priorities as well. Meters are useless if they are not maintained to a reasonable degree of accuracy. Ensure that key energy measurements are treated appropriately in the preventive maintenance schedule.

Accounting. In order to maintain and evaluate the effectiveness of an energy management program, energy data must be collected and analyzed. Collecting credible utility data in the plant data historian provides tracking capability at a low "people" cost. Ultimately, all energy use and energy-saving projects and actions should be collected in a database of some sort. Careful forethought will minimize the time required, rework, and irritation. It is certainly possible to start with a spreadsheet, but designing an energy database and thinking ahead about its contents and dimensions are worth the time. Again, it is more important to get started than to create the perfect database, so by all means go ahead and build the spreadsheet while you think about the database.

That said, an ideal energy database would pull historian data into a central repository for all sites. In many cases, however, that level of communication through various firewalls is difficult and human intervention is necessary. A benefit of intervention is that the energy lead can "smell test" the data and catch metering errors or even a significant issue in the plant. The downside is that it takes time for the energy lead to enter the data into a separate database, and data entry is not the best use of an engineer's time. Ideally, the site lead would initiate an automated data collection, give it a quick look, and then upload it into a combined company database. Discussing the functions needed in the database with the IT department should lead to a solution that achieves your needs in a way that is time efficient, easy to learn, and robust.

It is also useful to track what type of energy is being used. Again, complexity and accuracy must be in balance, but it is reasonable to separate electricity, fuels, and different levels of steam. These histories can also be used to optimize utility systems.

Chapters 17–19 discuss steam balances, steam pricing, and steam system optimization. Tracking the utility type can also allow the environmental benefits of energy savings to be tracked accurately.

Benefits are the more exciting—and the more complex—side of energy data collection. How will you track improvements? Again, thinking ahead will give you credible data that you can use to support longer term programs as well as report the results of your early efforts. Reporting is a balance between useful details and the time and complexity required from the reporters.

Experience provides some useful details to collect about each energy-saving action:

- Type of project
  - Operations change
    - Immediate or relatively quick
    - No or little spending required
    - May require a change in operating procedures or targets
    - Generally provides ongoing savings for a given operating mode
  - Maintenance spending (e.g., steam leak or steam trap programs)
    - Generally requires ongoing spending
    - Generates savings that exceed costs and are ongoing or have a sawtooth decay/repair pattern
  - Capital project
    - Requires justification and budgeting
    - Has a timing profile associated with both savings and spending
    - Generally provides ongoing savings
- Savings profile
  - Savings rate and timing
  - Savings duration (ongoing or one time)
- Spending profile
  - Spending rate and timing
- Source of savings
  - Steam and pressure level
  - Fuel
  - Electricity
- Environmental benefit of savings (dependent on energy source and site)
- Project status
  - Completed
  - In progress
  - Not currently active (useful to collect ideas or maintain "almost" projects)

The following are some hard-won insights into useful features to build into a database, should you decide to build one:

- Plan for easy uploading of data collected from plant historians.
- Design input forms that are as intuitively obvious as possible.
- Insist that each number in the database has an associated set of units. Energy could be entered in kWh or in Btu or in kJ, but the unit is a part of the value and the value is useless without the unit. This is obvious to engineers, but experience suggests it is not so obvious to programmers.
- Facilitate unit conversions so that people can enter data in the units most familiar to them.
- Provide reporting tools so that common reports can be generated easily.
- Consider the time increments for data gathering and for reporting. Should data be collected daily? Monthly? Quarterly? What roll-up increments are necessary, and how will you handle increments of varying length?
- Provide common analysis tools, but also allow the data to be dumped into a spreadsheet for a more personalized analysis. Pivot tables are very powerful.
- Finally, think about how to track actual energy costs. Costs are often markedly different from site to site and from utility to utility. An energy management program should focus on saving energy, which is controllable, rather than on money, which fluctuates with prices, but the program must be able to explain its worth. Consider the effort–accuracy balance for site and overall energy prices. Whether prices are internal to the energy database or tracked externally is your call.

Benchmarking. It is useful to compare each unit and site with its own past performance to determine energy savings, but the internal view is not very useful in determining how the process compares with its competitors. Several companies provide confidential benchmarking services, and benchmarking is discussed at greater length in Chapter 5. An understanding of benchmark energy use and any structural differences between your processes and competitors' processes is necessary to set realistic goals for each site and for the organization.

Another valuable source of "benchmarking" information is the original design data for each process, or the updated design after major revisions. Does the process operate at or near the specific energy inherent in its design?

Goal Setting

An energy goal focuses attention on improvement, and the goal can reflect a number of corporate concerns. If competitive position is important, the goal can be stated in terms of closing any gap between the company's performance and the benchmark energy performance of competitors. The company may have the capability and desire to be a pacesetter in energy performance, and the goal can reflect a commitment to change the energy profile of the company. Some companies spur employees into action with a "big, hairy, audacious goal," believing that a big vision produces big results. Different processes or sites will have different opportunities and different opportunity costs. Goals can be tailored to fit those different opportunities, but in that case they must be communicated very carefully so as not to create the belief that energy conservation is the responsibility of only some sites and not others. It may be better to set the same goal for all sites and excuse poorer performance in sites with fewer opportunities.

Whatever the chosen goal, it should include the expected change in performance and the time allowed to make the change.

Identification of Opportunities

The right people resources have been identified. Many will already have the skills and good ideas necessary to capture energy improvements, but they will be even more effective with a more specialized set of energy tools and methods. The second half of the book contains more detailed discussions of technologies and tools for saving energy. A very brief overview is given herein, with an emphasis on bringing resources together to use the tools and identify the opportunities.

Site Energy Reviews. Often the most effective way to find and begin implementing energy improvements is to get all the interested parties into one room for a structured discussion of their plant. Careful preparation and actively involved stakeholders are critical to success. A good energy review leaves a site with its own plan, developed by its own people and supported by its management.

Work with the plant energy lead to gather site energy data, benchmarking data, process flow diagrams (PFDs), and major equipment specifications. The U.S. Department of Energy (DOE) plant energy profiler (ePEP) tool's questions help gather all the relevant energy and equipment information for the plant and the profiler identifies many potential opportunity areas. The energy group and the site energy lead can work through the Profiler in preparation for the site energy review. At the time of this writing, the Plant Profiler is available for free download at https://ecenter.ee.doe.gov/EM/tools/Pages/ePEP.aspx.

The actual review can be led by knowledgeable internal resources (e.g., energy group members) or by an external consultant, and should include representatives from engineering, operations, and maintenance functions. The team should "walk" through the unit via the PFDs, taking note of changes in operation or in performance compared to design. Are major energy users operating at or near their design efficiencies? Is all the heat recovery equipment clean and in service? Have any short-term operating changes been overlooked and become long-term energy drains? The team can discuss any opportunities identified by the Plant Energy Profiler as well. Chapter 25 also discusses process reviews and improvements.

The team should also physically walk through the unit looking for opportunities such as steam leaks, failed steam traps, missing or damaged insulation in hot or cold areas, or compressed air leaks, and noting their findings.

The team, or a portion of the team, should then consolidate the notes and estimate the energy savings available from each opportunity identified, and the costs, time, and resources required to capture each opportunity. The estimates do not have to be perfect, but they should be reasonable and defensible. Group the opportunities into categories—things to do immediately with available resources, capital projects, turnaround projects, and maintenance programs—and put together a rough action plan for the site. Each review should end with a report out to the plant management, including value, costs, timing, and comparison of expected results with the energy goal.

Energy Tools Overview. A wealth of energy tools is available through both public sources and commercial vendors. Some are easy to use and others require training or even certification. We cannot begin to list the commercial tools for process simulation, furnace simulation, steam or cooling water network evaluation, or even energy management practice evaluation. Many vendors of equipment and services would be happy to survey your plants' version of their systems, either in anticipation of sales or for a fee. Steam trap surveys are a good example (Chapter 13).

We will highlight some offerings from a free, public source: The US Department of Energy's suite of tools is available for free either for on-line use or as downloads, and their entire tool suite is outlined at https://ecenter.ee.doe.gov/Pages/default.aspx.

The tools offered have changed over time in response to changing needs. Some of the tools available at the time of this writing are discussed below, with the DOE's descriptions:

1. *Steam System Modeler Tool (SSMT):* This new on-line tool allows you to create up to a 3-pressure-header basic model of your current steam system. It includes a series of adjustable characteristics simulating technical or input changes, thereby demonstrating how each component impacts the others and what changes may best promote overall efficiency and stability of the system. SSMT replaces an earlier spreadsheet-based modeling tool, *Steam System Assessment Tool (SSAT).*
2. *Process Heating Assessment and Survey Tool (PHAST)* introduces methods to improve the thermal efficiency of heating equipment. This tool helps industrial users survey process heating equipment that consumes fuel, steam, or electricity, and identifies the most energy-intensive equipment. The tool can be used to perform a heat balance that identifies major areas of energy use under various operating conditions and test "what-if" scenarios for various options to reduce energy use.
3. *Pumping System Assessment Tool (PSAT)* is a free online software tool to help industrial users assess the efficiency of pumping system operations. PSAT uses achievable pump performance data from Hydraulic Institute standards and motor performance data from the MotorMaster+ database to calculate potential energy and associated cost savings. The tool also enables users to save and retrieve log files, default values, and system curves for sharing analyses with other users.

4. *Other:* The DOE also offers tools for assessing fans and compressed air systems, motors, and solar power, as well as guides for energy management.

Another very useful tool available for free download is *3E Plus®*, which comes from the North American Insulation Manufacturers Association (NAIMA). The program calculates the most economical thickness of industrial insulation for user input operating conditions. The user can carry out calculations using the built-in thermal performance relationships of generic insulation materials or supply conductivity data for other materials.

Good Engineering Practices. Designing energy efficiency into new processes is the best way to ensure that they operate efficiently. Process engineers routinely optimize operating pressure, recoveries, and costs in distillation. They can utilize pinch analysis (Chapter 26) to optimize heat recovery and its costs and evaluate new exchanger technologies. They can also be aware of energy loss through pressure drop. When a process change is anticipated, process engineers should take the time to look at its energy impact and reoptimize the utility systems if necessary. A major energy efficiency change could also bring new process-side opportunities, particularly if the energy user was a bottleneck for process throughput or reaction conditions.

Likewise, robust mechanical engineering standards and practices should ensure that life cycle costs are considered in equipment choices and should set expectations for motor efficiencies, repair versus replacement decisions, preventive maintenance, and insulation and refractory standards.

Energy Optimization. Just as including energy costs and constraints in process-side advanced process control provides benefits, utility systems can benefit from advanced process control and optimization. A complex steam system can be operated more efficiently if boiler rates, steam rates at each pressure, pump drivers, and variable steam demands are optimized (Chapter 19).

Implementation of Opportunities

This section provides some recommendations for implementing different sorts of energy opportunities to achieve real energy reduction results. The most important recommendation for a new energy reduction program is go for the "low-hanging fruit." Early and free, or inexpensive, savings improve energy program credibility in several ways: They justify the use of people resources to look for savings, they prove that savings are available, they show that even good operations can be made better, and they show that the energy team is and will be good stewards of the resources assigned to the program. Look for waste and for easy operating changes first. Capital requests to capture energy benefits are easier to justify when all the no-capital opportunities have already been captured, and the low-hanging fruit can help fund longer term projects.

In general, energy opportunity projects should be implemented as if they were any other project, using the processes and procedures already in place in the plant. Since energy use is subject to "drift," some extra care should be taken to document and preserve the new equipment, program, or operating scheme in the collective consciousness.

Operating Changes. The site energy lead should engage operations and process engineering to implement changes in set points or procedures. In some cases, an operating change may take a process into an unknown territory. Use the best practice team to see if another plant can provide information on its impact. Ensure that the change is sustainable by recording it into operating orders or procedures.

Maintenance Programs. Once again, the site energy lead should engage the maintenance and operations staff to implement a change. Often several sites have similar maintenance needs and programs, and the energy group should consider advocating for a corporate policy and a set-aside budget for the work.

Capital. Usually energy projects are subject to the same capital limitations and procedures as any other project, and must compete for funding against all other projects. Sometimes energy projects are developed in an "energy silo." Ensure that other groups are informed and involved in the development of the energy project so that any associated improvements or alternative uses are considered in the project's justification, and also that energy impacts are considered in the justification of other projects. It is much easier to justify energy improvements in conjunction with process changes or debottlenecks. Also ensure that the risks associated with project earnings are outlined appropriately. As discussed earlier, energy projects generally have a lower risk of loss than many other types of projects. Project earnings are reduced if energy prices or throughputs fall, but energy costs never go away while the unit is running—therefore, for most energy projects, earnings cannot go to zero.

Reporting and Recognizing Achievements

Communicating energy-saving progress is critical to maintain support for the program and to recognize good work by the sites, as well as to hold each site accountable for its goals. Progress reports can be generated from the energy database or from separate record keeping. It is useful to be able to provide the same report format at a unit, site, division, and corporate level. Some sites may wish to express the same goal differently.

Often the energy-savings goal is expressed as a percentage of a baseline year's use. Tracking savings is relatively simple, and each site is held to the same standard, regardless of its base usage. Figure 1.3 shows an example format for three sites.

While sites cannot be held responsible for energy price fluctuations, costs are something they truly understand. It is inspirational to translate energy savings into cost savings for a time period. One meaningful way of tracking the value of the program is shown herein. If you have chosen to track energy prices (either by site or by average of the company), it is relatively easy to combine cumulative energy savings for any time period and the price for that time period. You can calculate the value of the savings to date for each period. You could also subtract the costs of the program in each period and

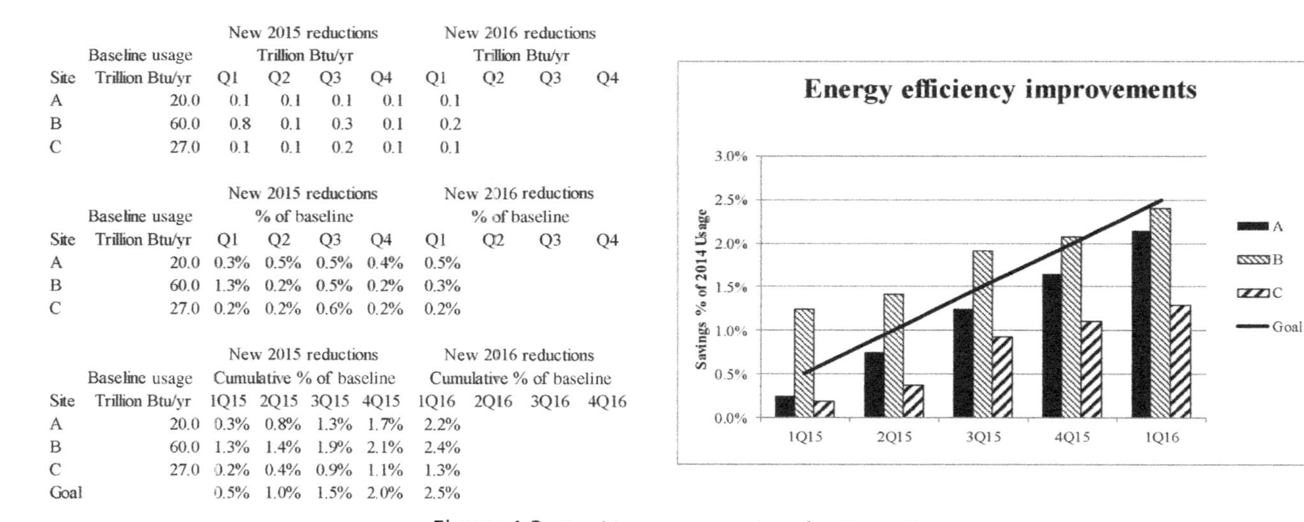

Site	Baseline usage Trillion Btu/yr	New 2015 reductions Trillion Btu/yr				New 2016 reductions Trillion Btu/yr			
		Q1	Q2	Q3	Q4	Q1	Q2	Q3	Q4
A	20.0	0.1	0.1	0.1	0.1	0.1			
B	60.0	0.8	0.1	0.3	0.1	0.2			
C	27.0	0.1	0.1	0.2	0.1	0.1			

Site	Baseline usage Trillion Btu/yr	New 2015 reductions % of baseline				New 2016 reductions % of baseline			
		Q1	Q2	Q3	Q4	Q1	Q2	Q3	Q4
A	20.0	0.3%	0.5%	0.5%	0.4%	0.5%			
B	60.0	1.3%	0.2%	0.5%	0.2%	0.3%			
C	27.0	0.2%	0.2%	0.6%	0.2%	0.2%			

Site	Baseline usage Trillion Btu/yr	New 2015 reductions Cumulative % of baseline				New 2016 reductions Cumulative % of baseline			
		1Q15	2Q15	3Q15	4Q15	1Q16	2Q16	3Q16	4Q16
A	20.0	0.3%	0.8%	1.3%	1.7%	2.2%			
B	60.0	1.3%	1.4%	1.9%	2.1%	2.4%			
C	27.0	0.2%	0.4%	0.9%	1.1%	1.3%			
Goal		0.5%	1.0%	1.5%	2.0%	2.5%			

Figure 1.3. Tracking energy savings for three sites.

show net savings, or simply report the total cost along with the accumulated savings (Figure 1.4).

Re-evaluation and Renewal

Just as data become clearer with analysis, the potential for energy savings at each site becomes clearer as it is explored, and the efficacy of each energy program component becomes clearer as it is put into use. In the spirit of continuous improvement, you should pause periodically to adjust methods and expectations and refocus the energy effort. Chapters 2 and 3 have good discussions of how this works even in mature and successful programs.

The following are some of the points to ponder:

How is the company performing versus its goals? Is the performance a function of the goal or of the actions in response to the goal? If most sites are exceeding the goal, maybe the goal was set too low. How much work has been done? Perhaps the company should be celebrating an exceptional effort, or perhaps it should set its sights higher. On the other hand, if most sites are not reaching the goal, there are other questions to ask. Is the goal unrealistic, or has the response been less than exceptional? Are the sites availing themselves of the services and tools available to them? Has the company supported the goal with both people and economic resources? Consider all these factors in continuing or adjusting the goal, and follow through with the discussions necessary to reward good efforts, align the resource support with the opportunities available, and refocus the laggards on expectations.

What is the potential for energy improvement going forward? Have justifiable projects and actions been proposed? What is the balance between potential savings at low cost and potential savings that will require more expensive changes to the process? Are the resources available to execute the projects and actions? Is it worthwhile to develop an energy plan for each unit or site that includes cost and ultimate energy benefit, both in savings value and in competitive position, and a realistic timeline for execution?

How is the best practice team performing? Are true best practices recorded, shared, and adopted? Are the team members engaged? Talk to the team to see what elements are working for the members and what they feel could be de-emphasized. What are their greatest needs to help them do a better job?

Are the right resources employed in finding and implementing energy savings? Do you have the right members on the energy team? Are there technical or managerial gaps, or new resources that could be applied to the effort? Are the capital and maintenance budgets sufficient to meet the goals?

ONWARD AND UPWARD!

Now that we have discussed the "hows" of starting and sustaining a successful energy management program, we will spend the rest of Section 1 examining successful corporate energy management programs, benchmarking techniques, and general energy

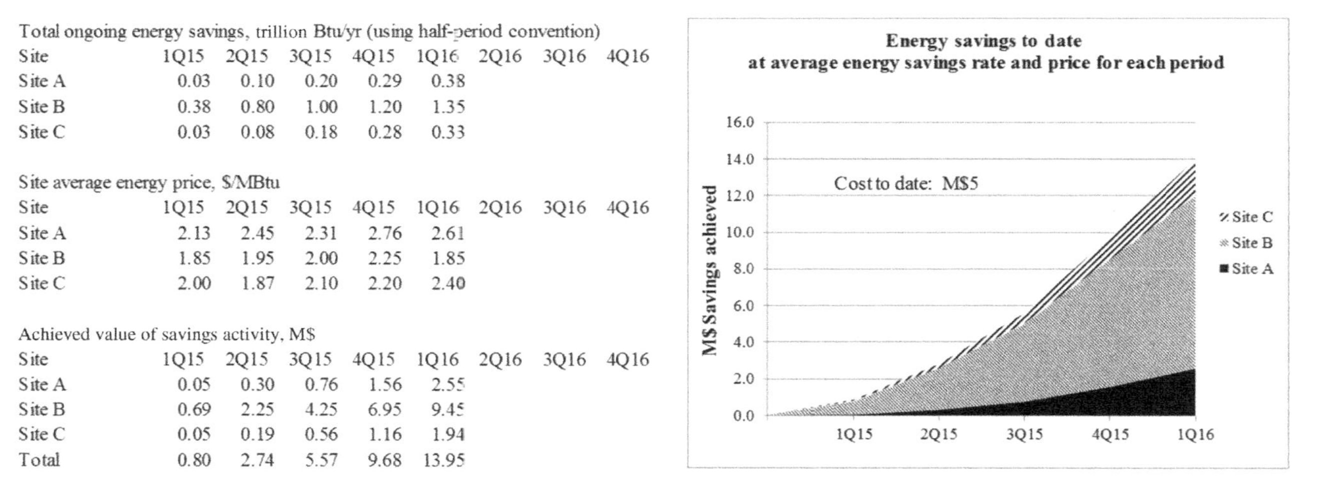

Total ongoing energy savings, trillion Btu/yr (using half-period convention)

Site	1Q15	2Q15	3Q15	4Q15	1Q16	2Q16	3Q16	4Q16
Site A	0.03	0.10	0.20	0.29	0.38			
Site B	0.38	0.80	1.00	1.20	1.35			
Site C	0.03	0.08	0.18	0.28	0.33			

Site average energy price, $/MBtu

Site	1Q15	2Q15	3Q15	4Q15	1Q16	2Q16	3Q16	4Q16
Site A	2.13	2.45	2.31	2.76	2.61			
Site B	1.85	1.95	2.00	2.25	1.85			
Site C	2.00	1.87	2.10	2.20	2.40			

Achieved value of savings activity, M$

Site	1Q15	2Q15	3Q15	4Q15	1Q16	2Q16	3Q16	4Q16
Site A	0.05	0.30	0.76	1.56	2.55			
Site B	0.69	2.25	4.25	6.95	9.45			
Site C	0.05	0.19	0.56	1.16	1.94			
Total	0.80	2.74	5.57	9.68	13.95			

Figure 1.4. Tracking accumulated cost savings for three sites.

management systems. In Section 2 we will concentrate on the "whats": expert technical help in techniques and process insight toward saving energy, with particular focus on the energy users most significant to the process industries.

REFERENCE

1. U.S. EPA (2004) Unit Conversions, Emissions Factors, and Other Reference Data. Available at http://www.epa.gov/appdstar/pdf/brochure.pdf (accessed November 2004).

2

THE DOW CHEMICAL COMPANY: ENERGY MANAGEMENT CASE STUDY*

Joe A. Almaguer

The Dow Chemical Co., League City, TX, USA

THE DOW CHEMICAL COMPANY: OVERVIEW

Headquarters: Midland, MI
CEO: Andrew Liveris
Revenues (2013): $57 billion
Energy-saving results: 40% reduction in energy used per pound of product, 1990–2013
Key efficiency strategy successes:

- *Saving energy:* Since 1990, Dow's energy efficiency strategy has saved 5800 trillion Btu of energy. These energy savings also led to $27 billion in cost savings and prevented 308 million metric tons of carbon dioxide equivalents (CO_2e) from entering the atmosphere. The energy savings and avoided greenhouse gas (GHG) emissions are equivalent to the average footprint of more than 48 million single-family homes.

* This chapter is adapted and updated with the permission of the Center for Climate and Energy Solutions, formerly the Pew Center on Global Climate Change [1].

Energy Management and Efficiency for the Process Industries, First Edition. Edited by Alan P. Rossiter and Beth P. Jones.

- Sustaining savings over a 10-year period, and then strengthening the target for the next 10 years.
- Finding significant energy savings in an already efficiency-oriented, energy-intensive primary manufacturer.
- Leveraging the company's energy business unit to provide a wide range of energy efficiency and related technology and operations services.

ENERGY EFFICIENCY STRATEGY OVERVIEW

Energy is a small cost for most companies, but for Dow Chemical, nothing about energy has ever been small. About half of every dollar the company spends goes toward energy, mostly in the form of natural gas and natural gas liquids, which are the energy feedstock for the company. Not all of that energy is used to power Dow's operations; in fact, two-thirds of the energy molecules Dow buys are used as feedstock, transformed via chemical processes into myriad products. That still leaves 30% of Dow's costs as energy to run its plants, making it one of the world's most energy-intensive companies. Dow's energy-intensive nature, coupled with its continuous-processing, 24-7 operating mode, makes its energy efficiency strategy somewhat different from companies using moderate amounts of energy. While certain aspects of Dow's strategy are more applicable to other energy-intensive companies, many elements will hold relevance for a broad range of companies. These include Dow's efforts to organize an effective program, set up a detailed reporting system, and gain cooperation across business units to meet ambitious energy-saving goals.

Everything about Dow's energy operations is large in scale. The company's energy purchases (feedstock included) roughly equal Australia's entire energy bill, and are equivalent to 10% of all the oil the United States imports. Dow's Freeport Texas site is nothing if not large, taking up several square miles along the Gulf Coast and accounting for over 1% of all the energy consumed in the state.

Energy is so big that it constitutes on its own a business unit at Dow. The energy business sells electricity, steam, and natural gas to other business units. It is also a major player in world, national, and state energy markets, selling as well as buying energy on a wholesale basis. At the Freeport Texas site, Dow operates some 1000 MW of electricity generation, as much as the largest utility power plants. From its high-tech control center with its panorama of brightly colored display screens and switch-studded control panels, Dow buys and sells power among its production units, and with the Texas power grid operator, the Electric Reliability Council of Texas (ERCOT). Dow's power generation is fully interconnected with ERCOT, allowing Dow to behave like other large generators on the system. Operators monitor market conditions and Dow operations minute by minute, making decisions on dispatching power flows. On relatively rare occasions, when ERCOT prices are high enough, Dow will elect to shut down one or more production units for short periods and sell power into the grid, because it is more profitable to sell electricity than to make chemicals under those conditions.

This kind of sophistication in its energy business, driven by the essential role of energy at Dow, flows to the end-use level as well. When energy is such a huge part of production costs, reducing the energy needed to make a pound of product is a matter of competitive survival. It is not surprising then that Dow was among the first companies to set quantitative, measured energy-saving goals. In 1995, the company set a goal of cutting energy use per pound of product 20% by 2005. Dow beat that goal, realizing 22% savings as of 2005. However, from 2002 to 2007, Dow's energy bill rose from $8 to $27 billion as natural gas prices skyrocketed. While these price effects offset the energy intensity improvement, Dow saved almost $8 billion compared with the energy bills it would have paid without the efficiency strategy.

Against the backdrop of sustained high energy prices, in 2006 CEO Andrew Liveris raised the bar, increasing the goal by pledging to slice another 25% off the energy needed to make a pound of product. Dow's energy efficiency goal is represented in Btu per pound of product produced, so natural gas used as feedstock (which does not emit greenhouse gases in the production process) is not included as part of the goal. However, Dow is exploring ways of becoming more efficient in its use of feedstock, primarily due to cost concerns. Figure 2.1 illustrates the company's goal and progress to date. The 2008 usage shows an increase in energy per pound, largely due to the economic slowdown that cut production and whose effects linger.

The way in which the 2015 energy goal was announced illustrates Dow's commitment to energy efficiency. Rather than simply issuing a press release from headquarters in Midland, MI, Liveris spoke at a special event in Washington, DC, delivering a speech, largely unscripted, that showed a detailed understanding of the numbers and the

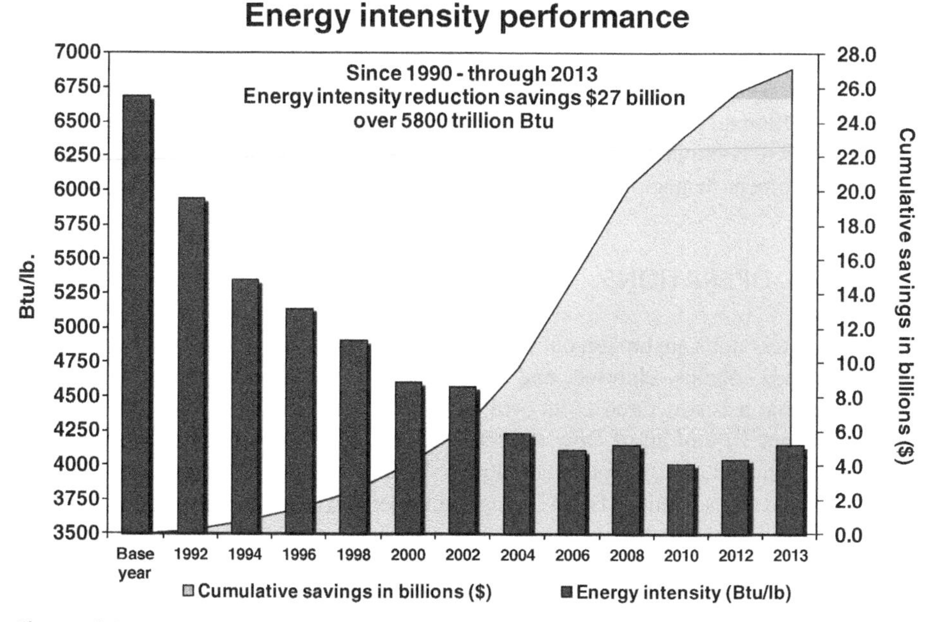

Figure 2.1. Dow's energy intensity performance 1990–2013. (*Source:* Courtesy of The Dow Chemical Company.)

technologies involved. Dow's extra effort and the CEO's personal and visible involvement indicate that this is a high priority for the company. And the energy-saving commitment, with the company's other sustainability goals, is measured and reported on the Dow website on an ongoing basis. The Dow home page includes "sustainability" as a top-level menu title, so visitors are able to easily track the company's progress toward meeting its goals.

Andrew Liveris continues a hands-on tradition of Dow leadership driving energy innovation dating back to Herbert H. Dow's 1897 launch of his bleach business. Even in those early days, energy was critical to Dow's business. As Herbert Dow envisioned today's bulk chemical production process and business model in the early twentieth century, he realized that the energy technologies of the day would not allow him to produce chemicals at the scale he needed. George Westinghouse, pioneer of the gas power generation turbine among other energy technologies, worked in partnership with Dow to create the specialized electricity and steam technologies that enabled the Dow business model to advance. The fact that Dow leadership and employees remember and tell this story a century later reflects a deep appreciation for the role that energy innovation plays at Dow.

ENERGY EFFICIENCY AND DOW'S CLIMATE STRATEGY

Because energy use is such a major factor in Dow's operations, energy efficiency and GHG reductions are closely linked. Dow's goal is to maintain GHG emissions below 2006 levels on an absolute basis for all GHGs, thereby growing the company without increasing its carbon footprint.

Dow's energy efficiency and chemicals management efforts have significantly reduced the company's GHG emissions footprint. As a result, Dow has prevented over 308 million metric tons of GHG emissions from entering into the atmosphere since 1990. Dow will continue to focus on managing Dow's footprint and delivering solutions to help customers manage theirs. Figure 2.2 illustrates the GHG emissions goal.

INTERNAL OPERATIONS

Dow's business units are broken out by product type, even as their operations are highly interconnected: olefins, chlorine, and so on. Energy is such a large part of Dow's operations that it is structured as its own business unit. The energy business owns and operates about 10% of all of Dow's assets, making it the second-largest business unit after the olefins business, followed closely by the chlorine business. The Energy Efficiency and Conservation (EE&C) program leadership team operates from the energy business, engaging some 40 roles overall: about 26 leads in the various business units and 14 site leaders, and some individuals at larger individual plants. Like many companies, Dow does not have many people with a full-time energy efficiency job title—that distinction goes to corporate energy efficiency manager Joe Almaguer. As Almaguer explains it, the matrixed nature of the EE&C team makes estimating total labor effort

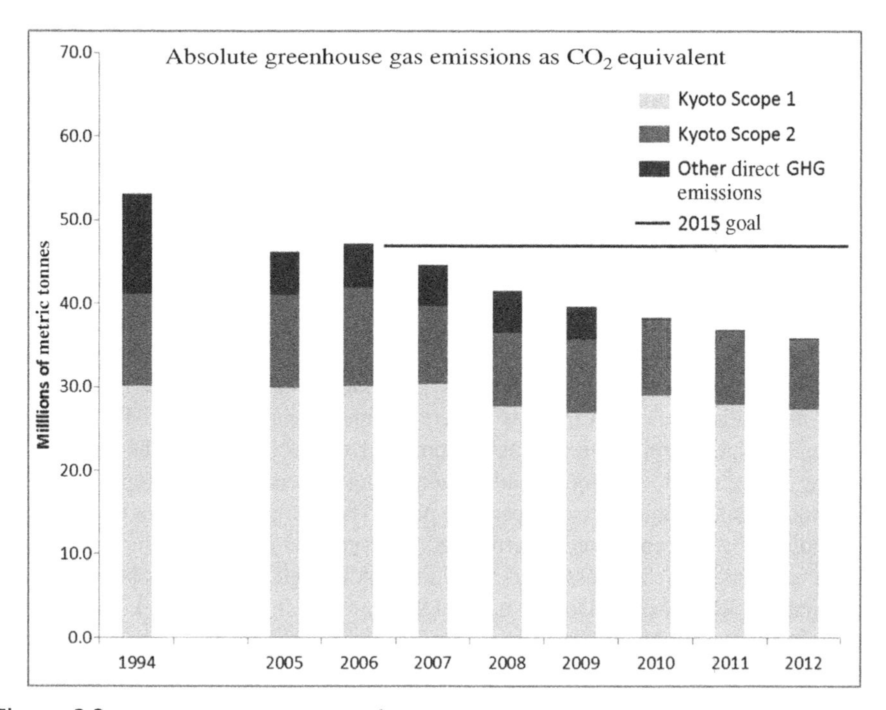

Figure 2.2. Dow's GHG emissions performance 2005–2012. (*Source:* Courtesy of The Dow Chemical Company.)

involved in the company's energy efficiency efforts an "all and none" paradox. Most EE&C members work on a number of issues including energy; energy might take up all their time some days and little or none on others. The amount of time spent on energy is driven partly by the EE&C team's regular meeting and reporting cycles, and partly by specific actions undertaken at given plants or sites in response to operational changes or project investments.

HOW THE EE&C PROGRAM OPERATES

A note on Dow's terminology for its operating units helps in understanding the EE&C operation. Because of the large-scale manufacturing of Dow, and the many different chemical products involved, its manufacturing centers are termed "sites," within which it operates several individual "plants." A plant is typically defined by its product; so a site might hold an olefin plant, a chlorine plant, a power plant, and others. The Texas site, for example, contains some 60 individual plants of the 310 total U.S. plants Dow operates.

Because 95% of Dow's energy use flows through its 14 largest sites, the 14 site leaders play an important role, especially when it comes to energy efficiency opportunities that cut across different plants. For example, all the major sites have extensive steam systems with miles of piping; at the Texas site, one pipe moves steam 4 miles from the power plant to a remote production plant. Steam systems need regular maintenance, especially for steam traps, the valve-like devices that remove condensed steam from the line while keeping the live steam flowing. A site EE&C lead will typically develop a site-wide steam trap maintenance contract. Plants can opt out, but typically do not because of the cost-effectiveness of the larger contract and the operating efficiency benefits. Site leaders also play an important role in the company culture, building relationships with plant operators, providing them help and information, and leaning on them as needed to achieve energy goals.

As in most successful energy efficiency programs, energy efficiency team members faced challenges in gaining the active cooperation of production plant operators and other site-based staff. Getting participation in the company's data reporting system was not as big a challenge, in that Dow already possessed in most cases the metering and billing information, based on the way the energy business is structured. The company had long used a centralized reporting system, and was able to build the new energy reporting elements into it. More detailed monitoring systems and data reporting occur within each business unit and each operating plant. The energy metrics thus act as a high-level indicator with centralized information; finer-grained information is kept decentralized at individual business units and sites. Although the energy business is a functional unit like others in Dow, it has had the core data it needs to track performance at a high level. Setting aggressive high-level energy performance targets motivated business unit and plant site staff to look harder at their finer-grained operating data to look for ways to meet the targets. Companies with more numerous and disparate operations may have to go through a longer and more complex process to establish and achieve compliance with the data reporting system. Each company must seek the right balance between centralized and decentralized data.

However, Dow shares with other companies the challenge of persuading production managers to consider changes in operating practices and technologies. Production staff members are focused on product quality, production volume, and reliability of equipment and systems. Energy improvements, be they changes in operating or maintenance procedures or new technologies, pose potential risks to these iron-clad principles. Dow addresses the potential conflicting interests of the energy team and the production plants partly through its Tech Centers. Each business unit has a Tech Center, with experts on its particular production technology, and with one or more EE&C experts. These Tech Centers each have a director reporting to the business leadership. As a result, the Tech Centers create focused teams that not only develop technology and operating solutions but also build trust relationships that help the EE&C team gain the cooperation of the product staff. The involvement of Tech Center staff in developing, testing, verifying, and then instituting new practices and technologies is key in this respect, as is the highly interactive nature of the EE&C team. Regular conference calls, webinars, email exchanges, and site visits build trust and help spread technical information.

Dow's EE&C team is part of the larger Dow Sustainability Team. EE&C includes team members from each business unit and each large site, as illustrated in Figure 2.3. Having an EE&C team member at each large site allows coordinated and collaborative efforts among the many plants operated by different business units within a given site.

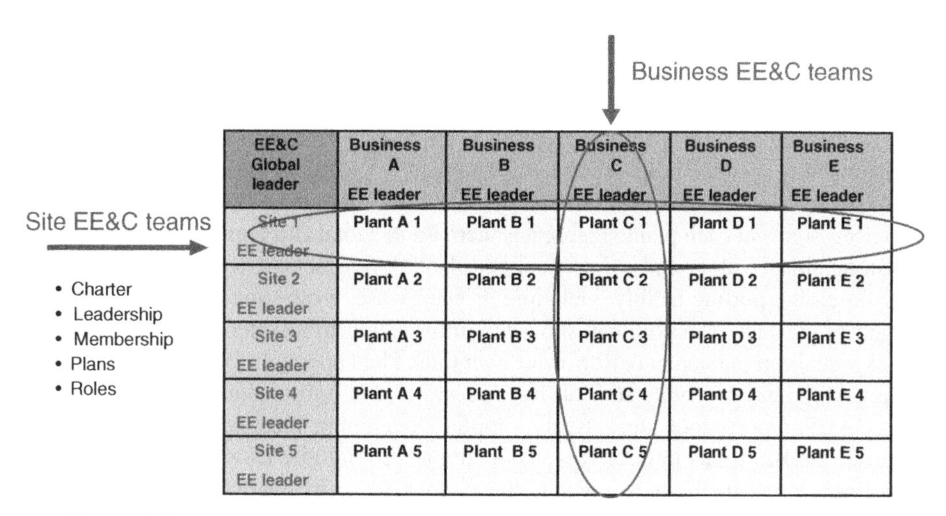

Figure 2.3. Dow's energy team. (*Source:* Courtesy of The Dow Chemical Company.)

Drilling deeper into the EE&C structure, Figure 2.4 shows how EE&C works within a given business unit. Each business's EE&C team leader is sponsored by the respective business Tech Center leader. The team leader works with people at the individual plants in the business. For example, each olefin plant has a person assigned to work with the olefin plant's EE&C team leader. The Tech Center's role is to provide engineering and technology support for the business, developing new specifications for production equipment and ancillary equipment, and working with plant operators to advance and sustain operating practices.

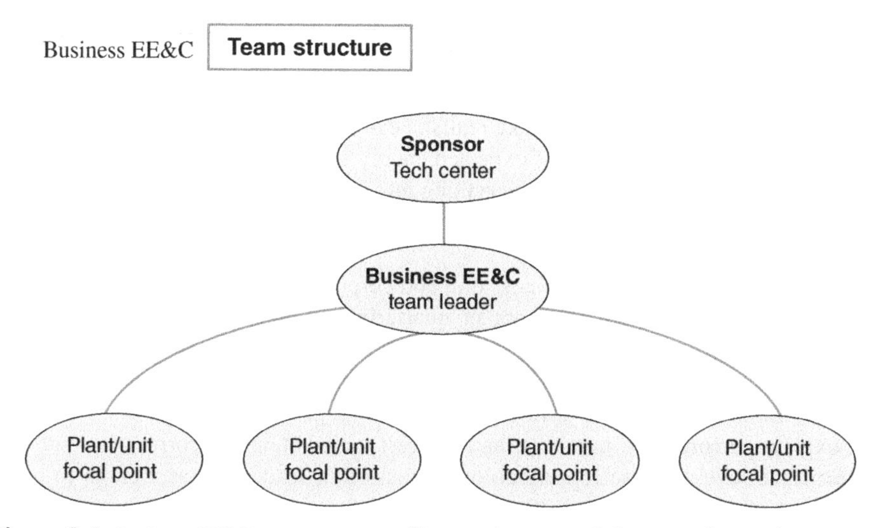

Figure 2.4. Business EE&C team structure. (*Source:* Courtesy of The Dow Chemical Company.)

DATA COLLECTION AND REPORTING: THE GLOBAL ASSET UTILIZATION REPORTING SYSTEM

A key element of Dow's EE&C program is the energy data collecting, reporting, and accountability system built around the company's global asset utilization reporting (GAUR) system. GAUR collects energy data from the numerous metering and sub-metering points at each plant, aggregates them up to a total Btu number, and reports the Btu total along with pounds of product per plant. A typical GAUR report consists of a row for each reporting facility, including its energy use, production, and energy use per pound of product. The columns show this information by quarter and year, allowing quick visual comparison of performance over time. Plant operators, business unit leaders, EE&C staff, and corporate staff all have access to this information.

GAUR reports look deceptively simple. Developing the calculations needed to produce the Btu totals in GAUR was not simple. A longtime Dow engineer painstakingly developed methods to convert the various flows of energy commodities, and intermediate energy flows in Dow's complicated operations, to a standardized set of Btu equivalents. Calculating GHG equivalents added another layer of computational complexity. Dow uses lots of primary (raw) energy, but also converts it into intermediate forms such as steam and electricity. Accounting for the conversion and distribution losses, and the energy content of various energy streams as they are delivered to each plant, involves complex calculations, which led Dow to institute a formal, documented procedure to ensure consistency and quality of information across the company.

Because natural gas is the dominant source in Dow's energy purchases, all energy forms consumed are translated into equivalent Btu of natural gas and adjusted to reflect real-world efficiencies and losses associated with energy conversion. This method allows staff across the company to use a common currency, in terms of how much natural gas has been or would have to be used to produce the energy consumed at a given plant. This methodology also reflects the ability of each energy form to do real work and not just theoretical work. It therefore estimates how many Btu of natural gas equivalent would be saved from a given energy-saving opportunity. This gives the EE&C team a consistent way to make realistic calculations of primary energy impacts and thus derive better estimates of expected energy cost savings. The company knows the price of natural gas, so turning everything into equivalent natural gas units makes it possible to compare energy performance and energy savings across plants, sites, and energy markets.

GAUR reports roll up to business unit and corporate levels, and form the basis of the energy and production numbers in quarterly sustainability reports, which are posted on the Dow website. However, GAUR numbers are also used to produce internal reports with much more detail. In addition to the basic energy and production numbers, the internal reports assess actual versus target performance, discuss issues that explain performance numbers, and suggest action items to correct any sub-par performance levels. These reports go to business unit leadership as well as to plant operators and site leaders, and often become the focus of extended discussions and follow-up actions.

HOW DOW ACHIEVES ENERGY SAVINGS

Dow achieves its energy-saving goals somewhat differently than many other companies. Dow is much more capital-intensive, and less labor-intensive, than most. One can drive around a Dow site for some time without seeing a human being, whereas in most companies people are everywhere. Dow also typically operates in a continuous processing mode—the chemical plants normally run 24-7. This reduces the opportunities to apply downtime shutdown strategies that some companies have used to great advantage to gain "free" energy savings.

Dow plant operators rely on the continuous flow of data they get on energy and other performance indicators to determine whether their operations are on track. This may occasionally include operating adjustments or maintenance actions, but most big efficiency improvements come from process technology changes. For example, Dow's Light Hydrocarbon (LHC)-7 plant at the Freeport Texas site won a 2008 American Chemistry Council (ACC) Responsible Care® Energy Efficiency Award in 2009 for replacement and upgrade of its ethylene furnace capacity. Dow replaced the existing 10 furnaces with 5 new, larger state-of-the-art furnaces with significantly higher energy efficiency. Interestingly, this investment was initiated for compliance with NO_x emission regulations, but the additional benefits in energy savings, yield improvements, corporate sustainability, and extended lifetime of the facility made the business case very strong. This investment has improved LHC-7 plant energy efficiency by more than 10%, producing energy savings of over 2,000,000 MBtu/year,[1] equivalent to the annual electricity use of more than 17,000 homes, and cutting CO_2 emissions by 105,000 metric tons annually.

Across the Atlantic in Antwerp, Belgium, Dow and its partner BASF in 2009 opened the world's first commercial-scale propylene oxide (PO) plant based on the innovative hydrogen peroxide to propylene oxide (HPPO) technology the two companies jointly developed. Producing up to 300,000 metric tons of PO per year for the polyurethane industry, the new Antwerp plant reduces energy use by 35% and wastewater by up to 80%.

FINDING THE FUNDING FOR ENERGY PROJECTS

While these projects represent major efficiency gains and provide several streams of business benefits, the capital for such projects is hard to come by, even for a large company like Dow, and especially in the current economy.

Dow's capital budget must serve many priorities, and meeting energy goals is just one of them. Dow applies various financial criteria to project investment opportunities, including discounted cash flow, net present value, and internal rate of return. Business units use their own analysis and prioritization methods, driven by competitive forces and the norms of their businesses. In many cases, investments that are seen as critical to maintaining or improving competitiveness are moved to the fore; this forces the EE&C

[1] 1 MBtu = one million Btu.

team to make a broader business case for energy-related investments, documenting the cobenefits in competitiveness terms. For example, a given efficiency investment might not meet a simple payback criterion, but if it keeps the company competitive in a specific market and avoids the loss of a specific customer or a certain share of the market, such considerations can help justify the investment.

In one sense, energy has long been a key issue for Dow, so energy efficiency competes relatively well when the numbers are compelling. Dow expects to ultimately factor in carbon price expectations in assessing energy projects. However, because it does not expect carbon prices to appear in energy markets for several years, except in Europe, carbon prices only have a material effect on larger capital projects, and these tend to be major process changes that involve more than energy as driving forces. Dow does have facilities that are regulated under the European Union's Emissions Trading System. The carbon price created by the trading system is an added cost that Dow incorporates into the evaluation of energy efficiency projects.

Dow has looked at trying to leverage outside capital through performance contracting, but it has found the performance contracting business model hard to match to its own operating models. Energy service companies typically require long-term agreements to make their financing structures work, and most corporations are not inclined to enter such long-term contracts. Performance contracts also tend to require long repayment terms to enable energy savings to exceed payments, because they must roll up all of the development costs, interest costs, and other "overhead" factors into debt service payments on top of direct capital costs. Moreover, in a complex industrial operation like Dow's, establishing the critical baseline energy use calculations on which savings are based can be too difficult and uncertain to build into long-term agreements. Dow uses companies like Johnson Controls, which has a large performance contracting business, but tends to use them for more specific technology and service contracts on a traditional sales or fee basis.

PURSUING OPERATIONAL IMPROVEMENTS VERSUS CAPITAL INVESTMENTS

The challenge of raising capital for efficiency investments turns the focus back to operating improvements. Dow works hard at this on an ongoing basis, but EE&C team members also know that the sheer scale of energy and product flows in a continuous processing environment limits the relative impact of operating changes. Using the engineering talent in the Tech Centers and other EE&C team members, Dow sometimes uses Pinch analysis, a methodology evolved in industrial circles in the last three decades. The Pinch method (described in more detail in Chapter 26) is designed to minimize energy consumption in chemical processes by calculating technically feasible energy targets, and then achieving them by optimizing heat recovery systems, energy supply methods, and process operating conditions. In a Pinch analysis, process data are typically represented as a set of energy flows as a function of heat load against temperature. These data are combined for all the energy flows in the plant to give composite curves, one for hot flow and one for cold flow. The point of closest approach between the hot and cold composite curves is called the Pinch temperature or Pinch point. By finding this point, various heat recovery or transfer solutions can be developed.

Dow engineers constantly use their creativity combined with the variety of energy and material streams in their plants to find new efficiency ideas. For example, at the Texas site, hydrogen occurs as a by-product in one process. EE&C staff have experimented with blending this available hydrogen with natural gas fuel in their power turbines, taking the mix up to the limits that the turbines will tolerate.

For its efforts, the EE&C team also seeks outside input to find new ways to save energy. It maintains a panel of outside energy and environmental experts to provide technical and strategic input. Dow also participates in national programs: the U.S. EPA's ENERGY STAR Partner Program, the U.S. Department of Energy's (DOE) Better Plants Partnership, the U.S. Council for Energy Efficient Manufacturing, the Alliance to Save Energy, and the American Council for Energy Efficient Economy (ACEEE), to name a few. Dow in turn supports various national and international energy initiatives such as the development of ISO 50001, the international standard for energy management systems.

SUPPLY CHAIN

Dow sits near the top of the energy supply chain: It purchases raw materials and transforms them into bulk materials, and in some business units produces products for business or consumer markets. In this context, there is little energy efficiency opportunity for Dow to exert with its suppliers: Its efficiency strategy focuses primarily on internal operations, where most of its footprint occurs. Dow has not conducted a formal analysis of its suppliers' energy or carbon footprint. In the energy arena, with 60% of the company's costs incurred as energy and most of that as natural gas, the supply footprint is relatively small compared to the footprint created by the conversion of that energy in Dow's processes. Dow does, however, focus on supply chain issues in its broader sustainability commitments. In its Sustainable Chemistry commitment, for example, the company works to reduce its impact on water resources, and seeks renewable sources of natural gas, such as landfill gas or sugarcane. The Antwerp HPPO plant described earlier, for example, saves even more water than energy on a percentage basis. The company is using landfill gas at its Dalton, GA, facility.

PRODUCTS AND SERVICES

Avoided emissions resulting from the use of Dow products is an important contribution to reduce the overall footprint of human activities. A life cycle assessment documented that emissions saved by Dow insulation products are about seven times greater than the company total direct and indirect Kyoto and non-Kyoto GHG emissions. Dow's insulation products, mostly sold under the STYROFOAM™ brand, are used to reduce energy use in homes and businesses. One square foot of 1 in. thick Dow STYROFOAM™ insulation will save 1 ton of CO_2 emissions over the average life of a house. Dow building insulation material saves hundreds of millions of metric tons of CO_2 each year. Dow actively supports programs and policies that encourage energy efficiency, in buildings, industry, transportation, and utilities, and continues to innovate efficient products and other solutions for its customers.

CONCLUSIONS

Dow is one of the few companies that has not only set and met a 10-year efficiency target but has also gone on to set another, more aggressive 10-year target. This "in it for the long haul" demonstration of commitment is one of the hallmarks of a successful sustainability program. Dow has also set and met its goals in a manufacturing environment where energy has long been a critical factor of production, so the company had been paying attention to energy issues since Herbert Dow pioneered bulk chemical manufacturing a century ago. This means energy savings were not as easily found as they might be in companies where energy has only recently become an issue. Dow has made effective use of its energy efficiency team from its energy business unit, reaching across business units and functional lines to create teams of experts and local site leaders to keep the efficiency program moving forward.

Key lessons learned from Dow's energy efficiency successes, include the importance of the following:

- Demonstrating corporate leadership commitment. In Dow's case, the CEO has made a very clear and public commitment to achieving the company's energy-saving goals.
- Building an appropriate organizational structure to institutionalize and lead the program. This includes cross-functional teams with clear lines of accountability.
- Establishing robust measurement, tracking, and reporting systems so that management can monitor progress and identify potential problems.
- Establishing clear goals and objectives, and then revising them over time as the initial targets are met.
- Communicating the importance of energy efficiency as a core company value to both internal and external stakeholders.
- Recognizing success by rewarding employees and business units for energy-saving innovations.

REFERENCE

1. Prindle, W. (2010) *From Shop Floor to Top Floor: Best Business Practices in Energy Efficiency*. Center for Climate and Energy Solutions. Available at http://www.c2es.org/energy-efficiency/corporate-energy-efficiency-report (last accessed February 27, 2014).

3

EASTMAN CHEMICAL'S ENERGY MANAGEMENT PROGRAM: WHEN GOOD IS NOT ENOUGH*

Sharon L. Nolen

Eastman Chemical Company, Kingsport, TN, USA

Eastman Chemical Company has a long history of promoting and implementing energy efficiency. It is widely known that energy costs are significant to the chemical industry. Due to cost pressures and a need to remain competitive, energy has been a common target for process and efficiency improvements for many years. In addition to cost pressures, Eastman has sustainability goals to create value through environmental stewardship, social responsibility, and economic growth. Energy efficiency addresses both these goals and cost concerns.

In 2010, the company set an aspirational goal to reduce energy intensity, defined as MBtu/kg of product, by 25% over a 10-year period, making a public commitment to the U.S. Department of Energy (DOE) Better Buildings, Better Plants program. Eastman recognized that changes were needed to take the energy program to a new level. Since that time, Eastman's energy program has seen significant advancements and received national recognition.

This chapter will discuss the actions taken by the company that proved significant in transforming the already good energy program to a great one that has received national

* Material in this chapter was presented at the Proceedings of the Thirty-Sixth Industrial Energy Technology Conference, New Orleans, LA, May 20–23, 2014 [1]. Adapted with permission.

Energy Management and Efficiency for the Process Industries, First Edition. Edited by Alan P. Rossiter and Beth P. Jones.

attention and recognition, including winning 2012 and 2013 ENERGY STAR® Partner of the Year awards and the 2014 ENERGY STAR Partner of the Year Sustained Excellence Award.[1] No other chemical company has ever won the award more than once and no others have won the highest award, Sustained Excellence.

INTRODUCTION

Many are familiar with Jim Collins' book, *Good to Great* [2]. Collins describes the transformation of 11 companies that, through various means, became great ones. At Eastman, our management inspires us by talking about the difference between good and great and encouraging us for the future. The same type of introspective evaluation can be applied to many things, including an energy program.

In 2010, by many measures, Eastman had a good energy program. There was a Corporate Energy Team and some site energy teams in place, an energy policy had existed for decades, and many energy projects had been completed. Some of these projects won American Chemistry Council (ACC) energy efficiency awards. In fact, by 2010, Eastman had won these awards for 17 consecutive years.

Eastman ran a concerted study in the early 2000s to generate project ideas for energy efficiency and capture these in a database. The study resulted in a long list of potential projects. The Department of Energy had conducted on-site training and assessments to increase skills and knowledge of DOE tools and identify additional opportunities.

There were also programs in place to improve insulation and lighting efficiency and to reduce steam leaks, with money allocated to address these issues at Eastman's two largest manufacturing sites. There was general management support, especially at these two sites that accounted for over 90% of the company's energy use. In particular, at the largest site in Kingsport, TN, the site manager took a long-term view and allocated millions of dollars over several years to improve metering of energy consumption with no direct evidence of immediate return. This foresight was instrumental in later years in providing valuable data.

However, in 2010, something changed. Eastman decided on an aspirational goal to inspire radical improvement and made a public pledge to the DOE Better Buildings, Better Plants Program to reduce energy intensity by 25% over 10 years with a baseline of 2008, the year Eastman became an ENERGY STAR Partner. Suddenly what had been good enough in the past was recognized as falling short of what was needed to meet this ambitious goal.

EXECUTIVE-LEVEL SUPPORT

It is widely recognized that executive-level support is needed for a successful energy program. As Eastman's program was revamped in 2010, an Executive Steering Team was formed. The team consists of 2 of the 11 Executive Team members and other senior

[1] The U.S. EPA's ENERGY STAR program is discussed in more detail in Chapter 7.

managers representing environmental, sustainability, and engineering organizations. The manager of the worldwide energy program meets quarterly with the Steering Team, which provides support and asks challenging questions. Two things in particular have resulted from this high level support: funding and data-based decision making.

Funding

In 2010, no capital money was allocated specifically for energy projects. While many good projects that saved energy and money had been identified, they often fell below the approval level as they competed with other projects, due to other priorities. When shown a list of unfunded projects and their projected returns, the Steering Team immediately agreed to fund $4.2M of energy projects. As projects were completed and more high return projects were identified, the budget grew to $8M per year within 2 years.

While the budget was essential for completing more projects, it also heightened interest in the energy program. Employees in manufacturing recognized an additional avenue for funding projects and approached the Energy Team with ideas. The dynamics changed, with the members of the Energy Team now becoming invited guests rather than their effort being seen as an additional burden.

Data-Based Decisions

Prior to 2010, most of the work dealing with energy efficiency was done at the site level. As the Executive Team members became increasingly involved, they challenged the Energy Team to develop quantitative information to identify the major factors affecting the energy intensity measure, assess the magnitude of each, and evaluate progress toward the goal. Intensive data analysis resulted in a tool that was used to set priorities for funding and to quantify the gap between current plans and the energy intensity goal. One of the significant learnings from this work was that repairing steam leaks played a much bigger role than expected. The value of this analysis was proven and the magnitude of each factor influencing energy intensity is now updated on an annual basis for the Steering Team's review to further guide the program.

Additional Linkage

Around this same time, the company's Sustainability Council was formed and the Steering Team became a subcouncil of this higher level team with additional Executive Team representation. Energy issues were frequent topics of discussion among the full council. Energy conservation was even added to an annual strategy discussion for the full Executive Team.

STRATEGY AND PRINCIPLES

While the energy program was based on basic strategies and principles, these were not written, agreed to, or generally understood. Strategy documents were therefore developed and then endorsed by the new Executive Steering Team. This has proved to be a vital communication tool for the energy program and a valuable check step as decisions were made. The strategy utilizes the following five key components in the program:

- Measures
- External resources
- Awareness
- Initiatives
- Projects

Each of these, including their development and implementation, will be discussed in more detail.

Guiding principles were developed as a reference to ensure that all decisions related to the energy program were, and remain, consistent with its intended direction. The principles are as follows:

- Ensure the accuracy of utility information
- Maximize operating efficiency
- Incorporate energy efficiency in capital investment decisions

As mentioned previously, site management at the Kingsport site proactively decided to add meters in strategic locations. This has proven extremely helpful in determining actual energy use. In addition, energy surveys check the accuracy of allocated costs and correct placement of meters. Recent modeling efforts have been able to predict energy use on a product level, with varying accuracy depending on the number of products and run days. The belief is that manufacturing managers are more than willing to make good energy decisions—they just need the right information to enable them to do so.

Rotating equipment is tested to ensure that each piece of equipment is operating at the best efficiency point on the operational curve. This equipment includes turbines, pumps, chillers, and compressors. Rotating equipment is tested annually and the results are compared with previous test results. Equipment that is not performing as designed is scheduled for maintenance to restore optimum performance.

It is recognized that the most opportunities for energy-efficient equipment and processes occur during the design stage rather than in later retrofits. According to the DOE's sourcebook for improving motor and drive system performance, electricity costs make up about 96% of the total life cycle cost of a motor, while capital costs (3%) and maintenance (1%) account for the rest [3]. As a result, energy efficiency considerations can have a large effect on the total ownership costs related to machine drives, including the use of optimally sized energy-efficient motors and proper motor maintenance.

Well Defined, Auditable Measure with Meaningful Goals

Eastman's primary goal was to reduce energy intensity by 25% over a 10-year period. Secondary goals were established as well, including maintaining minimum returns with the energy project portfolio and tracking spending. Although Eastman has tracked energy intensity for many years, the existing measure was found to be lacking in several respects:

Lack of Definition

While it was clear that an energy number was divided by a production mass, there were no clear definitions of what should be included in either value. The following definitions were developed:

- The energy measure includes any source of fuel or energy that is purchased by Eastman. Energy sources at Kingsport include coal, natural gas, and electricity. Energy produced from waste heat is not included in the measure. An example is steam that is generated from waste heat from chemical processes. While renewable energy is viewed positively from a sustainability standpoint, its use is irrelevant to the energy intensity measure.
- Production is tied to a specific hierarchical level of the business software used. Eastman's very large sites were found to provide information that was not comparable to other chemical companies that may have many small sites, which count production every time a product leaves one site for processing at another. Using a specific level within Eastman's business management system provided consistency in the measure across the corporation regardless of the size of the site.

DOE guidelines assisted in determining what should and should not be included for both energy and production.

Inconsistent Frequency

Corporate data were collected on an annual basis prior to 2010. It was recognized that the limited data available were not sufficient to track trends, provide any kind of analysis, or allow intervention should the energy intensity increase.

Opportunity for Errors

All data were originally generated by an individual at each site. There were many opportunities to make mistakes. Not only was the definition vague, but it was also unclear where the data were obtained. There were no clear processes for transitioning to a new individual when jobs changed. While there are a few exceptions at some of the smaller sites, the majority of the data for the energy intensity measure is now obtained directly from online business software, thus minimizing the potential for error.

As sustainability has become a higher priority, germane to the success of a company, and influential to shareholders, it is increasingly important that the measure is sound. Accordingly, Eastman's energy intensity measure has been subject to internal auditing and found to be satisfactory. It is now reported externally in Eastman's Sustainability Report [4].

Use of External Resources

Many resources are available to assist companies in developing a new program or enhancing an existing one. An excellent resource is the U.S. EPA ENERGY STAR Guidelines for Energy Management [5]. Following the restructuring of the energy program in 2010, Eastman utilized the guidelines to identify gaps in the existing program. In working to address the gaps, Eastman found that other ENERGY STAR partners were often willing to share ideas through benchmarking and by providing best practices on the ENERGY STAR website and at ENERGY STAR meetings and webinars.

In addition, ENERGY STAR provided both a mentor (an energy manager at another company) and a technical advisor; both proved very helpful. Each reviewed the existing worldwide energy program and provided specific suggestions for improvement. ENERGY STAR and DOE hold meetings where companies share information through both formal presentations and networking opportunities. In addition, DOE provides onsite training and assessments. Eastman takes advantage of all of these opportunities, recognizing that the value far exceeds the costs.

Employee Awareness

While energy efforts previously focused on completing projects and involved people in procurement, engineering, maintenance, manufacturing, and technology, it was recognized that additional engagement and awareness could yield benefits. Resources were available to assist in this effort. ENERGY STAR is an excellent resource for employee awareness campaigns. Posters, brochures for energy efficiency at home and at work, children's activity books, and displays are available for the asking for ENERGY STAR partners. ENERGY STAR has the benefit of being a well-recognized and positively perceived brand that instantly garners favorable attention when used in company awareness campaigns.

Eastman learned that other companies held employee energy fairs to promote energy awareness at home. After visiting another company's fair, the value became apparent and the first annual energy fair was held in 2011 at the company's largest manufacturing site, colocated with the corporate headquarters. The fair used ENERGY STAR resources with booths manned by local utilities and retail stores showcasing home energy-efficiency products. This has truly been a win–win with utilities and local companies now readily participating as they also receive value from opportunities to proactively provide information to their customers.

Eastman also learned that other companies have Green Teams geared toward sharing information with employees who have personal interests in preserving the environment. ENERGY STAR provides a Green Team Checklist [6] that provides a framework for development of these teams. At Eastman, anyone who is interested can join a Green Team. Monthly newsletters, which are appropriate to share on company bulletin boards, are developed at the site level. These are e-mailed to Green Team members who are asked to share the information with their coworkers and post them within their work areas. Topics typically include energy efficiency ideas, local opportunities for recycling or other activities, and suggestions for family activities. Out of consideration for workload, communications to Green Team members are limited to twice per month and are typically sent only once per month. The site newsletters are shared among the developers at all sites. While some of the newsletter topics promote local activities, others are applicable to everyone, regardless of location.

Eastman utilized the ENERGY STAR Pledge [7] to raise awareness. This pledge is designed for individuals to pledge to save energy at home. It includes installing more energy-efficient light bulbs or appliances or simply turning off lights when not needed. The pledge forms were distributed via e-mail to Green Team members, included in the company newsletter, and provided at manned tables in cafeterias and lobbies as a promotion during October, Energy Awareness Month. Receiving national recognition for the campaign as a Top 5 pledge driver in 2012 and 2013 contributed to employee interest and enthusiasm for the energy program.

While the majority of Eastman's energy use is consumed in manufacturing, office building energy use was targeted to demonstrate savings and to gain awareness and recognition. Eastman used ENERGY STAR's Portfolio Manager [8] to track energy use intensity. This program is simple to use and requires only a limited number of inputs: area, number of occupants and computers, and monthly energy use. Eastman has initially focused on buildings outside the plant fence that receive an independent energy bill. Portfolio Manager provides a monthly score, a percentile, which indicates building performance compared with that of other similar buildings across the United States.

Work began in the small office building that houses the Worldwide Energy Manager and the energy engineers. A Portfolio Manager score of 75 or better is required for certification, demonstrating that the building performs in the top quartile of similar buildings. Knowing this, an initial score of 39 was discouraging. However, implementing projects and asking for employees' help to turn off unneeded lights and equipment proved worthwhile, with the building becoming an ENERGY STAR Certified Building in 2013 with a current Portfolio Manager score of 96 and a 57% energy reduction during that period. In total, Eastman received three ENERGY STAR building certifications in 2013, and occupants of other buildings are now asking to be involved. While some buildings are not independently metered, energy-saving tips can be shared and implemented, although certification will not be possible. Participation in ENERGY STAR's Battle of the Buildings provides additional opportunities for communications and reinforcement. An internal competition between two buildings of the same size at Eastman increased employee enthusiasm. In fact, significant improvements in the performance of these buildings resulted in two of the three being named number 9 and number 10 in the 2013 National Building Competition [9].

While these examples have focused on engaging general employees, it is also worthwhile to consider whether specific organizations should be targeted for participation in the energy program. For example, Marketing Communications has proven a valuable ally. They developed a campaign with a unifying graphic to promote the ENERGY STAR Partner of the Year win and have incorporated the news and logo in company communications. Production Planning agreed to consider energy costs as well as the cost for inventory after discussing the possibility that some products can be run more efficiently as a campaign rather than at a lower level to minimize inventory.

Even groups who were previously involved were found to have opportunities for greater contribution. A Process Efficiency Team was chartered in Eastman's Technology organization to look at existing chemical processes as if they were being designed from scratch to foster breakthrough ideas. Distillation was identified as a very energy-intensive process and a task force was assembled to develop a list of best practices. Finally, Technology developed "Energy Briefs," short descriptions of energy-saving ideas identified early in the project that would follow it through the design process for consideration.

In addition to improving employee awareness, further buy-in can be obtained by celebrating success. Team celebrations are held to recognize ACC award winners, employees in ENERGY STAR certified buildings, and energy team results. Letters signed by Executive Team members were personalized for each site energy manager recognizing their contributions to the award-winning energy program. Eastman also utilizes an existing performance award system to recognize individual's contributions through timely cash bonuses.

Energy Initiatives

Prior to the revitalization of the Energy Team, it was recognized that certain energy efficiency efforts were not related to manufacturing and could best be handled centrally rather than by each manufacturing area. These included replacing lighting, repairing leaks, and adding insulation. The two largest sites had programs in place for addressing these issues, with manufacturing personnel happy to let maintenance take the lead. While the programs were good, there were opportunities for improvement. The programs were limited to the two sites and sharing of best practices was limited. Several improvements were made.

Identification and Sharing of Best Practices

An excellent example of identifying best practices is related to steam leak repair. By analyzing steam leak repair data during a six sigma project to improve the plant steam system, one area of the Kingsport plant was found to have a minimal number of leaks compared with other areas. Several key learnings were identified from this area. When leaks occurred, an entire length of tracing was replaced rather than just repairing the leak. And, when tracing was replaced, the material of construction was changed from copper to stainless steel. Finally, a single maintenance coordinator was dedicated to leak repair. Over 10 years, this area decreased the number of leaks by 98%. These learning are now disseminated as best practices and are incorporated into the program.

Identification of Other Initiatives

While seeing the value of a centralized, standardized approach to leak repair, it was determined that there were opportunities for other initiatives to be managed the same way. The list was expanded to include steam traps, motors, and HVAC.

Evaluation

After identifying the full list of initiatives, a questionnaire was developed to assess the progress of each site in each area. The results serve to identify common areas of concern, needs for improvements, and best practices at individual sites, which can then be shared more broadly through the Worldwide Energy Team.

Energy Projects

A database of potential projects is updated as ideas are identified. The database is regularly mined and the best projects are selected as circumstances and energy prices change. Typical projects include upgrades to more energy efficient equipment and taking advantage of heat recovery opportunities. While initiatives are broad and applicable to many sites, the projects are more specific and limited to certain process areas. However, similar processes at other sites benefit from being aware of successful projects for possible future consideration.

LINKS WITH OTHER INITIATIVES

Energy projects often have tangential benefits. Highlighting these benefits and linking them to other corporate initiatives may gain additional support for the energy program. Safety is extremely important to Eastman, as indicated by its strong "ALL IN FOR SAFETY" program. Some energy projects improve safety. For example, changing to more efficient lighting provides improved light, making for safer working conditions. These upgraded lights also have to be changed less often, so employees will be working fewer hours at elevated heights to change bulbs. The strategy to do group relamping rather than replacing lights one at a time also links nicely with the company's efforts to improve productivity. The labor required for changing bulbs, including required permits, obtaining tools and supplies, and lock out/tag out for safe operation can be reduced by a factor of 6 for each light by replacing a group of lights at one time rather than waiting for each individual light to fail.

In addition, at least for Eastman, linking with the fairly new Sustainability organization is a beneficial alliance. The Sustainability organization has stronger linkage to the businesses, while the energy program is more closely tied to manufacturing. The connections with and insights into the other organizations have proven valuable for both. The director of Sustainability is an advocate for the energy program and includes the Energy Program manager in team meetings and discussions.

CONSISTENCY

The gas shortages of the 1970s and the threat of an interrupted supply of petroleum captured Eastman's attention, along with that of many other companies. There was a significant response that changed the future of the company. Eastman invested in coal gasification, reducing its dependence on foreign oil. The company also responded by naming the first of a long line of energy managers and developing the first energy policy.

Over the next several decades, however, interest in and attention to energy efficiency varied with the availability and cost of energy. As we take a look back, it is clear that the previous energy programs had many good things to offer. At times, the energy manager was a full-time position reporting to site management. There were also periods when designated capital funding was available for energy conservation projects, but many of these good practices were not sustained consistently over time.

It is more likely that there will be consistency in the future as energy efficiency is no longer just about cost. It is a key element of Eastman's sustainability commitment. Hopefully, this lesson has been learned and the future program will be consistently strong, at least as strong as it is today.

GROWTH

While Eastman is proud to display its three ENERGY STAR Partner of the Year Awards, it is recognized that an effort must be made to continue to improve and grow. Eastman expects to grow the energy program with more acquired sites to be added. After all, as the company grows, so does the opportunity for energy efficiency. Eastman has a goal to add acquired manufacturing sites to the program within 3 years of acquisition. With the Solutia acquisition in 2012, the largest in Eastman's history, there was an opportunity to meld the two energy programs, and this resulted in an overall better program. While Solutia had only a limited number of sites in their program, others were eager to come on board when they recognized the value of the program. An integration strategy was developed to facilitate the transition and to be available for use with any future acquisitions.

Eastman is also considering expanding the energy program to include other natural resources such as water. Many program elements lend themselves to being more efficient with any natural resource, not just energy.

RESULTS

Eastman's efforts are paying off. Energy intensity has improved 8% since the baseline year of 2008. If energy intensity had remained the same since then, we would have spent $29M more in 2014 for energy based on current production and energy prices.

CONCLUSIONS

Beginning in 2010, Eastman deliberately took steps to transform an already good energy program into a great one. While energy efficiency in industry may be perceived as purely a technical challenge, a robust energy program must broadly seek support and input from many internal organizations with engagement of employees at all levels. Many of the items discussed in this chapter may not appear groundbreaking in scope and may even seem easy to implement. However, a great deal of effort and discipline is required to maintain focus and influence progress. And whether you choose to have a good program or a great one, almost any program can benefit in some way by learning from others.

ACKNOWLEDGMENTS

The author would like to acknowledge that the success of Eastman's corporate energy program is due to the efforts of many employees, representing all levels and all sites. The author would especially like to recognize Lisa Lambert, Tennessee Site Energy Coordinator, for her significant influence on the development of the corporate program.

REFERENCES

1. Nolen, S. (2014) Energy management: when good is not enough. *Proceedings of the Thirty-Sixth Industrial Energy Technology Conference*, New Orleans, LA, May 20–23, 2014.
2. Collins, J. (2001) *Good to Great*, HarperCollins, New York.
3. Lawrence Berkeley National Laboratory and U.S. Department of Energy Office of Energy Efficiency and Renewable Energy Industrial Technologies Program (2008) *Improving Motor and Drive System Performance: A Sourcebook for Industry*. National Renewable Energy Laboratory, Golden CO, p. 39. Available at http://www1.eere.energy.gov/manufacturing/tech_assistance/pdfs/motor.pdf (accessed June 3, 2014).
4. Eastman Chemical Company (2013) *Sustainability Report: Science and Sustainability—Positive Progress*. Eastman Chemical Company, pp. 8–10. Available at http://www.eastman.com/Literature_Center/Misc/2013ProgressReport.pdf (accessed June 3, 2014).
5. ENERGY STAR *The ENERGY STAR Guidelines for Energy Management*. U.S. Environmental Protection Agency. Available at http://www.energystar.gov/buildings/about-us/how-can-we-help-you/build-energy-program/guidelines (accessed June 3, 2014).
6. ENERGY STAR *ENERGY STAR® Green Team Checklist*. U.S. Environmental Protection Agency. Available at http://www.energystar.gov/ia/business/challenge/bygtw/Green_team_checklist_FINAL_4.pdf (accessed June 3, 2014).
7. ENERGY STAR *Pledge to Save Energy*. U.S. Environmental Protection Agency. Available at https://www.energystar.gov/campaign/takeThePledge (accessed June 2, 2014).
8. ENERGY STAR *Use Portfolio Manager*. U.S. Environmental Protection Agency. Available at http://www.energystar.gov/buildings/facility-owners-and-managers/existing-buildings/use-portfolio-manager (accessed June 2, 2014).
9. ENERGY STAR *Battle of the Buildings: 2013 Wrap-up Report on Trends & Best Practices*. U.S. Environmental Protection Agency, p. 4. Available at http://www.energystar.gov/buildings/sites/default/uploads/tools/2013_NBC_report_FINAL.pdf?63bd-346a (accessed June 3, 2014).

4

GENERAL MILLS' ENERGY MANAGEMENT SUCCESS STORY

Graham Thorsteinson

General Mills, Atlanta, GA, USA

OVERVIEW

General Mills has grown its energy program from the responsibility of one corporate manager to a thriving program with dedicated plant-based energy engineers. This new energy team has provided a competitive advantage for General Mills and continues to grow rapidly. General Mills has

- delivered $6.7 million in annual energy savings in the Cereal Division in 2013 and 2014,
- reduced energy use by 6% on a per pound of product basis in the Cereal Division for 2013,
- developed a plan to deliver more than $20 million of annual savings in all divisions,
- executed hundreds of energy projects,
- developed an internal proprietary energy continuous improvement (CI) process and tools,

Energy Management and Efficiency for the Process Industries, First Edition. Edited by Alan P. Rossiter and Beth P. Jones.

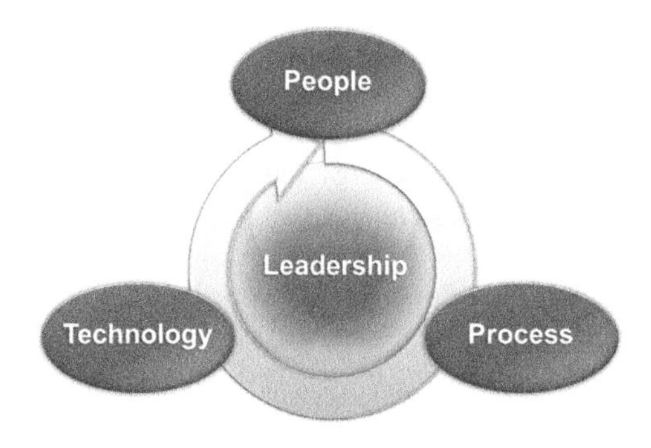

Figure 4.1. General Mills' energy program strategy: Strong leadership drives results with the right people, a focused process, and an innovative technology.

- won the 2012 and 2013 Association of Energy Engineers International Young Energy Professional of the Year awards, and
- lowered energy use by 29% on a per pound basis in the Georgia plant from 2008 to 2013, with an additional $5 million in savings.

How did the program grow so fast? The short answer is results. Expanding on this question offers many interesting insights into starting a successful energy program. As Figure 4.1 illustrates, when strong leadership interacts with the right people, a focused process, and innovative technology, extraordinary results occur. This chapter will discuss meaningful strategies for each area to develop a successful energy management program.

BACKGROUND

General Mills, one of the world's leading food companies, operates in more than 100 countries and 6 continents. With more than 41,000 employees, General Mills produces 100+ food and snack brands, including Cheerios, Yoplait, Nature Valley, FiberOne, Green Giant and more, available to consumers around the world.

General Mills' goal is to continually reduce the company's environmental footprint from agriculture practices to global operations, particularly focusing on reducing natural resource consumption and sustainably sourcing the raw materials used for General Mills' products.

With rising energy prices in a mature, cost-competitive food industry, the need to aggressively reduce energy usage has increased dramatically. General Mills continues to show itself as a global manufacturing leader as it delivers significant innovation in the energy space.

COMMON CHALLENGES WITH ENERGY MANAGEMENT PROGRAMS

While plant-based people usually have accountability for the energy budget, it is generally a small percentage of their overall objectives. Others in the plant may support energy reduction with a small percentage of their time, and everyone's actions in the plant affect energy use. This resourcing model leads to no one's objectives truly being tied to energy performance and very slow progress. Some prior gains are even lost with no leadership to sustain the short-term results of focused energy reduction initiatives. This lack of plant energy resourcing makes it very difficult for the corporate energy manager to make consistent progress.

For corporate energy managers without plant-based energy engineers, pushing new energy practices on the plant is like pushing a rope uphill. Some will gain traction, but many will not. If only one person is responsible for reducing energy in dozens of plants, progress will be slow. The scope is far too large to understand the details of the energy usage in each plant to the point of driving sustained reductions. Data collection for reporting and energy contracts can fill the time very quickly.

The corporate energy manager is forced to pay for numerous energy audits at the plants. Without a plant owner, the list of ideas generated will never be executed. The energy manager can fund capital projects to be executed by external engineers, which will deliver great short-term results. However, without a plant owner, the results will not be sustained. The energy manager will be in a cycle of doing the same improvements every couple of years. The other issue with external energy audits is that they generally focus on lighting, (heating, ventilating, and air conditioning) HVAC, variable frequency drives, and hot water. This covers the majority of energy spend for a commercial building, but a small percentage of the load for an industrial user. Most of the energy is used in the process, and no one understands the process better than internal resources. When it comes to supplementing an energy program with external engineering, General Mills has found more value in hiring specialists that are deep technically in specific equipment areas as opposed to generalists that tell trained energy engineers what they already know.

Energy inefficiency is a loss that is not highly visible, and the energy bill is paid each month without question or analysis. Plant personnel typically think that the plant uses only the amount of energy that it needs to make product, and that their actions do not significantly affect the use. Utility buildings (steam, compressed air, refrigeration, hot water, etc.) have high reliability, and operations start to take for granted that those utilities are available in infinite amounts for no cost. One of the most common examples is tapping into compressed air lines for inefficient uses such as cooling or conveying product. This leads to the last major issue with many energy management programs: They focus on the utility building and efficient generation. The resources in utility generation are typically less comfortable driving change in the operation. Generation is important, but most of the energy loss is in the use of the utilities in the plant's operation.

PEOPLE

General Mills has found dedicated plant energy engineers to be the solution to the above issues. The plant energy engineer should have the plant's energy performance prioritized on his or her yearly objectives. The engineer will oversee both the utility generation efficiency and operations use of the energy. It may seem overwhelming to hire energy resources. Food manufacturing is not nearly as energy intensive as many other industries, but these resources still deliver savings that are multiples of their salaries. Even 5% energy savings per year leads to fairly significant cost savings.

The good news is that these resources are probably already on the payroll somewhere. General Mills hired no new energy engineers. Instead, priorities for an engineer at each plant were realigned. The program targeted people who excelled as drivers of change, innovators, and technical leaders. In fact, most of the new energy engineers did not have any prior energy, environmental, or utility experience. It is far more important that they are passionate about finding inefficiencies and exploring innovative solutions. They should be motivated to navigate around roadblocks, especially the most common: "We have been doing it like this for 30 years." General Mills developed a strong training program along with the standardized tools that allowed new energy engineers to deliver savings quickly. External organizations, such as the Association of Energy Engineers, have training options as well.

As General Mills has a team of dedicated energy engineers, the corporate energy manager does not have to execute projects in dozens of plants, and instead can focus on strategy, breakthrough innovation, securing funding and support, and best practice standardization across different plants. General Mills has found that it is a best practice to share top–down program objectives with each energy engineer. Energy team members support each other to identify and share solutions. In summary, dedicated, trained energy resources that are managed as a team will deliver significant savings.

PROCESS

Assigning people to the problem is not enough: They need to follow a standardized reduction process to maximize results. There are very good standards available for energy management (e.g., ISO50001 (Chapter 6) and ENERGY STAR guidelines (Chapter 7)). However, the approach that General Mills used gained even more traction. It combined the best practices and concepts from those management programs with its existing continuous improvement program. Using the standard CI tools and language enabled anyone from an operator to a vice president to understand the process.

Like all programs, General Mills' process begins with corporate and plant commitment to energy reduction. General Mills will not begin the focused energy reduction process until the plant is committed to the following:

- Providing a dedicated site resource to lead the effort
- Allocating dollars for innovative energy tests, submetering, and energy efficiency projects
- Placing energy reduction on all leadership's objectives

After a plant has demonstrated the above prerequisites, they will follow a systematic process to

- develop plant energy allocation of usage for every unit operation,
- develop technical solutions for significant energy users,
- execute improvements, and
- sustain reductions through real time energy monitoring.

The plant energy allocation exercise accounts for every Btu being consumed. For example, on the electrical side, the tool would include energy usage for lighting, compressed air, refrigeration, pumps and fans, processing motors broken out by line, and HVAC. On the thermal side, gas usage would be broken out into HVAC, hot water, and each processing unit's steam or gas usage. Within these areas, the tools contain more than a thousand subpoints for unit operations that are either measured with a meter or estimated based on nameplate data.

The energy allocation exercise clearly identifies the areas of high energy consumption. General Mills developed its own improvement tools for high usage areas. Examples include boiler and steam systems, refrigeration systems, compressed air, industrial drying, lighting, motors, pumps, fans, and HVAC. Each tool allows the energy engineer to input data from his or her system, controls, and operation. The tools automatically identify key opportunity areas and even calculate savings for each area. This makes it easy for a plant to prioritize its largest opportunity losses without having to fully engineer each idea. The tools were built on known solutions and best practices from other plants.

These tools allow best practices to be easily spread from plant to plant and the energy team to track progress on the execution of all known improvements. Having defined the energy opportunity with these tools, the energy engineer puts together a multiyear execution improvement plan and leads the site in the execution of this plan.

The last step of the process is to sustain savings already realized. The goal of this step is to move away from only looking at monthly utility bills to managing energy in real time. However, monitoring energy usage data without overlaying production data provides little clarity on how the plant is actually performing, since production mix has the largest impact on usage. To solve this issue, General Mills developed an energy reporting solution that identifies a target energy usage based on real-time production data for each meter (from submeter to billing meter). This allows us to calculate volume and weather-independent energy reduction targets for each plant. The target helps the operation view "energy as an ingredient" to the pounds of each product produced: Material inputs and energy yield a useful product. Using this language allows operations to manage energy usage in a similar manner to its focus on ingredient waste.

These daily management systems alarm if a unit operation is operating outside its optimized energy efficiency. They also engage production operators in energy management, allowing energy performance for each unit operation to be summarized quickly at shift production meetings. Operator engagement is critical to sustaining energy success. The operators can also have the most innovative energy savings ideas, since they know

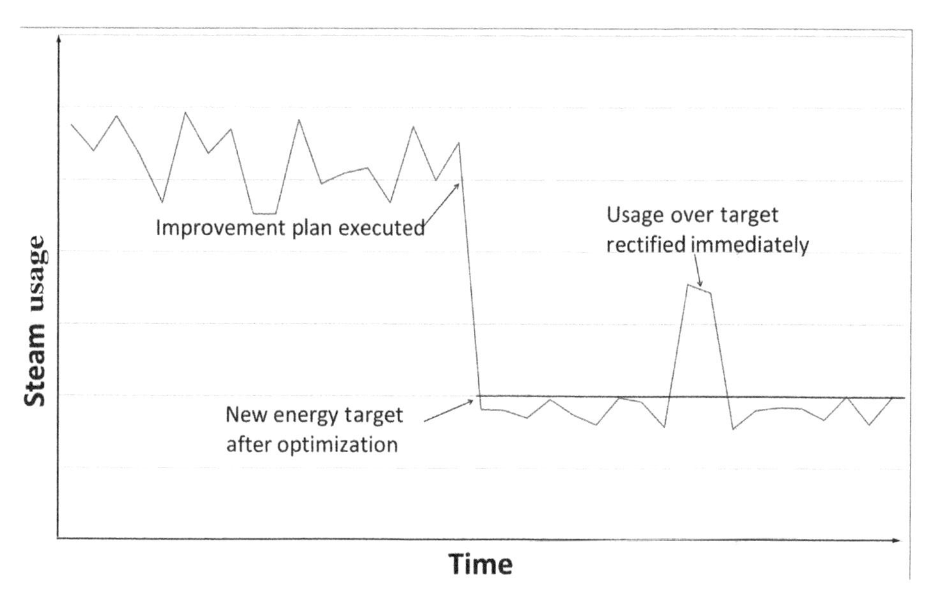

Figure 4.2. The steam usage on an industrial dryer illustrates the process working. Improvements executed from loss tools for large energy users drastically decreased usage. Submetering was then utilized to sustain savings.

the operation very well. The energy engineer needs to excel at building strong relationships with operators in addition to seeking engagement with senior leaders to develop holistic support of the program.

Figure 4.2 shows the result of this energy reduction process used successfully on an industrial dryer's steam usage. The General Mills dryer improvement plan was developed and executed. These improvement plans are holistic and include all four M's (Man, Machine, Material, and Method). In other words, the maintenance program, operation, operator training, new equipment, and controls were all optimized for energy efficiency. The steam meter for that individual unit operation (not the header meter) verified the savings (Figure 4.2). A new energy baseline for this unit operation was established for each product. Then the dryer's usage was monitored and the focus was sustained, driving almost $100,000 in savings.

TECHNOLOGY

The formalized structure of driving known solutions from tools does not limit innovation but instead encourages it. It prevents time waste resulting from multiple engineers investigating solutions to the same problem. At General Mills, there is an annual innovation pipeline meeting, where energy engineers brainstorm new and creative ideas, controls, and technologies. Even the fundamental design of the processing systems and products is challenged. Engineers are encouraged to benchmark externally and to be aware of new transformational technologies where applicable. However, some energy programs depend excessively on external innovation. General Mills funds innovative internal solutions for identified energy losses.

General Mills uses a rigorous process to prioritize these innovative ideas. Each plant is assigned an idea to investigate along with appropriate funding and resources for the investigation. There are regular meetings to check on innovation pipeline progress. If an idea is successful, the solution and calculations are added to the official improvement tools, which add the idea to the execution responsibility of all the other energy engineers in similar operations. The value of a new idea can be multiplied by implementing it in several similar operations within the company. To ensure that efforts are not being wasted on redundant reduction ideas, General Mills focuses its time and money into innovative ways of optimizing one pilot unit's operation. Once it is determined which solutions work and which do not, the successful unit's optimized example can be replicated many times. In the past, different plants would have different controls for similar unit operations and utilities. Now, with focus on developing the one best way, the optimal solution can be deployed to all plants. This innovate/replicate strategy is the engine that allows the energy team to continue to drive savings in multiples of their salaries, allowing the company to continue to invest in the team.

FUTURE CHALLENGES

Even after the energy management program's people, process, and technology are aligned for success, there are still potential roadblocks to a successful program. Capital funding is an important concern for ongoing success in energy reduction. Many energy projects' cost savings fall short of the corporate hurdle rate of return on investment (ROI). For example, with current low natural gas prices, advanced low-grade heat recovery solutions are less attractive than they were several years ago. However, with rising energy costs, it is important to reevaluate nonfunded initiatives every year. General Mills updates energy costs in its energy loss tools in order to identify past ideas that are worthy of funding with new energy rates. Because of this project log, the pipeline of future projects waiting for energy inflation is strong.

It is also important to understand how the rate of return is calculated for the company's projects. Many times, accelerated depreciation, rebates, energy inflation, and incentives are not properly considered and cause the calculated ROI to be lower than the real return. Other ROI calculators may limit the savings to 10 years despite the much longer useful asset life of many energy solutions, such as combined heat and power. It will be challenging to change the ROI calculator for the company. However, these are important added benefits of a more accurate calculation, and the finance team needs to be informed about them, especially for projects that are just short of funding requirements. Another added benefit to energy projects is that the confidence level of delivering savings is very high compared to new products or other cost-saving ideas. Because of this confidence and in order to reach corporate sustainability goals, General Mills continues to prioritize energy projects.

In addition to funding energy projects, educating employees at all levels of the company about critical energy losses is another opportunity area for most companies. It is important to build energy training for operators, so that they truly understand how their daily decisions impact costs. The maintenance organization also influences many energy losses such as those due to poor insulation, air leaks, and steam trap failures. These items can take lower priority during shutdowns, so ensuring that the organization is properly incentivized to support energy reduction is critical. In every plant, some operators and mechanics are more passionate about energy reduction than others. Finding ways to involve them as informal leaders on the floor will pay big dividends.

The new product research and development team is another important organization to educate on energy reduction. This team develops the new products that the plants will run for years. Energy-intensive practices are easier to build into new processes and practices, but more challenging to change once the product identity is finalized. Regular energy meetings can engage and energize the new product team.

Educating the project engineering community is also essential, so that they install new capital projects with the latest energy standards. It is common to see a new production line installed with an energy-intensive application that was just redesigned on the line directly next to it. This forces the energy team to find funding to improve a new, nondepreciated asset. Submetering is very inexpensive for a new project and should be included at the time of installation, as it is harder to justify later. To overcome these challenges, General Mills has built evaluation of critical energy losses into their standard project engineering tools.

SUMMARY

- The return on investment of resourcing energy management is significant.
- A focused process will drive sustainable results.
- Innovation will fuel the people and the process.

A successful energy management program requires all three aspects to be successful. Leadership at the plant and corporate levels turn this model into results. As General Mills has found, when the energy program starts saving millions of dollars, momentum will drive program growth quickly.

5

ENERGY BENCHMARKING

Mark Eggleston

Phillip Townsend Associates, Houston, TX, USA

WHAT IS BENCHMARKING?

Benchmarking means the comparison of different types of assets or practices to find the most effective or economical. Many different types of analyses have been called benchmarking, but standard practice has evolved based on statistical comparison of actual operating data, not estimates.

One use of the term benchmarking has been to describe a study of business operating practices, sometimes even comparing different industries. Much can be learned, for instance, in comparing high-level parameters such as inventory-to-sales ratio, delivery time, or customer satisfaction in very different businesses. Often these studies are more anecdotal than statistical, but they can show differences between industries.

Another use of the term benchmarking has been to describe the use of estimates and public assessments of operating company details. For instance, a given type of plant may be estimated to use eight shift operators in the Texas Gulf Coast at a given cost per worker. The same type of plant may be estimated to have eight shift operators in Europe with a different cost per worker and a shorter work year. More operators at lower cost per worker may be assumed in Middle East or in Asia, and so forth. The

Energy Management and Efficiency for the Process Industries, First Edition. Edited by Alan P. Rossiter and Beth P. Jones.

concern with estimating such factors of cost and operations is accuracy. With insufficient accuracy, benchmark estimates do not provide reliable information for business decisions.

Some benchmark assessments use public data with widely varying base years and very high-level comparisons. Sometimes estimates as described above can be combined with old and new public data, gathered from individual operating companies or even from the public press. Such studies do not provide a strong basis for business decisions. However, comparing data from within a given industry can provide producers in that industry with valuable and "real" information to drive results and improve competitive position, without compromising confidentiality.

Confidentiality

In the past, side-by-side analysis of operating industrial companies with two or three very similar plants was done by comparing technical and economic details. These types of studies can be very useful and show precise differences, but the issue of confidentiality can arise. Government competition authorities can become concerned if direct competitors in a given region are comparing operating data. These types of studies were phased out in Europe during the 1990s for this reason, and are seldom seen anymore.

Now antitrust guidelines are similar throughout the developed world, and often local joint ventures of Western companies are still expected to comply with American or European competition law. Current practice usually requires that at least four companies be included in any comparison group, to avoid revealing any one company's data points. The thinking behind this guideline is that any operating company will first extract their own data from the comparison set, leaving only three companies' data, but it is difficult to break the data down further. In contrast, if only two or three companies are included in a comparison data set, once the company's own data are extracted the remaining data can be broken down to reveal exact data points of their competitors.

Confidential side-by-side comparison of actual operating data (not estimates) following precise definitions and using validated input data from at least four similar company plants, is the current best practice for industrial benchmarking. Results are usually reported back to individual producers as a set of averages and statistical measures, compared to that producer's own data. Such comparisons, especially when the identity of the competitors can be revealed, can provide a strong basis for key business decisions and ongoing improvement efforts.

Several for-profit providers have established well-respected and heavily subscribed benchmarking services for different industries around the world. Energy benchmarking experience in several key industry sectors is discussed at the end of this chapter.

Data Sensitivity

Different types of operating data may give rise to varying levels of confidentiality concerns. Comparison of pricing, capital investment plans, and future operating plans are obviously of the highest concern to government competition authorities, and the strictest standards apply.

Business confidentiality concerns always exist between competitors, and economic data such as fixed and variable costs, margins, staffing, and operating practices can be quite sensitive. Even when complying with antitrust guidelines, usual industry practice precludes benchmark studies that reveal the exact data of any competitor. If producers feel that their input data will not be properly protected, they will not consent to join a benchmark study. Proper confidentiality protection is usually insisted upon as a condition of participation, and if confidentiality is compromised, participation in the study will evaporate. Typically, no type of general publication of actual benchmark results to the public will be allowed by the participants.

Energy benchmarking may be considered less sensitive by many companies, especially where the parameters considered consist of a limited set- for instance, only fossil fuel, electricity, and steam consumption. Government authorities in various regions have demanded transparency on energy consumption figures to drive improvement efforts since the late twentieth century. In fact, energy consumption figures for various processes have been published publicly by government and industry groups, as guidelines for implementing control programs.

WHY BENCHMARK?

The most important reason for benchmarking is to drive business improvement. A reliable benchmark study can show a producer's standing versus competitors, and indicate areas where further efforts will pay off. Every business is faced with limited resources, which must be deployed in the best possible way.

Good business practice usually includes annual budgeting, including the costs of improvement initiatives that should be implemented during the current business cycle. A Strategic Improvement Plan can include longer term targets, typically on a 3- or 5-year horizon, with reviews on some frequency—perhaps quarterly or annually. The most reliable way to set business performance goals includes both the assessment of current performance and reliable comparisons with the leading competitors either on a regional or a global basis.

Reliable, verified benchmark results, for both the producer's own business and the important competitors, provide the most reliable basis for establishing the Improvement Plan.

Precise benchmark results for discrete parameters, along with a monetary gap calculation against a leadership standard, provide economic justification for improvement efforts. A good gap calculation can provide monetary justification for resource allocation or for capital investment. With a realistic estimate of resources and time required, these figures can be incorporated into the annual budget cycle and the longer term business plan.

Another reason for energy benchmarking can be a government mandate. Various government agencies around the world have been mandating energy benchmark programs to drive conservation and energy consumption improvement efforts. Although the simple economic justification of energy improvement projects should make the projects sufficiently attractive to producers, energy has become politicized globally for various reasons. Some governments consider that energy reductions, even beyond those dictated by normal economic returns, are an important goal for society. Mandated industrial energy benchmark programs are the result.

USE OF ENERGY BENCHMARKING TO DRIVE THE IMPROVEMENT PLAN

Energy consumption and costs are important parts of any industry benchmark in the energy-intensive industries. Such industries may consume 5 or 10% or more of total operating costs as energy inputs.

Even though both fixed and variable components exist for energy consumption, the most straightforward benchmark approach considers energy as a variable input. This simplification assumes that most plants in an industry are operating well above 50% capacity utilization, where energy consumption typically varies with production rate in a linear way. It is true that a typical production plant will still consume substantial energy even at very low capacity utilization. The most efficient energy use is usually when the plant is running at a very high rate, just as with many other operating parameters.

Typical energy performance indicators usually result in a simple measure such as gigajoule per ton for consumption figures, and dollars (or pounds or euro or yen) per ton for cost figures.

The Improvement Plan can indicate a certain improvement goal per year in gigajoule or in dollars per ton of product. Target consumptions can be established for each year, and then combined with estimated costs for the various energy components as inputs to a longer term Business Plan. Reductions in operating costs due to energy savings and the costs of capital improvements can be estimated and included in the Improvement Plan at the appropriate time.

Whether the energy benchmark parameters are derived either from a comprehensive benchmark study covering all aspects of a business or from a mandated government program, energy improvement goals should be derived. Especially if the energy performance of a given producer is poor, investigation of causes and improvement options should be pursued. The following are the typical areas for investigation:

- Modernization of a plant design if the current plant uses an earlier generation of technology
- Energy optimization if the plant was designed based on a previous lower energy cost scenario that is no longer relevant
- Straightforward improvement initiatives such as insulation, leak control, and so on
- Inattention to energy optimization in day-to-day operations

Investigating root causes and corrections as part of the Improvement Plan can often provide substantial savings in energy consumption and operating costs.

POSSIBLE BENCHMARKING METHODOLOGIES

Several types of benchmarking methodologies have been used:

- Global or regional surveys
- Best practice reports
- Energy audits

Of the above, energy audits of operating practices and installations are not properly considered a benchmark approach, even though they are included in various government-mandated energy benchmark programs, so they will not be considered here.

Energy Survey

A global or regional survey is the most rigorous approach to energy benchmarking and is always preferred when it can be properly implemented. A global survey would include production plants from anywhere in the world, of different ages and technologies. Since energy costs vary substantially around the world, good energy performance in one region may be average or worse performance in other places. Europe and Japan have had high energy prices for some decades, and typically plants in these regions have better energy performance than plants in lower energy cost regions. Additional energy savings equipment would have been installed, and careful operating procedures to minimize energy use would be in place.

A regional energy survey avoids somewhat the issue of historical energy pricing and so may be considered a fairer basis for comparisons in some situations. The opportunity to compare with the best global standard is of course lost.

The first step of a global or regional survey is to compile an appropriate listing of operating plants for that product. Sometimes a set of operating data will exist already, and this can be used as a starting point. In other cases, the plants will have to be contacted to encourage them to submit operating data. Unless all of the plants in the selected region are mandated to participate by a government authority, participation will be voluntary and some companies will elect not to participate. Old and small plants, which the owners know to be noncompetitive, often choose not to participate. This voluntary study may even exclude some regions that would be interesting but where the producers do not want to participate. Communication or other issues can also arise, especially if local authorities have already made public comparisons that discourage certain producers from participating. As a result, the plants included in a survey can often end up being mainly in certain regions, with only the larger and more modern plants represented. Technology differences, plant perimeter, and local factors may have to be considered, using sound normalization practices and engineering judgment.

Best Practice Reports

In certain types of government-mandated energy benchmark programs, too few plants may agree to participate to execute a confidential benchmark survey. In this situation, usually a special approach is needed to gather public and licensor information in order to make a reasonable assessment of the plant energy performance. Special guidelines must be agreed with the government authority. In these cases, a report on the best practices in the plants covered by the program is sometimes issued in lieu of a benchmarking study.

COMPARABLE TECHNOLOGY

Decisions may be required on comparable technology for similar processes. Typically, the minimum criteria are that processes to be compared should have similar raw materials, products, co-products, and by-products.

Even when the input and output streams are similar, issues in technology can arise and must be reconciled between the producers involved and with any regulating authority. Sometimes even if a co-product is present in some plants and not in others, if the stakeholders agree on a benchmark basis, the comparison can proceed.

For instance, if some plants make a simple series of products, and other similar plants have a more complex, specialized product slate, the more complex products may require additional energy inputs. Potential differences must be resolved.

In some cases, a generational jump in technology may have occurred, and the treatment methodology must be resolved. If an older energy-intensive process has become obsolete, those plants may insist on separate treatment or adjustment to compensate.

Other processes may by their nature be different in some way in nearly every implementation. These situations must be carefully handled if a reasonable benchmark is to be executed. Simply because the same product is produced in a number of plants, it does not necessarily follow that similar technologies are being utilized. For instance, some technologies may be more energy efficient, at higher capital cost. In other cases, an older plant may use a different configuration that makes technical comparisons of specific energy inappropriate, since a revamp to reduce specific energy is not practical.

ESTABLISHING THE PERIMETER

One of the most important steps for consistent benchmarking is establishing the plant perimeter. For a consistent result, process steps that are included in some plants but not in others should be excluded. Typical plant areas to be excluded could be raw material or feedstock preparation, utility areas, product packaging or downstream processing, and by-product or co-product capture or further processing. Every effort must be made to ensure that similar production facilities are compared on a like-for-like basis.

A good practice is to agree upon a high-level process sketch of the plant area to be benchmarked, perhaps by reviewing the sketch with a few major producers before the data collection instrument is distributed. Such a sketch would show what equipment is included inside the perimeter to be benchmarked, and what processes and operations are meant to be excluded.

Since feedstock and raw material qualities can vary based on the source, it may be necessary to exclude preliminary purification, grinding, heating or cooling operations, and so on to get the most consistent set of input conditions. In some cases, a minor change in conditions may not affect energy performance much and can be ignored. In a similar way, some kind of industry standard condition for product purity and conditions should be agreed upon.

By-product and co-product processes must be carefully considered, especially if different for the various participating plants. For instance, if some plants are liquefying or compressing a co-product for separate sale, these typically energy-intensive processes should probably be excluded since they are not present in all plants.

Utility production areas are often totally excluded, since some plants will have the areas included in their battery limits and other plants will use purchased or centrally supplied utilities. Usually the goal is to evaluate the energy performance of the product facility and not that of utility generation. For this reason, the main utility components of fossil fuel, steam, and electricity would be gathered separately. In some processes, chilled water or refrigeration and stream-to-stream heat transfer process should also be considered. Fossil fuel used as boiler fuel or as process feed or off-gas must be carefully checked in the data validation process to avoid double counting.

LOCAL FACTORS

Local factors often must be considered. However, the number of corrections or adjustments should be kept to a minimum to keep the benchmark analysis and result as simple as possible. For instance, climate factors such as extreme heat or cold can increase energy consumption. Unusual utility conditions, such as interruptions in supply, can be a factor. Poor economic conditions during the benchmark base period can result in low operating rates and thus inefficient energy performance. Raw material or other supply interruptions could cause starts and stops, or intermittent low rate operation, leading to inefficiencies.

The risk is in giving special consideration for every possible technology, product or local difference, which can result in great complexity of the benchmarking activity, obscuring the actual differences in plant performance. Benchmarking is always a trade-off of the effort consumed in executing the study, compared with the benefit to be derived in more complex analysis.

VALIDATION AND VERIFICATION OF INPUT DATA

A critical issue in consistent benchmarking is the definition of the input data, and then the validation and verification of the data during the collection process. Effort should be made to provide consistent and detailed definitions of every input variable, even if it seems self-explanatory. While producers are collecting the data, rapid response to questions on the definitions can expedite the data collection process and minimize errors. Simple error checks can be built into the data collection instrument.

Once the data are submitted, careful validation and verification of the inputs is required. The first series of checks are based upon examination of the data set for internal consistency. Do reasonable material and energy balances apply? Do steam conditions of temperature, pressure, and flow match with enthalpy and total energy figures?

Once a producer's data are shown to be internally consistent to a reasonable standard, it can be loaded into a database with the other producers, so that outlier checks can be executed. For instance, questions such as "Can you justify the high steam consumption reported, or is it possible that double counting has occurred?" or "Are you using a large number of steam turbines as rotating drivers instead of electric motors?" may be necessary.

Once the internal consistency and the database outlier checks have been executed successfully, a preliminary data publication can be made to the participating producers as a final check. Then any final comments and changes can be incorporated for the final report.

If for some reason, some input data in the database cannot be justified or explained by the producer, and the data are reliably regarded as incorrect, these data must be excluded until properly validated.

REPORTING OF RESULTS

Reporting of individual company energy consumptions has been used at times, for instance, to show the "best demonstrated performance." However, use of statistical tools to protect confidentiality of individual producer performance is strongly preferred.

If sufficient data exist in the database to protect confidentiality, display of energy benchmark results in quartiles can be useful. Sometimes a special approach is mandated by a government regulatory authority, but two standardized approaches are commonly used:

1. A simple grouping by the number of plants is sometimes used. For example, if there are 20 plants in the database, 5 plants would be represented in each quartile. This approach is suitable when all plants are of almost the same size, because no adjustment is made for very large or very small plants.
2. A more rigorous approach uses the weighted average and weighted standard deviation to calculate quartiles, with the weighting factor as tons of annual production (see Figure 5.1 and the statistical formulas in Table 5.1). This approach compensates for differing plant size.

ENERGY-INTENSIVE INDUSTRY BENCHMARKING EXPERIENCE

Metals

Steel industry processes are widely practiced in a standardized way, and much experience in energy benchmarking exists. Other metals are more variable with far

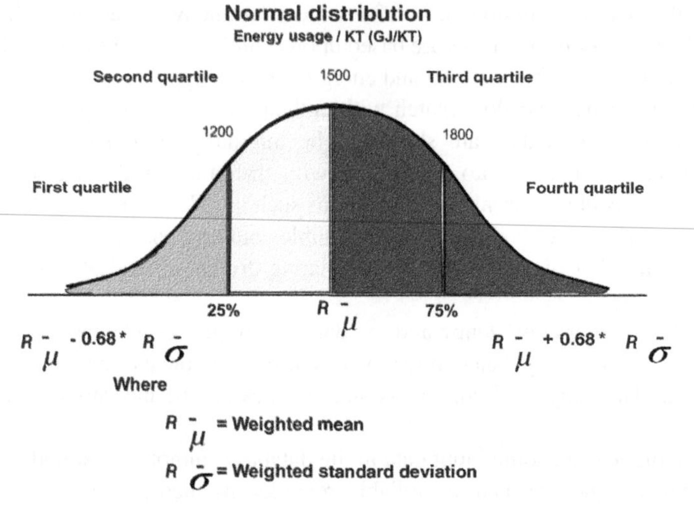

Figure 5.1. Weighted averages and weighted standard deviations compensates for differing plant sizes when calculating quartiles.

fewer facilities, and may be more dependent on the basic ore utilized. Benchmarking experience is therefore scattered.

Petroleum Refining

Energy benchmarking has been practiced for many years in global refineries. Nearly every refinery has different processes and facilities, so a factor-based approach has become well accepted, based on the basic crude distillation process. Much experience and data already exist.

TABLE 5.1. Statistical Formulas Used for Weighted Averages and Weighted Standard Deviations

Statistical Formulas

$\mathbf{R}_{\bar{\mu}}$ = Weighted mean

$$R_{\bar{\mu}} = \frac{\sum_{i=1}^{N} W_i^* R_i}{\sum_{i=1}^{N} W_i}$$

R_{σ} = Weighted standard deviation

$$R_{\sigma} = \sqrt{\frac{\sum_{i=1}^{N} W_i^* (R_i - R_{\bar{\mu}})^2}{\sum_{i=1}^{N} W_i}}$$

$R_{25\%} = R_{\bar{\mu}} - (0.68^* R_{\sigma})$

$R_{75\%} = R_{\bar{\mu}} + (0.68^* R_{\sigma})$

N = Number of plants in sample

W_i = Weighting factor of plant i

R_i = Metric ratio for plant i

Petrochemicals and Plastics

The more widely produced petrochemicals and plastics have been benchmarked for some years, and reliable databases exist for the more common products. Products made in fewer locations or using widely varied technologies have required more specialized analysis such as a Best Practice report.

Industrial Chemicals and Fertilizer

Many large volume industrial chemicals and fertilizers have been benchmarked reliably for some years and reliable databases exist. Similar to petrochemicals, some more specialized materials have required a Best Practice report.

Cement, Glass, and Pulp and Paper

These quite energy-intensive processes are standardized, at least in the commodity processes, and have been benchmarked reliably in different regional studies.

Food Products

Efforts to benchmark food products encounter process differences and a wide variety of products manufactured. Since food products are less energy intensive in general, fewer databases exist.

FUTURE DEVELOPMENTS

It can only be assumed that benchmarking for energy-intensive industrial processes will be more widely practiced in future. Energy consumption has been politicized, and many governments feel pressure to show their constituents that efforts are being made to improve and economize. Individual producers also feel pressure to compare their performance with others, in search of efficiencies and cost savings.

In an atmosphere where capital is tight, and energy is considered only as another economic input, energy conservation projects may have a different hurdle rate than capacity increase, environmental safety, and product quality projects. Sometimes during the final capital review for a new plant, reductions are made in energy-saving equipment, since these can be retrofitted later if their cost is justified. Energy benchmarking helps to provide economic justification for a design basis that ensures competitive performance. Mandated energy benchmark programs have grown, as governments attempt to manage demand for this scarce commodity. Benchmarking spurs improvements, which raise the energy efficiency and lower the production costs of the industries that participate.

6

ENERGY MANAGEMENT STANDARDS

Kathey Ferland,[1] Paul E. Scheihing,[2] and Graziella F. Siciliano[3]

[1]*Texas Industries of the Future, The University of Texas, Austin, TX, USA*
[2]*Advanced Manufacturing Office, U.S. Department of Energy, Washington, DC, USA*
[3]*Office of International Affairs, U.S. Department of Energy, Washington, DC, USA*

INTRODUCTION

ISO (International Organization for Standardization) is the principal organization that develops and publishes voluntary standards that impact specifications for products, services, and good practices globally. ISO 50001:2011, Energy Management Systems—Requirements with Guidance for Use, adopted in June 2011, is the international standard for energy management systems. There were national or regional standards for energy management prior to the adoption of ISO 50001 [1]. However, the adoption of ISO 50001 eliminated this patchwork of national or regional energy management standards, making it more straightforward for large companies with sites in many countries to consider implementation across their organization. This chapter will mainly focus on the ISO 50001 standard and related programs based on this ISO standard.

Energy Management and Efficiency for the Process Industries, First Edition. Edited by Alan P. Rossiter and Beth P. Jones.

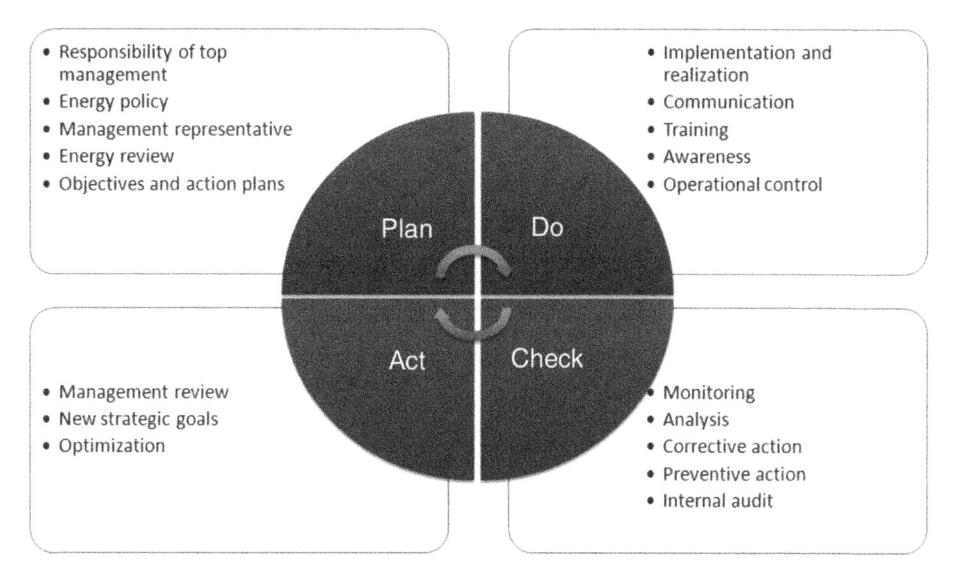

Figure 6.1. Plan–Do–Check–Act cycle. (Copyright © 2012, FW8100—CC-BY-SA-3.0 (http://creativecommons.org/licenses/by-sa/3.0)—via Wikimedia Commons, unaltered.)

ISO management systems have been implemented in the operations of many organizations since the mid-1980s with the publication of ISO 9001 for quality systems in 1986, followed by the publication of the ISO 14001 standard for environmental management systems in 1996 [2]. The ISO standards are voluntary; thus, the drivers for adoption can vary from standard to standard and from company to company.

ISO 50001, ISO 9001, and ISO 14001 are all based on the Plan–Do–Check–Act (PDCA) cycle, also known as the Deming cycle. Figure 6.1 illustrates the PDCA cycle in the context of energy management.

Figure 6.2 compares the number of certifications with ISO 9001 and 14001, based on years since standard publication. Factors cited in the growth in adoption of ISO 9001 include customer pressure, quality improvement, cost reduction, corporate image, and government requirement [2]. Even as early as 1991, it was noted that ISO 9001 had become a "de facto requirement of many firms" [2]. Six years after their publication, the adoption paths of ISO 9001 and 14001 diverged. With only 2 years of data, it is simply too early in the ISO 50001 adoption process to forecast the slope of the line. Will it take off, similar to a hockey stick (ISO 9001), or follow a more gradual slope (ISO 14001)?

Table 6.1 lists the number of certifications by geographical region for ISO 50001 based on January 2014 data from Reinhard Peglau, Senior Scientific Officer on Environmental Management in the German Federal Environment Agency, time series data (via personal communication with DOE EERE Advanced Manufacturing Office).

Germany clearly leads in the adoption of ISO 50001. Overall, adoption is being driven in large part by European Union countries, with those countries accounting for 85% of all certifications issued. As of January 2014, Peglau reported a total of more than

[1] Reinhard Peglau, Senior Scientific Officer on Environmental Management, German Federal Environment Agency, time series data (via personal communication with DOE EERE Advanced Manufacturing Office).

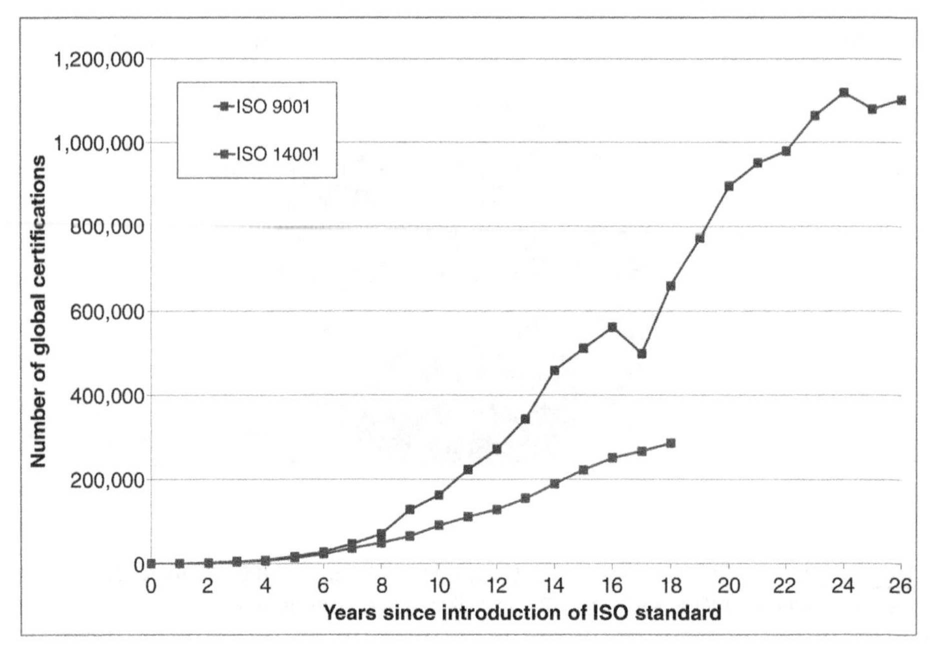

Figure 6.2. Number of ISO certifications by year after publication of standard.

2870 certifications worldwide to ISO 50001.[1] For a discussion of how various countries are encouraging the adoption of ISO 50001, see Section "International Context."

ISO MANAGEMENT SYSTEM BASICS

As already described, ISO 50001 is based on the PDCA cycle also found in the ISO 9001 and ISO 14001 management system standards. The three standards share a common framework with similar components, such as the requirements for document control, internal audits, and management review. This similarity gives organizations

TABLE 6.1. Global Certifications to ISO 50001 by Region as of January 2014

Region or Country	% of Certifications
North America	2
Africa	0
Central and South America	1
Europe (excluding Germany)	33
Germany	52
East Asia and Pacific	7
Central and South Asia	4
Middle East	1

Source: U.S. Department of Energy [3].

adopting the standard the option of integrating the ISO 50001 management system into an existing ISO management system, and early adopters have generally followed this path. However, the management systems can also be implemented separately. The question of whether and how to integrate ISO 50001 into existing management systems is one of the first decisions that an organization deciding to implement ISO 50001 must address.

The three ISO standards also share the following characteristics:

- They specify the *process* for accomplishing the goals of the organization. They do not provide specifications for the product or service itself. For example, an energy-efficient manufacturing process may produce an energy-inefficient product—the manufacturing of incandescent light bulbs.
- They do *not* include thresholds for *goals* or *targets* to be achieved, except that an organization commits to being in compliance with legal requirements in the subject addressed by the standard (i.e., energy, environment, or quality). This gets to the heart of the continual improvement process. Through the iterative process of the PDCA cycle, an organization identifies opportunities for improving its management system, which may be addressed immediately or in subsequent implementation cycles. Here, however, the similarity with the other standards ends, as ISO 50001 also requires that an organization continually improve the outcome of the energy management system, which is the resulting energy performance [4]. Improvement in the outcome of the management system is not required in ISO 9001 or ISO 14001.
- The three standards can be applied to organizations of *any size* and *in any sector*. For example, the energy management team required by ISO 50001 can be a team of 1 or 10, depending upon the complexity of the organization ([4], Section 3.10, p. 3). If the need for a sector-specific standard becomes apparent, sectors can develop standards more specific to their industry. For example, TS 16949 is a quality management standard that applies to suppliers in the automotive industry and is based on ISO 9001.
- Certification is implemented through a distributed system of independent accreditation bodies and auditors. This has led to one of the major criticisms of the ISO management standards: that of a perceived difference in the requirements of different accreditation bodies ([2], Appendix). This issue has been addressed directly in the United States by the development of testing requirements for ISO 50001 auditors, to be discussed later in this chapter.
- Organizations can choose to have a third party certify their performance to these management standards, using an independent accreditation body (already discussed). However, organizations can also choose to implement the standard and forego the costs of a third-party audit. In this case, an organization is said to conform to the standard.

The ISO 50001 standard contains a comparison of the sections of ISO 50001, ISO 9001, ISO 14001, and ISO 22001, a food safety management system standard ([4], Annex B). ISO announced in 2012 that it had completed work on Annex SL, which harmonizes the format of all ISO management system standards to be revised or published in the future. This harmonization will impact the structure, text, and terms and conditions [5].

ISO 50001

With this background on ISO management system standards, we now turn to the energy management standard itself. The purpose of the standard is to

> "enable organizations to establish the systems and processes necessary to improve energy performance, including energy efficiency, use and consumption. Implementation of this International Standard is intended to lead to reductions in greenhouse gas emissions and other related environmental impacts and energy cost through systematic management of energy" ([4], p. v).

As already described, ISO 50001 differs from ISO 9001 and ISO 14001 in one significant element. It requires that organizations improve the effectiveness of their management system and their energy performance as part of the PDCA cycle. However, the standard does not specify what constitutes an improvement in energy performance. Instead, it is up to the organization to define its own objectives and targets, consistent with its energy policy. National programs based on ISO 50001, such as the U.S. Department of Energy's (DOE) Superior Energy Performance® (SEP™), have sought to address this by requiring a minimum percent improvement over a period of 3–10 years. See Section "US Context."

ISO 50001 contains a number of key concepts central to understanding the standard. These will not be foreign concepts to those versed in energy management. However, it is important to understand specifically how the standard addresses these concepts. These include the following.

Significant Energy Use

The standard defines Significant Energy Use (SEU) as an "energy use accounting for substantial energy consumption and/or offering considerable potential for energy performance improvement." Again, it is up to the organization to set the criteria for determining significance ([4], Section 3.27, p. 5). This provides organizations with implementation flexibility, because designation of an energy system, a process, or a production line as an SEU triggers the following: identifying and planning for effective operational control and maintenance activities; evaluating the impact of the SEU on the ability of the organization to meet its objectives and targets; outlining the procurement process for related energy services, products, or equipment; ensuring the competence, training, and awareness of persons associated with SEUs; and monitoring, measurement, and analysis. As you would expect, developing, implementing, and documenting (as necessary) these processes could require significant organizational resources. Organizations with a number of energy systems and limited resources will find it useful to begin with what is achievable, and, as necessary, add to or even designate different SEUs in subsequent ISO 50001 cycles, as objectives, targets, or policy changes. For an example and discussion of how other considerations might be incorporated into the SEU selection, see Level 2 Step 2.5.3 of the U.S. DOE's e-Guide 2.0, which describes how legal requirements (reduction of greenhouse gas emissions) or business plan priorities (reducing monthly energy expenditures) can be used as additional criteria for the selection of SEUs [6].

Although the concept of SEUs makes sense and may have been implemented by organizations with mature energy management programs under other names, ISO 50001 takes this concept one step further. The standard formally links the SEU designation to five other components of the energy management system already described: operational control and maintenance activities; objectives and targets; the procurement process; competency, training, and awareness; and monitoring, measurement, and analysis. This delivers the benefit of a systematic approach to the management of SEUs while providing flexibility in the designation of SEUs.

Energy Baseline

An energy baseline is established by the organization based on a data period that represents the organization's energy use and consumption [7]. Annex A of the standard elaborates on this. The data period is to account for variables that impact energy use and consumption, such as weather, seasons, business activity cycles, and other conditions ([7], Section A.4.4, pg. 17). The organization compares this baseline with the performance in later years in order to evaluate its energy performance. The baseline is adjusted under a set of defined conditions, such as a change in the processes, operational patterns, or energy systems of the organization [7].

Energy Performance Indicators

The comparison of change in energy performance indicators (EnPIs) from the baseline to the reporting year is at the heart of the standard. The EnPIs are quantitative measures of performance, ranging from a simple metric to a complex model with more than one variable. This is how organizations measure their continual improvement, compared with the baseline. An organization can have many EnPIs depending upon its complexity and needs. EnPIs can be chosen at different organizational levels, such as the facility, production line, process, or a business unit level. Again, this flexibility in designating EnPIs allows an organization to select EnPIs that, taken together, will meet the varied needs of different levels of the organization. Some EnPIs will be of interest to the management (business unit annual energy consumption), while other EnPIs provide utility managers with timely feedback so that they can ensure an energy system is operating as expected (e.g., kW/N m^3 for a compressed air system). EnPIs can answer at least two questions: How well did the organization do compared with its target (retrospective) and how well is this system presently performing compared with a target. The latter EnPI provides information to operators who can correct system performance as soon as it is noticed, and thus contribute toward achieving the larger organizational targets. This is a very complex topic and data collection and processing can be expensive to implement, if an organization does not have existing equipment in place to measure, record, and analyze performance at the appropriate line or process. On the other hand, if an organization develops its EnPIs with an eye on outcomes, it can use this information to manage its performance where it will have impact and savings.

Other Related ISO Energy Standards

Other standards developed throughout 2014 in the 50001 family pertain to energy auditing (ISO 50002:2014): requirements for bodies providing audit and certification of energy management systems (ISO 50003:2014); guidance for the implementation, maintenance, and improvement of an energy management system (ISO 50006:2014); measuring energy performance using energy baselines and energy performance indicators—general principles and guidance; and measurement and verification of organizational energy performance—general principles and guidance (ISO 50015:2014) [8]. Although these standards may simply provide general principles or guidance, they can be very helpful to organizations implementing the standard for the first time.

U.S. CONTEXT

Superior Energy Performance is an energy management program managed by the U.S. DOE, which extends beyond ISO 50001 by adding a verification component to ensure energy savings in industrial facilities. SEP is a voluntary certification that industrial facilities earn by demonstrating continual improvement in energy efficiency and conformance to ISO 50001. Organizations can use the SEP framework as a roadmap to achieve ongoing energy improvements, even if they are not yet ready to pursue SEP or ISO 50001 certification.

SEP builds on ISO 50001 to analyze and prioritize energy use and consumption by tracking progress with energy performance metrics. SEP is accredited by the American National Standards Institute (ANSI) and the ANSI–American Society of Quality (ANSI–ASQ) National Accreditation Board (ANAB). SEP certification requires independent verification of the two requirements by an ANSI–ANAB accredited verification body:

- ISO 50001 conformance
- Energy performance improvement levels corresponding to the ANSI/MSE 50021 standard for SEP

SEP provides a robust protocol for energy performance measurement and third-party verification and also generates reliable data for company management and validation to external stakeholders of continual energy performance improvement. Figure 6.3 shows the relationship of the requirements of ISO 50001 with SEP.

SEP Results to Date

As of January 2014, 17 facilities have achieved SEP certification. Nine of the 17 facilities collaborated with DOE and shared their energy-saving results and SEP implementation

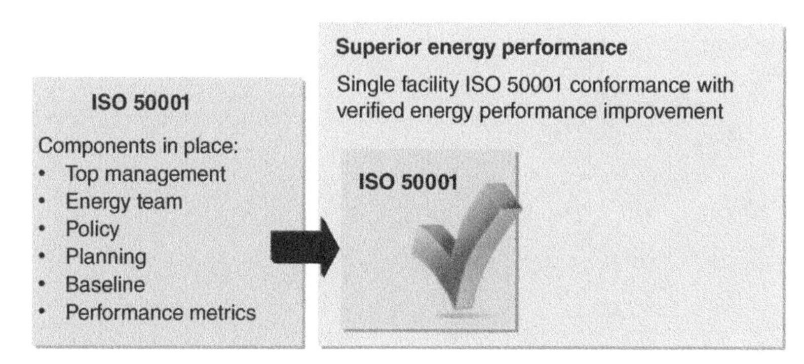

Figure 6.3. ISO 50001 and Superior Energy Performance. (*Source:* U.S. Department of Energy.)

costs. Figure 6.4 shows average quarterly percentage energy savings as a function of average quarterly baseline energy consumption for all nine facilities. Therkelsen et al. [9] report results aligned across facilities so that the first quarter starts when facilities received their first SEP training. Prior to the first SEP training (−Q4 to −Q1) business

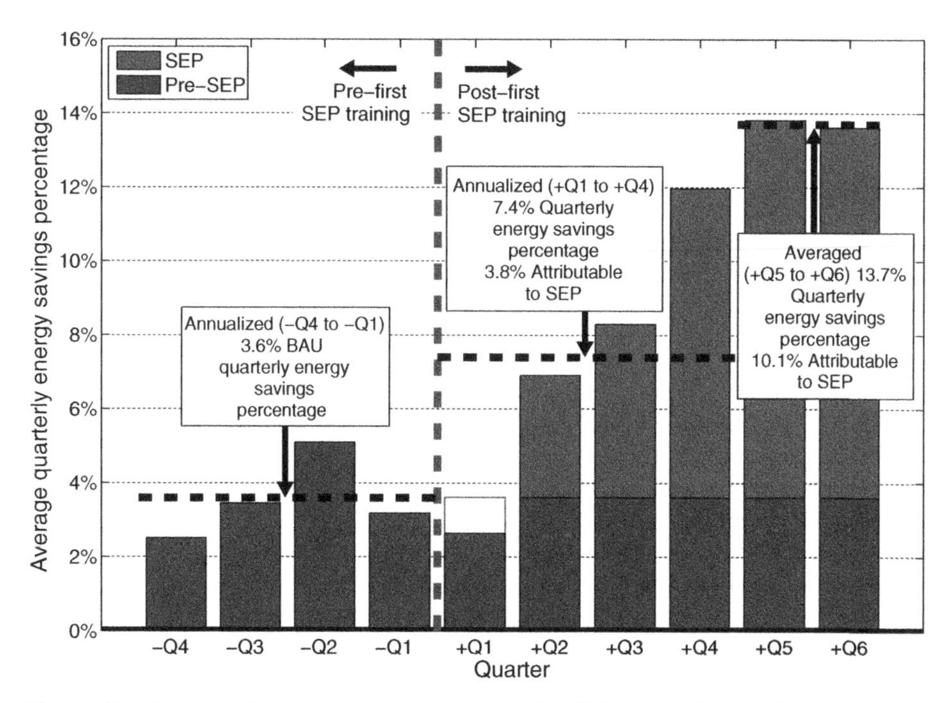

Figure 6.4. Energy savings percentages pre- and post-first SEP training. (*Source:* U.S. Department of Energy [9].)

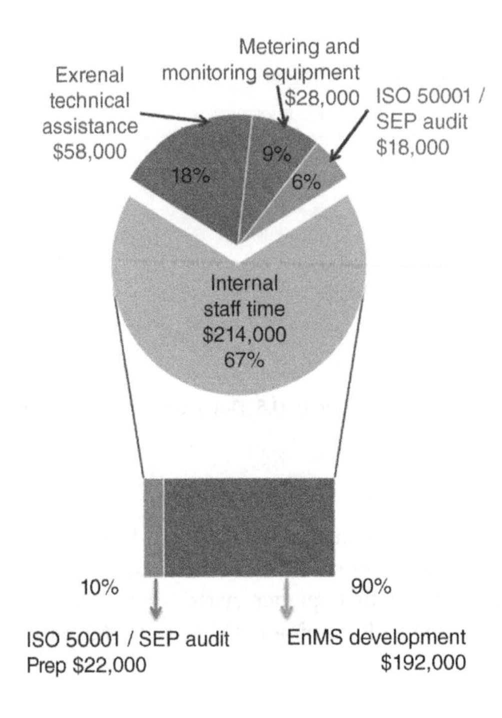

Figure 6.5. Average costs for SEP implementation in demonstration project plants. (*Source:* U.S. Department of Energy [9].)

as usual (BAU) energy performance improved by an average of 3.6% annually against the four quarters that constitute the baseline. Energy-saving percentages increased to an average of 7.4% for the year during quarters +Q1 to +Q4 and an average of 13.7% during quarters +Q5 and +Q6. The increase in percentage energy savings from the first year to the second year after SEP training coincides with the time facilities required to design and implement their energy management system (EnMS). There may be further benefit from maintaining energy savings realized from previously implemented energy performance improvement actions. This is a feature of a fully functional EnMS.

Figure 6.5 shows the average SEP costs for implementation, based on data from the nine facilities included in the analysis. The average cost per facility was $318,000. Cost reduction is anticipated in subsequent implementations when companies transfer the SEP best practices and lessons learned from their first SEP-certified facility to other facilities or enterprise-wide.

Figure 6.6 shows the SEP payback period versus the plant energy consumption. Plants with an energy consumption greater than approximately 0.3 trillion Btu per year will typically experience a payback on the overall SEP cost of 2 years or better. U.S. DOE is working with industry to develop an enterprise-wide SEP certification, which will reduce costs further. The silver, gold, and platinum designation on the chart indicates plant energy improvement as a percent reduction from its baseline, ranging from silver at 5% (the minimum) to gold at 10% and platinum at 15% or greater.

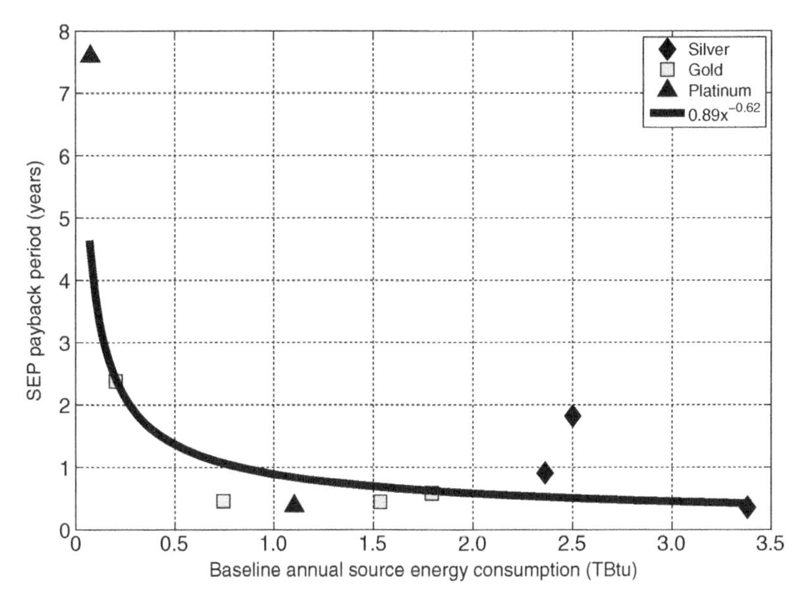

Figure 6.6. SEP payback period versus plant energy consumption. (*Source:* U.S. Department of Energy [9].)

A quantitative analysis of the costs and benefits of ISO 50001 is not yet available. However, this analysis of the SEP program provides some insight into this question. The SEP program goes beyond ISO 50001 requirements in two areas:

- Auditing energy performance data in order to demonstrate achievement of a minimum performance improvement
- Developing a facility-wide model of energy performance that passes specific statistical tests.

These additional costs beyond ISO 50001 are included in the data collected from the nine facilities. For the same facility, the costs of implementing SEP will be higher than achieving ISO 50001 compliance alone. However, this study represents the best analysis of the costs and energy savings attributed to implementing an energy management system that meets the international standard.

Sixty-seven percent of the costs are associated with internal staff time to implement the system. This cost estimate is consistent with the time estimates for implementing previous ISO standards, ranging from 0.5 to 2.0 full time equivalents, depending upon the complexity of the organization.

Workforce Qualifications

With the development of the SEP program, the U.S. DOE and other stakeholders concluded that successful implementation of the program would require a skill set not typically found in industry or at consulting firms providing energy-efficiency services. Implementers needed to combine knowledge of industrial energy use and consumption characteristics with the ability to apply this knowledge in meeting the requirements of the ISO 50001 energy management system standard continual improvement approach. The Institute for Energy Management Professionals (IEnMP) was established at the Georgia Institute of Technology to provide personnel certification services in support of the SEP initiative and other energy management system-related activities [10]. The certifications offered by IEnMP encompass the following areas:

- *Certified practitioner in energy management systems (CP EnMS) for the industrial sector:* CP EnMS certification holders are qualified to assist facilities in implementing ISO 50001 and the additional requirements of the SEP program. As of February 2015, 91 CP EnMS certifications have been awarded. The CP EnMS certification is ANSI/ISO/IEC 17024 accredited.
- *SEP lead auditor (SEP LA):* SEP LA leads the SEP audit team in conducting an audit to determine an organization's conformance to ISO 50001 and a facility's conformance to additional SEP requirements. These additional requirements are documented in *ANSI MSE 50021: Superior Energy Performance—Additional Requirements for Energy Management Systems*. As of February 2015, 11 SEP Lead Auditor Certifications have been awarded. SEP LA is in the ANSI/ISI/IEC 17024 accreditation review process.
- *SEP Performance Verifier (SEP PV):* The SEP PV is a required member of the SEP Audit Team whose responsibility is to determine whether a candidate facility has met or exceeded minimum energy performance improvement levels as defined by the SEP program and documented in *ANSI MSE 50021: Superior Energy Performance—Additional Requirements for Energy Management Systems*. The stated energy performance improvement level must be documented using the *SEP Measurement and Verification Protocol for Industry*. As of February 2015, 20 SEP Performance Verifier Certifications have been awarded. SEP PV is in the ANSI/ISI/IEC 17024 accreditation review process.

INTERNATIONAL CONTEXT

Government policies and programs have historically been significant drivers of EnMS implementation worldwide [11]. In fact, in the few short years since ISO 50001 was published, a number of countries have launched new energy efficiency policies and programs that reference and promote ISO 50001. These include the United States, South Korea, Germany, and South Africa, among many others. United Nations Industrial Development Organization (UNIDO) works with a number of emerging and developing countries to develop national energy efficiency strategies, with ISO 50001 as a cornerstone.

Government Efforts to Promote Energy Management

Government efforts to promote EnMS implementation are structured to meet the domestic needs of industry and to fit within the local policy context, which could include an array of policy goals that address industrial competitiveness, energy, and sustainability. Governments can position adoption of EnMS as a path to industry compliance with mandatory policies or regulations, or can design incentives to encourage voluntary adoption. In addition to creating key drivers for adoption such as regulations or incentives, governments recognize the need to address industry capacity to successfully implement EnMS by providing technical assistance, designing and disseminating tools and resources, and supporting broader workforce development programs to increase the number of qualified energy management professionals in the market.

Countries are promoting adoption of energy management systems at industrial plants through mandatory and voluntary programs. Table 6.2 compares examples of programs and policies implemented in five countries.

TABLE 6.2. Historical Energy Management Programs of Select Countries

Country	Program	Participation	Description
China	Top 10,000 Enterprises Program	Mandatory	Energy management is a key pathway for compliance for this program, which requires industry to set energy-saving targets. To meet targets, local authorities are required to establish programs to support industry implementation of China's national EnMS standard, GB/T 233331, which was revised in 2012 to align closely with ISO 50001
Denmark	Agreement on Industry Energy Efficiency	Voluntary	Companies participating in the voluntary program are required to implement ISO 50001 and identify and implement energy-saving measures in return for tax incentives
Germany	Renewable Energy Sources Act	Voluntary	Qualifying companies that have been certified to ISO 50001 may be eligible for a tax incentive
Sweden	Program for Improving Energy Efficiency (PFE)	Voluntary	Eligible energy-intensive companies may receive a tax incentive in return for implementing ISO 50001 and identifying, implementing, and verifying performance of energy-saving measures
The United States	Superior Energy Performance (SEP)	Voluntary	Recognition and certification program requiring participating companies to become certified to the ISO 50001 standard and verify energy performance improvements

China has an evolving mandatory program that requires industry to work closely with local authorities to set energy reduction targets at both the facility and provincial levels. The second phase of this program—the Top 10,000 Enterprises program—will run through 2015 as a part of China's 12th Five-Year Plan (FYP). The 12th FPY has an absolute energy-saving goal of 670 metric tons carbon equivalent (MTCE). The target of the Top 10,000 Program is to achieve 37% of that goal, or absolute savings of 250 MTCE, by 2015. Energy management is a key component of this expanded program, which now requires local authorities to establish programs to support industry implementation of China's national EnMS standard, GB/T 233331, which was revised in 2012 to align closely with ISO 50001 [12].

Until recently, Sweden had a voluntary program to promote energy management systems that was established in 2005 and designed to position energy management as a strategy to reduce the costs of industry compliance with directives of the European Union, and the resulting energy taxes on electric power for major energy-using facilities and new energy efficiency laws that were being implemented by Sweden. The program for Energy Efficiency (PFE) in energy-intensive industries offered attractive tax incentives to energy-intensive industries that joined the 5-year program, provided they met requirements such as certification to an EnMS standard. Since 2011, Sweden has ISO 50001 as its national standard and integrated certification to ISO 50001 as a core requirement in the PFE. As of January 2014, companies that had volunteered to participate in the PFE represented over 90% of energy use in the Swedish energy-intensive sector [13].

In Germany, companies certified to ISO 50001 could access tax incentives under Germany's Renewable Energy Sources Act [1]. With approximately 52% of all certifications globally as of 2014, Germany's substantial tax incentives have been very effective in encouraging industry to achieve certification to ISO 50001.

International Collaboration to Increase Worldwide Implementation

While countries develop and expand their national approaches to promoting EnMS standards such as ISO 50001, they continue to seek out best practices by engaging in international fora. The ISO process that produced ISO 50001 in 2011 is still very active and is developing a range of accompanying standards and guidance that will support international harmonization and dissemination of best practices for energy management system implementation. ISO's Technical Committee 242 totals 55 participating countries and another 16 observer countries.

In another international forum, the member countries of the Clean Energy Ministerial's Global Superior Energy Performance Partnership (GSEP) share their knowledge and expertise to identify and evaluate energy management system activities, opportunities, strategies, and best practices—working closely with industry and other stakeholders. Current activities include facilitating technical exchange between members on program design, implementation, and evaluation; identifying knowledge and skills needed to boost workforce capacity; collecting and disseminating EnMS implementation data; and building a publicly available toolbox of free resources from around the world to facilitate industry implementation [14].

Workforce Qualifications

Organizations seeking certification require auditors skilled in both business processes and the technical requirements of energy efficiency. The GSEP working group governments are working to increase the number of qualified auditors in the marketplace by creating a consensus-based international certification process for ISO 50001 Lead Auditors. This program will be launched in mid- 2015 [10].

In addition, GSEP has recognized that inadequate workforce knowledge and training are potential barriers to successful EnMS implementation. GSEP has prepared a document that describes the skills, knowledge, and types of personnel needed by organizations for successful energy management system implementation [15]. This document is an excellent resource to an organization as it considers implementing an energy management system. It includes listing of the skills and qualifications an organization will need to draw upon, as well as links to documents and tools on the web.

CONCLUSIONS

It is too early to tell how widely ISO 50001 will be adopted. At present, the drivers for third-party certification to an internationally recognized energy management standard are mixed. Some countries are using the energy management standard as a path to industry compliance with mandatory policies or regulations, while others have successfully provided incentives to encourage voluntary adoption. Countries that provide financial incentives have been successful in increasing the rate of certification of sites within their boundaries. In a few cases, successful experiences implementing ISO 50001 in one facility have led to broader corporate adoption of the standard. Irrespective of whether certification becomes a business or government requirement, the ISO 50001 family of standards will serve as a valuable blueprint for the energy manager seeking to put in place best practices in site or corporate energy management.

ACKNOWLEDGMENTS

Numerous people have contributed to this chapter. The authors particularly acknowledge Robert Auerbach of Robert Auerbach Associates for his insights, review, and comments.

REFERENCES

1. Kahlenborn, W., Kabisch, S., Klein, J., Richter, I., and Schürmann, S. (2012) *Energy Management Systems in Practice, ISO 50001: A Guide for Companies and Organisations.* Federal Ministry for the Environment, Nature Conservation and Nuclear Safety (BMU), Berlin, June 2012, p. 12.
2. Corbett, C.J. and Kirsch, D.A. (2001) International diffusion of ISO 14000 certification. *Productions and Operations Management*, 10(3), 327–342.

3. Reinhard Peglau, Senior Scientific Officer on Environmental Management, German Federal Environment Agency (2014), Global Certifications to ISO 50001 (via personal communication with DOE EERE Advanced Manufacturing Office).
4. ISO (2011) *ISO 50001 Energy Management Systems: Requirements with Guidance for Use.* International Organization for Standardization, June 9, Section 4.2.1c, p. 5.
5. Warris, A.M. and Tangen, T. (2012) Management makeover: new format for future ISO management system standards. *ISO News.* Available at http://www.iso.org/iso/home/news_index/news_archive/news.htm?refid=Ref1621 (accessed on Jan. 30, 2014).
6. U.S. Department of Energy (2015) *DOE eGuide for ISO 50001.* Available at https://ecenter.ee.doe.gov/Pages/default.aspx (accessed February 17, 2015).
7. ISO 50001 *Energy Management Systems: Requirements with Guidance for Use*, Section 4.4.4, pp. 6–7.
8. ISO Standards Catalog (2014) Available at http://www.iso.org/iso/home/store/catalogue_tc/catalogue_tc_browse.htm?commid=558632&published=on&development=on (accessed Feb, 6, 2015).
9. Therkelsen, P., McKane, A., Sabouni, R., Evans, T., and Scheihing, P. (2013) Assessing the costs and benefits of the superior energy performance program. *2013 ACEEE Summer Study on Energy Efficiency in Industry*, July 23–26, 2013, Niagara Falls, NY. Available at http://aceee.org/files/proceedings/2013/data/papers/5_030.pdf
10. Institute for Energy Management Professionals (2014) *IEnMP Overview 2014*, February 13, 2014.
11. Goldberg, A., Reinaud, J., and Rozite, V. (2012) *Energy Management Programmes for Industry: Gaining Through Saving.* International Energy Agency (IEA) Policy Pathway Series, IEA and Institute for Industrial Productivity, April, p. 5.
12. Institute for Industrial Productivity (2011) *CN-3b:Top-10,000 Energy-Consuming Enterprises Program.* Available at http://iepd.iipnetwork.org/policy/top-10000-energy-consuming-enterprises-program.
13. Global Superior Energy Performance Partnership (2013) Models for driving energy efficiency nationally using energy management. *ACEEE Industrial Summer Study 2013* (finalized April 2013, published June 2013), p. 8.
14. Clean Energy Ministerial Available (2015) at http://www.cleanenergyministerial.org/.
15. Energy Management Working Group (2013) *Knowledge and Skills Needed to Implement Energy Management Systems in Industry and Commercial Buildings.* Global Superior Energy Performance Partnership, November 2013.

7

PROTECTING THE ENVIRONMENT AND INFLUENCING ENERGY PERFORMANCE WITHIN PROCESS INDUSTRIES

Elizabeth Dutrow

ENERGY STAR Industrial Sector Partnerships,
U.S. Environmental Protection Agency, Washington, DC, USA

The U.S. Environmental Protection Agency (EPA) operates the ENERGY STAR program, which has worked and continues to work in partnership with businesses to provide tools and best practices to improve the energy efficiency of industrial processes in a way that returns economic benefits to the businesses and environmental benefits to the country. This chapter provides a history of ENERGY STAR's development and explains how the program offers a wide variety of useful tools and strategies to process industries. Additional information on the specific energy management tools developed within the ENERGY STAR program is available at www.energystar.gov/industry.

INTRODUCTION TO ENERGY STAR

The early 1990s began a new focus in environmental protection: a time for preventing pollution before it could occur. Environmentalists and businesses recognized that economic progress and environmental protection must be coordinated for greater gains.

In response to growing concern that greenhouse gas emissions were damaging the earth's climate system, the EPA built and managed market-based programs to reduce

Energy Management and Efficiency for the Process Industries, First Edition. Edited by Alan P. Rossiter and Beth P. Jones.

impacts on the climate. These voluntary programs were designed based on a consideration of the sources of greenhouse gas emissions such as carbon dioxide from fossil fuel combustion and methane from coal mining, natural gas pipelines, and landfills. The EPA evaluated the unique conditions that surround the sources of emissions and then built a program to work within the market by overcoming barriers to adoption of cost-effective solutions for emission reductions and control.

Carbon dioxide emissions span a large part of the U.S. economy. A cost-effective method for addressing carbon dioxide emissions from fossil fuel combustion is to improve energy efficiency through the adoption of more efficient technology and improved energy management practices. A barrier to adopting energy-efficient technology is its lack of widespread availability to consumers in the marketplace and a lack of understanding of the benefits and risks of a technology.

From the early to mid-1990s, EPA operated Green Lights [1], a successful market transformation program that worked directly with businesses to create demand for energy-efficient lighting. At the time, lighting was estimated to account for 20–25% of the electricity sold in the United States. New energy-efficient lighting technology was available but was not widely purchased, nor was it present in great volume in the market. EPA asked corporations large and small, governments, and other organizations to make a voluntary commitment to survey their facilities' lighting and upgrade 90% of the square footage that could be improved profitably without compromising lighting quality. At Green Lights' peak, over a thousand organizations had committed to install energy-efficient lighting. Through Green Lights, EPA moved businesses forward in selecting energy-efficient lighting and helped transform the U.S. lighting market by bringing affordable, energy-efficient lighting to store shelves for consumers.

Building on the success of Green Lights, in 1992 EPA designed ENERGY STAR to help businesses and consumers make good selections of a new product, the personal computer, based on its energy performance. Personal computers were not common in 1992; however, projections of growing usage in businesses and households were large as were the electricity use and related carbon dioxide emissions associated with powering these devices. A technology was available, but not widely installed by manufacturers at the time, for managing the power consumption of personal computers.

EPA faced a challenge in motivating and mobilizing the U.S. marketplace to offer an energy-efficient personal computer. Working with computer manufacturers, EPA helped build the business case for producing and purchasing an energy-efficient product, identified energy performance specifications for the personal computer, designed the recognizable ENERGY STAR label for products that complied with the specifications, and informed the market of the availability of a new ENERGY STAR compliant computer that performed well while saving energy and protecting the environment.

The market responded. Computer manufacturers accepted the performance specification and applied the ENERGY STAR label to their product. EPA, computer manufacturers, retailers, utilities, states, and others educated consumers and businesses about the advantages of purchasing an ENERGY STAR compliant computer. Businesses that were buying large stocks of computers selected products identified by the ENERGY STAR and were able to avoid higher energy bills. Individual consumers responded by asking for the ENERGY STAR when making a computer purchase.

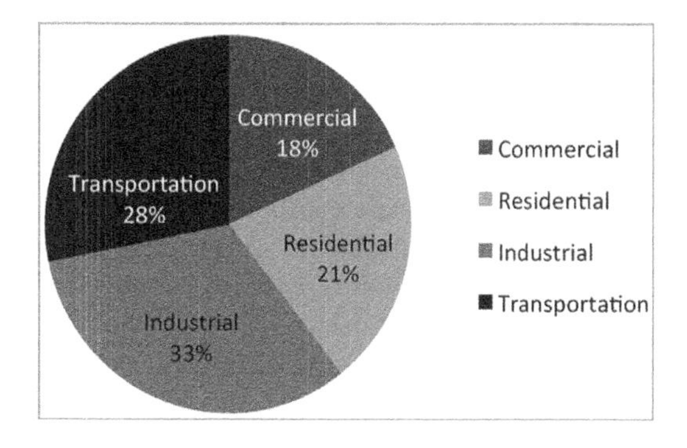

Figure 7.1. Percent share of total energy consumed by major sectors of the economy, 2012 (includes electricity consumption). (*Source:* U.S. Energy Information Administration [2].)

Within the personal computer market, ENERGY STAR became a symbol of savings and environmental protection by providing credible, objective information upon which consumers could make better informed decisions. As a result, purchasing an energy-efficient computer became easier and less risky.

Since 1992, EPA has leveraged ENERGY STAR and transformed it into a national symbol of energy efficiency, environmental protection, and savings. The ENERGY STAR logo is now recognized by over 80% of the U.S. public. EPA continues to apply the ENERGY STAR to products that meet strict energy performance specifications; more than 70 now bear the familiar logo. Further alliances with product manufacturers, retailers, utilities, state energy offices, public benefits fund administrators, and others expand the advantages that this national brand provides.

NOT JUST PRODUCTS

EPA has not limited ENERGY STAR's influence to products. The building and industrial sectors of the U.S. economy are major energy consumers and contributors to carbon dioxide emissions. The U.S. Energy Information Administration (EIA) reports [2] that in 2012 energy consumption in the building sector (commercial and residential) was 37.5 quad,[1] and 30.8 quad for industry, while their energy-related carbon dioxide emissions accounted for 38 and 28% of U.S. emissions, respectively [3]. Figure 7.1 shows the percent share of total energy consumed by major sectors of the U.S. economy in 2012.

Improving energy performance in these sectors depends on the availability of cost-effective means for reducing energy consumption. Studies show significant reduction potential in commercial and industrial enterprises. For instance, McKinsey & Company examined the opportunities for energy efficiency and concluded that efficiency potential is present "with substantial opportunities" [4].

[1] 1 quad, or quadrillion Btu, is 10^{15} Btu.

A STRATEGY FOR IMPROVING ENERGY EFFICIENCY IN THE PROCESS INDUSTRIAL AND COMMERCIAL SECTORS

Achieving improved energy efficiency depends on identifying good energy performance. Within the commercial and industrial (C&I) sectors, the main consumer of energy is a building or manufacturing plant and the operations within the facility. Just as consumers need help in selecting energy-efficient electronics, EPA has found in the C&I sectors that many companies struggle to identify good energy performance. Working through ENERGY STAR, EPA helps businesses make sound energy decisions and improve how energy is consumed.

EPA's strategy focuses on removing three limitations that affect most organizations that seek to manage energy in buildings and plants:

1. a lack of information on how to improve facilities' energy efficiency;
2. a lack of energy performance measurement tools that answer how well a facility should perform; and
3. a lack of understanding of the connection between managing energy well and creating value for an organization, often displayed as an absence of high-level corporate commitment to manage energy across an organization.

EPA promotes a straightforward pathway to manage energy across all operations and achieve the cost savings and environmental benefits that can result, starting with development of an energy management program. Through a proven energy management strategy that turns a basic interest in managing energy into action [5], ENERGY STAR is focused on improving energy efficiency in the C&I markets by

- engaging top-level corporate attention and creating a public commitment to secure resources for sustained improvement;
- building a credible and objective scoring system to assess the energy performance of facilities, validate savings, and recognize top performance;
- identifying actions to take in facilities through sector-specific guides on how to reduce energy consumption;
- providing visibility and recognition of a business' achievements; and
- offering access to a broad network of ENERGY STAR C&I partners who bring fresh, diverse, and creative approaches to managing energy.

An important component of EPA's strategy is a basic energy management program framework, the ENERGY STAR Guidelines for Energy Management. The guidelines mirror the best practices of EPA's ENERGY STAR C&I partners in managing energy (Figure 7.2). They promote a system for obtaining the commitment of senior management, setting goals and corporate policies, building action plans for the corporation as well as individual sites, ensuring progress according to plans, engaging senior management in program reviews, and recognizing the achievements of sites and individuals. Through EPA's work in the C&I sectors, the ENERGY STAR Guidelines for Energy

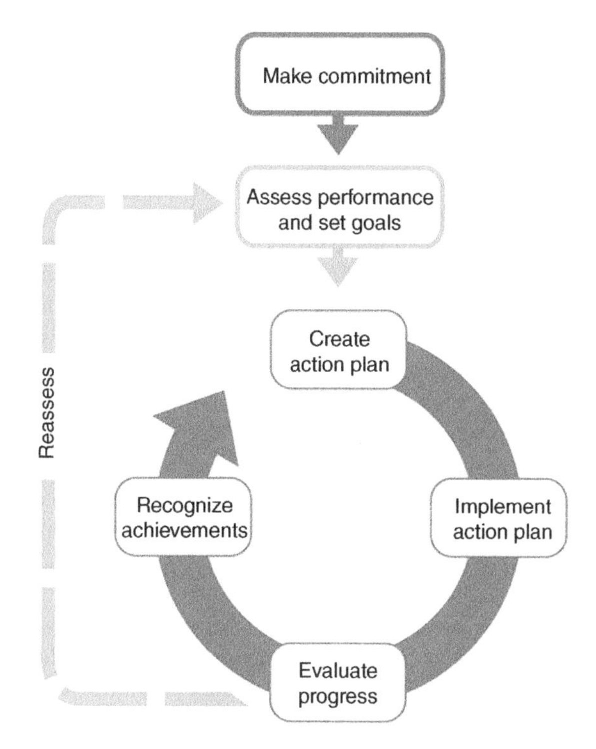

Figure 7.2. ENERGY STAR Guidelines for Energy Management. (*Source:* www.energystar.gov/guidelines.)

Management have become the foundation for energy programs in organizations throughout the United States and abroad and more recently have informed the development of an international energy management standard, ISO 50001.

Unique tools and resources are also available to move businesses through each of the steps of the ENERGY STAR Guidelines for Energy Management. For example, an industrial company seeking to identify actions to take in its facilities can locate industry-specific guides on the website that explain how to reduce energy consumption in both the manufacturing process and common plant systems such as compressed air, motors, and steam.

Many accept the premise that what cannot be measured cannot be managed. Until 1999, there was no reliable measure of energy performance or efficiency for U.S. buildings and industrial plants. Then EPA, using national building performance data, modeled energy performance for office buildings in the United States, producing a mathematical model that rated the energy performance of buildings. For the first time, a building owner or manager could use the model to make decisions about how well a specific building should perform on energy consumption relative to similar buildings [6]. To make facility benchmarking easier, EPA produced ENERGY STAR Portfolio Manager® for building managers to use to measure and track energy and water consumption and benchmark the performance of one building or a whole portfolio of buildings in a secure, online environment. For manufacturing plants, EPA works directly with manufacturing industries to construct ENERGY STAR plant energy performance indicators (EPIs) for various manufacturing sectors to benchmark the performance of a whole plant.

EPA's national system of evaluating plant and building energy performance enables users to manage energy at a level not previously possible. Businesses use ENERGY STAR to identify individual facilities for improvement, set energy performance goals, learn from efficient facilities, and earn positive recognition.

EPA's strategy in the C&I sectors breaks down three barriers to energy performance improvement: how to improve, how well a facility should perform, and a lack of corporate commitment. ENERGY STAR's guidance and its energy performance benchmarking system address the need within C&I markets for energy management tools. Breaking down the third barrier requires a different set of tools.

Corporate commitment implies energy management at all levels of an organization across all units of the business. EPA has long recognized that people are the most critical part of successful energy management and performance improvement. EPA's goal is to influence energy management from an organization's leadership to workers on the plant floor. Regardless of the level, people must be motivated to adopt and maintain good energy management practices. A clear behavior change is necessary.

From a comprehensive viewpoint, EPA uses the ENERGY STAR brand to engage companies in energy management. Many have built their energy programs on the ENERGY STAR platform by using its guidance and extended it worldwide throughout their organizations.

At the plant or building level, EPA enables organizations to identify good energy performance with the familiar ENERGY STAR logo and accompanying energy management tools. Plants and buildings where energy performance is achieved within the top quartile nationally for the plant or building type are eligible to earn ENERGY STAR certification and display the highly recognized ENERGY STAR banner or flag (Figure 7.3). Building performance must be verified using EPA's Portfolio Manager tool, while the relevant ENERGY STAR plant energy performance indicator must be employed for manufacturing plants to be eligible.

At the senior levels of business, EPA uses ENERGY STAR to assist executives in understanding the role they need to fulfill in assisting their organizations to successfully manage energy risk. Often senior leaders do not recognize the strategic value of an

Figure 7.3. ENERGY STAR certification banner for plants. (*Source:* U.S. Environmental Protection Agency.)

Figure 7.4. ENERGY STAR Partner of the Year logo. (*Source:* U.S. Environmental Protection Agency.)

energy management program, nor do they understand that the risk of implementing an energy project is relatively low and can have solid returns when compared with other investments. *Energy Strategy for the Road Ahead* [7] is tailored specifically to guide executives. They are encouraged to plan for their business' energy future, empower their organizations to take action, push for energy performance across the business, enable energy investments to take place by valuing them differently from other projects, and involve themselves on a continuing basis.

EPA's highest form of recognition is the ENERGY STAR Partner of the Year award that recognizes excellence in corporate energy management [8]. Organizations must demonstrate that energy management spans all levels of the company and that the energy management program covers all aspects of energy management as described in the ENERGY STAR Guidelines for Energy Management. EPA Partner of the Year awardees are permitted to display the Partner of the Year logo (Figure 7.4).

IMPROVING PERFORMANCE AND CHANGING BEHAVIOR

Can a national brand change behavior and motivate improvement in energy performance? EPA has seen evidence that it does.

At the plant level, occasional feedback from industrial participants has indicated that manufacturing industries highly value their ENERGY STAR plant certifications. To illustrate, a corporate energy manager reported his efforts to motivate plants to improve by earning ENERGY STAR certification (i.e., energy performance within the top quartile nationally for the industry). The energy manager lamented that each plant had achieved ENERGY STAR certification and he was finding it difficult to continue pushing for the plants to achieve higher goals. EPA conducted research into this industry's energy performance and found that the entire industry had shifted its energy performance curve, requiring an update of the plant energy performance indicator [9]. The model was rebaselined with new data, reviewed by industry, and provided again to the industry for use in energy management.

From a corporate perspective, EPA often hears comments from corporate energy managers that their executives have set goals for achieving Partner of the Year status. Clearly, the value of positive recognition cannot be overlooked in influencing all levels of an organization.

UNIQUE RESOURCES FOR PROCESS INDUSTRIES, INCLUDING PETROLEUM REFINING AND PETROCHEMICAL SECTORS

According to the U.S. EIA, a large portion of U.S. manufacturing sector energy consumption occurs within the process industry sectors. (While some energy serves as a feedstock for production, a large amount fuels the operation of plants.) Figure 7.5 shows that these sectors consumed over 50% of manufacturing energy use in 2010.

According to Figure 7.6, refinery energy consumption has not declined significantly during the 1995–2011 period. Worrell [11] reports that refinery energy consumption in 2011 was 3.1 quad.

EPA's work with the petroleum refining and petrochemical industries occurs through an ENERGY STAR Industrial Focus. Industrial Focuses are designed to engage manufacturers within an industrial sector to look at the barriers to energy efficiency and build tools and strategies to help reduce the impact of barriers.

To manage informational barriers, Energy Guides are available for both sectors to identify opportunities to improve energy management. Energy Guides profile reported energy savings, costs, and details of actual energy efficiency opportunities in each category. Companies seeking ways to manage energy use the guides to identify potential energy efficiency options for their plants. Examples of the types of categories of improvement measures included for refineries are listed in Table 7.1.

To address the lack of energy performance measurement tools that answer how well a facility should perform, EPA, when working with a new industry, will produce a statistical model, known as an ENERGY STAR plant EPI, that companies may use to judge the energy efficiency of individual plants. Using actual plant data, EPA models the plant energy performance for the entire industry to produce an energy performance scoring tool for plant energy performance within the industry nationally. The resulting tool is tested and reviewed by industry, finalized as an ENERGY STAR EPI, and released for use.

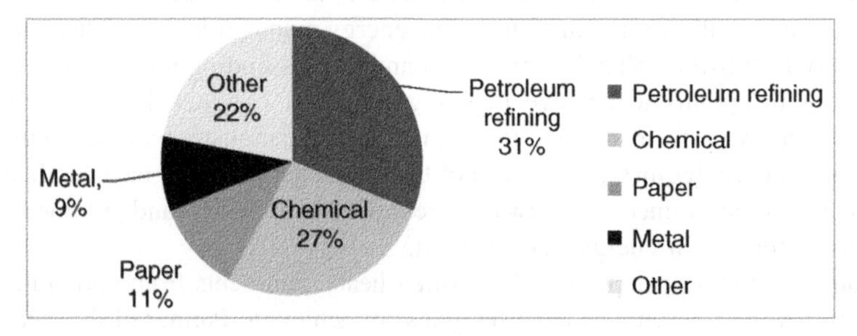

Figure 7.5. Percent energy use by type of industry, 2010. (*Source:* U.S. Energy Information Administration [10].)

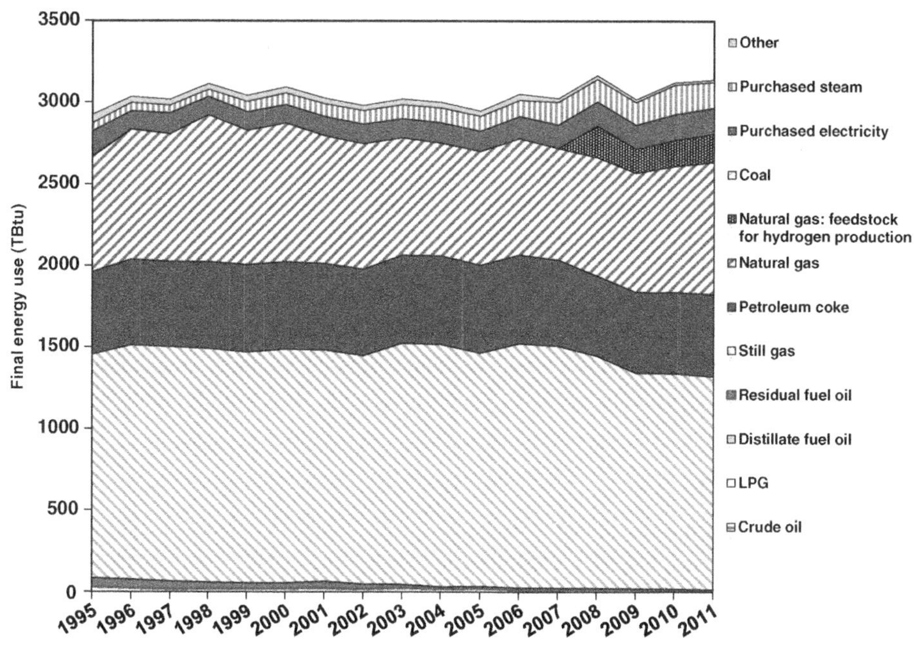

Figure 7.6. Annual final energy consumption of U.S. petroleum refineries for the period 1995–2011. (*Source:* U.S. Environmental Protection Agency [11].)

Within the petroleum refining sector, a private system (operated by HSB Solomon Associates LLC) exists where refineries are scored for their energy performance. Hundreds of refineries around the world participate in the system and receive an Energy Intensity Index (EII®) rating similar to the output of the ENERGY STAR EPI. Thus, it was unnecessary for EPA to produce a refinery EPI. (To date, EPA has produced EPIs for a

TABLE 7.1. Improvement Measures for Petroleum Refineries

Cross-Cutting Technologies	Process-Related Options
Motors	Flare gas recovery and management
Pumps	Power recovery
Fans	Hydrogen management
Lighting	Distillation
Cogeneration	Process heaters
Boilers and steam systems	Heat exchangers
Compressed air	Energy management

Source: U.S. Environmental Protection Agency [11].

variety of plant types in the glass, food, cement, pulp and paper, steel, metal casting, pharmaceutical, and automobile sectors.). Benchmarking is discussed further in Chapter 5.

To motivate improved refinery energy performance, EPA recognizes refineries[2] that achieve energy performance within the top quartile for their Solomon Associates-assigned size class. Since the inception of offering ENERGY STAR certification for refineries, 39 ENERGY STAR certifications have been issued for U.S. petroleum refineries. Among those, Marathon Petroleum Corporation has earned the most, having achieved 31 certifications shared among six of its refineries.[3]

To motivate corporate-level energy awareness, the ENERGY STAR Partner of the Year award recognizes excellence in corporate energy management. From 2009 to 2011, Sunoco, Inc. consistently achieved ENERGY STAR Partner of the Year for energy management across the entire company. Sunoco's energy management program spanned its refining, chemical, retail/marketing, and coking operations (see Box 7.1).

Box 7.1. Sunoco, Inc., ENERGY STAR Partner of the Year, 2009, 2010, 2011

- Consistently achieved energy intensity reductions in its refining and chemical businesses.
- Embedded energy management into a corporate-wide business improvement initiative.
- Created an internal plant energy assessment process that identified 300 capital and 140 noncapital energy projects in the refining business within 1 year.
- Built a transformative energy management work process system for plant operators and shift organizations that defined measurement and review, enabling employees to make greater contributions to energy management.
- Adopted a "total cost of ownership" approach for retail outlets.

NETWORKING: ANOTHER FORM OF BENCHMARKING

Early on in EPA's work with petroleum refiners and petrochemical producers, a regular network of energy managers was established to enable best practice sharing in a noncompetitive environment. Ground rules for networking were clear: no talk of

[2] A draft EPI was produced for ethylene plants but has not yet progressed to completion. ENERGY STAR certification is only offered where an EPI has been satisfactorily completed and offered for use in an industrial sector.

[3] ENERGY STAR certification is awarded to plants that achieve top quartile energy performance using an approved plant energy performance indicator and that have passed an environmental compliance screen.

confidential information, just discussion of energy management practices and technologies that have been useful in improving the performance of facilities.

Energy managers embrace the ENERGY STAR networking forum and each year have participated actively in discussions of tactical, strategic, and technical energy management topics relating to their industry. Networking is successful because the companies recognize that, while specific details of any project may not be shared, the fact that the project succeeded for one organization shows what might be possible for another. In addition, companies often focus just on project-level, short-term, technical activities. Learning about how others are implementing energy management practices within the mid-term tactical and long-term strategic realms enables energy managers to add new dimension to their energy programs. Topics are varied and have included the issues identified in Table 7.2.

TABLE 7.2. Variety of Energy Management Measures Discussed in ENERGY STAR Focuses on Energy Efficiency in Petroleum Refining and Petrochemical Manufacturing

Technical	Tactical	Strategic
Low-level heat recovery	Sustaining savings despite personnel changes	Securing the buy-in of senior management
Boiler optimization/fired heater optimization	Plant-level internal energy teams	Sustainability, including energy, water, and waste management
Waste heat recovery	Securing funding for projects	Building energy efficiency into project design
Variable-frequency drives	Corporate-wide energy teams	Lengthening payback periods
Flare optimization	Working better with operators	Allocating capital for energy projects
Insulate and reinsulate	Plant assessment techniques	Reporting of key energy performance indicators to upper management
Crude preheat	Steam system management	Evaluating capital projects for energy and carbon impacts
Use of energy score cards at the plant to drive improvement	Plant- and process-level benchmarking	Long-term dollar value of a corporate energy program
Operator engagement in energy management	Role of plant-level energy program	Value of a corporate-wide energy program
Energy management in cooling towers	Use of internal experts to assist in energy evaluations	Multiyear planning for energy program
Communicating performance expectations	Networking among plants	Building an energy culture
Reducing energy consumption during plant downtime	Sustaining energy management through routine measures	Managing energy across all businesses, even if diverse

CONCLUSION

EPA applies the ENERGY STAR to large parts of the U.S. market and uses it to identify superior energy performance in homes, buildings, and manufacturing plants, helping consumers and businesses to make good choices that protect the environment and save money. The power of the ENERGY STAR brand draws the participation of industry, and through it, EPA offers industries such as petroleum refiners and petrochemical manufacturers a place to learn and to manage energy better. In all applications, education and a choice in the marketplace for energy-efficient products, energy management improvement options, best practices, and energy efficiency services support the meaning of this impactful brand.

REFERENCES

1. U.S. Environmental Protection Agency (1993) *Introducing . . . the Green Lights Program.* EPA 430-F-93-050.
2. U.S. Energy Information Administration (2014) *Monthly Energy Review.* DOE/EIA-0035. Table 2.1: Energy Consumption by Sector. Available at http://www.eia.gov/totalenergy/data/monthly/archive/00351402.pdf (accessed February 2014).
3. U.S. Energy Information Administration (2014) *Energy Kids: Energy Sources, Recent Statistics.* Available at http://www.eia.gov/kids/energy.cfm?page=stats (accessed February 18, 2014).
4. McKinsey & Company (2009) *Unlocking Energy Efficiency in the U.S. Economy.*
5. U.S. Environmental Protection Agency (2003) *ENERGY STAR: The Power to Protect the Environment Through Energy Efficiency.* EPA-430-R-03-008.
6. U.S. Environmental Protection Agency (2010) *Celebrating a Decade of ENERGY STAR Buildings, 1999–2009.*
7. Global Business Network (2007) *Energy Strategy for the Road Ahead*, Global Business Network, San Francisco, CA.
8. U.S. Environmental Protection Agency (2014) *Partner of the Year Profiles in Leadership.* Available at http://www.energystar.gov/buildings/facility-owners-and-managers/industrial-plants/earn-recognition/energy-star-partner-year-award-0 (accessed in February, 2015).
9. Nicholas Institute for Environmental Policy Solutions (2010) *Assessing improvement in the energy efficiency of U.S. auto assembly plants.* Duke Environmental Economics Working Paper Series, Working Paper EE 10-01. Available at http://sites.nicholasinstitute.duke.edu/environmentaleconomics/files/2013/01/Duke-EE-WP-1.pdf (accessed March 11, 2014).
10. U.S. Energy Information Administration (2013) *Manufacturing Energy Consumption Survey 2010*, Table 1.2. Available at http://www.eia.gov/consumption/manufacturing/data/2010/pdf/Table1_2.pdf (accessed March 2013).
11. Worrell, E. (2015) *Energy Efficiency Improvement and Cost Saving Opportunities for Petroleum Refineries. An ENERGY STAR Guide for Energy and Plant Managers.* U.S. Environmental Protection Agency.

SECTION 2

ENERGY MANAGEMENT TECHNOLOGIES

8

THE TECHNOLOGIES OF INDUSTRIAL ENERGY EFFICIENCY

Alan P. Rossiter

Rossiter & Associates, Bellaire, TX, USA

Law number 1 of industrial energy efficiency: There is no silver bullet—no single method or technology that ensures that energy use will be optimized. Any comprehensive energy management program incorporates many different components that typically require a wide range of expertise, software, and equipment, and it is not practical for an energy manager to have full mastery of every aspect. Rather, the role of most energy managers is to manage resources—mostly human resources—embodying the expertise needed to accomplish corporate energy efficiency goals.

We can view the opportunities for improving energy efficiency in several different ways. For example, the opportunities can be organized by:

- department,
- time frame,
- equipment, utility systems, and process.

Many companies prefer the departmental classification, as it provides a focus on the administrative area within their organization where effort is needed to achieve specific energy efficiency improvements. The time frame classification, on the other hand, is helpful in deciding the order in which opportunities should be addressed. Finally, the

Energy Management and Efficiency for the Process Industries, First Edition. Edited by Alan P. Rossiter and Beth P. Jones.

equipment and systems classification highlights the physical facilities, software, and human resources that are needed to identify, develop, and implement each type of energy-efficiency activity, program, or project.

It is useful to consider all three forms of classification when developing an energy efficiency program and deciding which elements to include and which personnel to engage. This chapter starts with a fairly detailed discussion of the departmental classification. This is followed by shorter descriptions of the time frame and equipment and systems classifications, together with a consideration of the relationships between the different classification systems. The chapter ends with an overview of the other chapters included in Section 2, each of which addresses one or more of the equipment types or systems that are important in energy management.

DEPARTMENTAL CLASSIFICATION

In general, energy efficiency can be improved through four different types of changes: operational improvements, effective maintenance, engineered improvements, and new technologies. In most large companies, each of these corresponds to a different department or administrative area—operations, maintenance, engineering, and research and development. The types of improvements in each department are discussed below.

Operational Improvements

Before committing to projects that require capital expenditure, it is prudent to ensure that existing equipment is being used to its full advantage. This is illustrated in the following example that relates to a fluid catalytic cracking (FCC) unit at an oil refinery [1].

Piping exists to send a feed to the FCC fractionator either "hot direct" from an upstream process or to route it through a cooler and tank, creating a "cold feed" stream. The feed is used to remove heat from a pumparound circuit on the fractionator, and there has to be an adequate temperature difference between the pumparound stream and the feed stream to remove the pumparound heat. In order to ensure that the temperature difference was large enough to meet the pumparound duty, the feed was routinely routed through the cooler and the tank, and supplied as cold feed (see Figure 8.1).

As part of an energy efficiency study, several projects were scoped to recover the heat from the feed and minimize the heat rejection in the cooler. One option was selected and progressed to detailed engineering. At this point, the existing pump-around-to-feed heat exchanger came under scrutiny. Design specifications were reviewed and compared with current operating data, and it became apparent that the required pumparound duty could be achieved with the existing heat exchanger even if the feed was supplied "hot direct"—that is, without passing through the cooler and tank. Plant trials confirmed this conclusion. As a result, the operating procedures were changed, and "hot direct" routing is now considered the normal operating

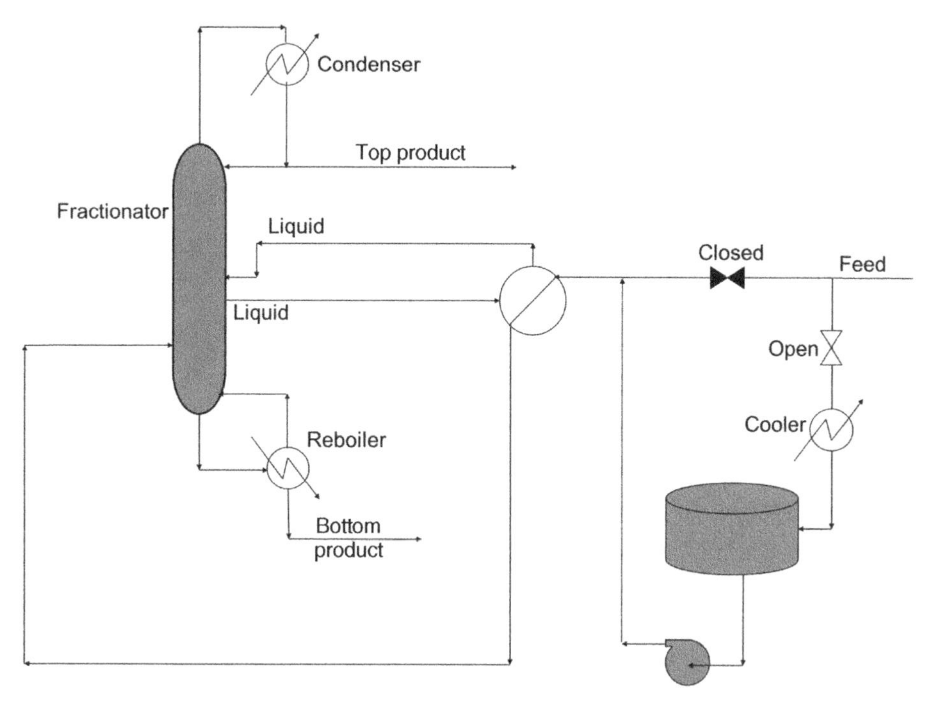

Figure 8.1. Hot feed is routed through the cooler and tank to provide cold feed to remove heat from the pumparound.

procedure (Figure 8.2). This change increases the temperature of the feed as it enters the fractionator. This in turn reduces the reboiler duty, saving about $600,000/year, with no investment required by the facility.

This example highlights the fact that operating practices tend to become ingrained over time. Operators may religiously follow procedures that were developed when the plant first started up, even though the requirements of the plant change over time due to new feedstocks, different product slates, changes in throughput, or a host of other factors. Another way that operating practices can become skewed is through responses to specific problems. For example, a valve position may be changed during an upset or unusual weather condition, and it may be left in the new position long after the abnormal situation has passed. The modified valve position becomes the new normal, and it can remain unchanged for years, until someone questions it. Further examples of this kind are discussed in Chapter 25.

When we become aware that operating conditions are suboptimal, our natural response is to adjust the process (e.g., change the valve position in the example above) in order to optimize the operation. While this may be a good first step, it is only a short-term fix. The next operating shift will very likely reverse the change and restore the status quo. Moreover, the optimum valve position today may not be optimum tomorrow, when plant throughput, feed purity, product slate, and ambient conditions may all be different from

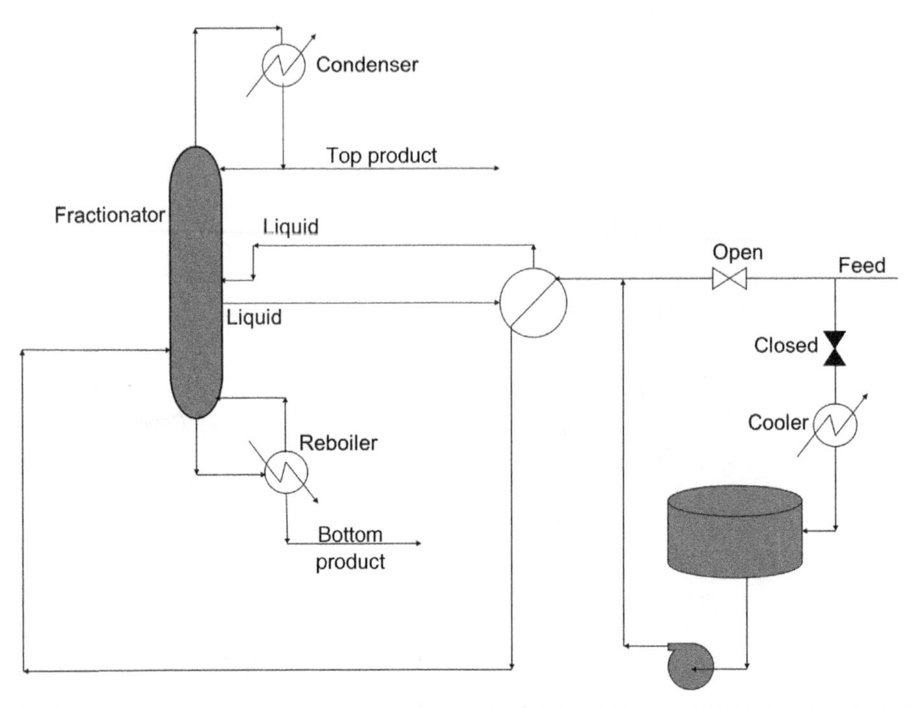

Figure 8.2. Feed is routed "hot direct," bypassing the cooler. Heat can still be removed from the pumparound and the reboiler duty is reduced, saving $600,000/year in energy costs.

their current values. Thus, we need more than just a one-time change to ensure that we get the most from our existing facilities.

The following are a few additional steps:

i. Modify operating procedure documentation
ii. Carry out additional operator training
iii. Add control valves
iv. Implement real-time optimization systems
v. Implement performance monitoring systems and key performance indicators (KPIs)

Items (iv) and (v) on this list are specific technologies that are discussed in more detail in Chapters 19 and 27, respectively.

Effective Maintenance

Another part of getting the most out of existing facilities is to ensure that the equipment is properly maintained. As the current theme is energy efficiency, our primary focus here is maintenance of the equipment and systems that have the largest impact on energy use.

One such equipment type is heat exchangers, especially when those heat exchangers are part of a preheat system. Heat exchanger cleaning programs are important in maintaining the performance of heat recovery systems, notably crude unit preheat trains in refineries. Refinery management is becoming increasingly sophisticated in this area, with both improved cleaning techniques and better tools for assessing appropriate cleaning intervals for the heat exchangers in the circuit, as discussed in Chapters 11 and 12.

However, the best cleaning methods and the most elegant optimization of cleaning intervals are of little use when communication fails. The following example illustrates how maintenance can go wrong [1].

During the course of a crude unit preheat train study in a refinery, one of the heat exchangers was found to be out of service, which is not unusual, as heat exchangers often require maintenance. However, the records showed that this particular heat exchanger had been idle for more than 3 months. Further investigation showed that the heat exchanger had been cleaned, and the work had been completed within a couple of weeks. The maintenance supervisor notified the shift supervisor that the cleaning was complete but shift personnel were busy with other activities, and the heat exchanger could not be brought back in service before the shift ended. Unfortunately, the shift supervisor failed to instruct the next shift to bring the heat exchanger back into service. There was no follow-up, and the cleaned exchanger remained out of service for 2.5 months.

When the unit manager was informed of this situation, it required only a few hours to bring the heat exchanger back in service. The energy loss during the period that the heat exchanger had been left idle after the cleaning was worth over $100,000.

In this case, the key problem was a breakdown in communication. Better systems were needed for tracking the status of maintenance jobs on the unit. A simple electronic reminder system, for example, could have alerted the operators to bring the heat exchanger back online.

In addition to heat exchangers, there are several other key systems and types of equipment that need careful attention to maintain energy-efficient operations. These include furnaces and boilers (Chapter 9), steam traps (Chapter 13), and insulation (Chapter 16), as well as compressors, pumps, and turbines.

Engineered Improvements

Additions and upgrades to plant facilities can lead to significant improvements in energy efficiency. Invariably these projects require significant input from engineering personnel to identify, evaluate, and design the projects. Opportunities can cover a wide range in scale and type, for example,

- simple piping changes,
- localized insulation projects,
- replacements of electric driver systems (e.g., installing variable frequency electric drives), and

- adding heat exchangers, steam turbines, distillation columns, or other major equipment items.

For some types of changes, there are established methods for identifying and screening options. For example, for heat recovery, pinch analysis has become the tool of choice (Chapter 26). Evaluation of steam balances is also a good way to identify inefficiencies in steam systems and develop projects to address them (Chapter 18). However, in many cases, the best ways to find opportunities involve various types of process review, knowledge sharing, and brainstorming (Chapter 25).

New Technologies

Engineered improvements typically use established equipment types and apply proven solutions to identified problems. In contrast, solutions that incorporate new technologies generally require some amount of validation through research and/or development. Thus, the amount of time required to implement opportunities involving new technologies and the degree of technical and financial risk are higher than for engineered improvements.

However, some of the largest energy efficiency improvements in the process industries have come through technological advances. For example, improvements in catalysts have allowed some processes to operate at less extreme temperatures and pressures, and with improved conversion and selectivity—all changes that tend to reduce energy usage. Technological advances have also improved some of the key equipment items that impact energy usage, such as heat exchangers (Chapter 10) and distillation columns, and incorporating some of the new types of equipment that are now commercially available can often lead to improved engineering solutions.

In most companies, new technology development resides within a research and development department. Most energy managers at a facility or a site level focus on the current performance of existing facilities and engineered improvements, and therefore do not play a large role in developing new technologies. However, some companies within the process industries do have strong research and development programs in energy efficiency, and their corporate energy managers typically have a strong interface with corporate research and development.

TIME FRAME CLASSIFICATION

Most successful energy management programs progress through a series of implementation phases (see Chapter 1). These tend to start with the relatively simple, low-cost, low-risk activities, and then progress to more challenging activities as the culture of energy efficiency matures within the company. This leads to the following simple three-stage "time frame" classification:

Short-Term Stage: "Quick Hit" Activities

In the early stages of an energy management program, it is important to gain buy-in from the site personnel, garner support, generate some early successes, and gain credibility.

Communication of goals and expectations is important, and so is the identification and implementation of readily achievable efficiency improvements. These are best obtained through programs that have a high probability of success with relatively minor technical support. These items should have high reward to cost ratios and be highly visible, and they might include repairing large steam leaks, shutting down unneeded equipment, improving heater efficiency, installing lighting projects, and so on. These largely overlap with the "operational improvement" and "effective maintenance" categories in the departmental classification. Very little capital investment is required, although there may be significant expense costs, especially associated with maintenance activities.

Medium-Term Stage: Core Program Components and Projects

Once the energy management program has been established and has gained some credibility, it is time to initiate some more challenging components:

- Initiate site-wide maintenance programs (steam leaks, steam traps, compressed air, etc.).
- Implement performance monitoring systems and key performance indicators.
- Conduct energy surveys using non-site personnel to generate new ideas and perspectives. The survey team can be comprised of personnel from other sites within the same company, consultants/specialists, vendors, and even network peers from outside companies.
- Systematically evaluate and implement potential capital project ideas.
- Incorporate energy efficiency concepts into maintenance planning discussions, and into all capital projects.

The activities during this phase can use a wide range of personnel, and they can correspond to all of the categories in the departmental classification.

Long-Term Stage: Sustainment

Many energy management programs fail through lack of follow through. Sustainment activities are needed to secure the gains that have been made in the earlier stages of the program and to promote a culture of continuous improvement. The following are some of the key features:

- Maintain and improve performance monitoring systems and key performance indicators, with more rigorous tools and more aggressive targets.
- Ensure continued funding for energy-related maintenance work.
- Reevaluate items that were dropped during earlier phases, and continue seeking new ideas from all sources.
- Explore large-scale "game-changing" projects such as cogeneration, importing and exporting fuel or steam between neighboring sites, and joint projects with other companies.
- Evaluate and where appropriate implement new technologies.

Once again, the activities during this phase potentially correspond to all of the categories in the departmental classification.

EQUIPMENT, UTILITY SYSTEMS, AND PROCESS CLASSIFICATION

The departmental classification is useful when considering how energy management activities interface with existing organizations within a corporation or a facility, and the time frame classification provides a good way to visualize the sequence of activities within the overall program. However, when the focus turns to individual component activities, we need to look at the equipment, utility systems, and process classification.

Any piece of equipment (e.g., heat exchanger, pump and furnace), utility system (e.g., steam system and fuel gas system), or process system (e.g., distillation train and reactor and recycle system) can be subject to operational optimization, maintenance, and physical improvements through engineering or technological advances. In most cases, the technical knowledge base that drives all the different types of improvements for a given item of equipment or system is similar, and it often resides with the same subject matter expert (SME) or group of specialists. In other words, if you are looking for resources for an energy management program—especially human resources—the key question is "What types of equipment or systems will the program address?"

Thus if, for example, an energy manager wishes to check the current performance of a furnace, improve its maintenance, and explore options to upgrade its physical design, most likely the expertise to drive all of these improvements resides in the same individual or group. An evaluation by a fired heater SME may find, for example, that the fired heater may only be able to safely achieve 3.5% excess O_2 due to a sticking stack damper. As a result of this finding, the interim operating target for excess O_2 may be set at 3.5%. However, if the sticking damper can be rectified, the SME may determine that the excess O_2 could be reduced to 2%. This triggers maintenance activities. Finally, the SME may determine that an additional convection bank can be added to the furnace. This would lower the stack temperature, and has significant energy-saving benefits. This leads to a capital project to upgrade to furnace.

SECTION OVERVIEW

The remaining chapters of Section 2 are broadly organized into three categories: types of equipment, utility systems, and process improvements. Given the diversity of the process industries, it is not possible to cover all types of facilities, but the most common ones are addressed; and many types of equipment and technology that are not named in the chapter titles (e.g., distillation columns, motors, control) are nevertheless discussed in examples within the text.

The chapters look at key factors to consider when evaluating existing facilities, and they provide some systematic and not-so-systematic ways of identifying potential improvements. They also illustrate some enabling technologies to ensure efficient long-term operation of equipment. The main focus is on the equipment, systems, and supporting utilities that are required to run the manufacturing processes themselves, although some consideration is also given to non-process issues such as lighting and space heating, ventilation, and air-conditioning (HVAC). The full list of chapters is as follows:

Equipment

Chapter 9: Energy Efficiency in Furnaces and Boilers

Chapter 10: Enhanced Heat Transfer and Energy Efficiency

Chapter 11: Heat Exchanger Cleaning Methods

Chapter 12: Monitoring of Heat Exchanger Fouling and Cleaning Analysis

Chapter 13: Successful Implementation of a Sustainable Steam Trap Management Program

Chapter 14: Managing Steam Leaks

Chapter 15: Rotating Equipment: Centrifugal Pumps and Fans

Chapter 16: Industrial Insulation

Utility Systems

Chapter 17: Heat, Power, and the Price of Steam

Chapter 18: Balancing Steam Headers and Managing Steam/Power System Operations

Chapter 19: Real-Time Optimization of Steam and Power Systems

Chapter 20: Fuel Gas Management and Energy Efficiency in Oil Refineries

Chapter 21: Refrigeration, Chillers, and Cooling Water

Chapter 22: Compressed Air System Efficiency

Chapter 23: Lighting Systems

Chapter 24: Heating, Ventilation, and Air-Conditioning Systems

Process

Chapter 25: Identifying Process Improvements for Energy Efficiency

Chapter 26: Pinch Analysis and Process Heat Integration

Chapter 27: Energy Management Key Performance Indicators (EnPIs) and Energy Dashboards

We trust that, whatever your current level of experience in energy management, you will find the information in these chapters useful and that it will help you to improve the planning and execution of your energy management programs and to achieve your energy efficiency goals.

REFERENCE

1. Rossiter, A.P. (2007) Back to the basics. *Hydrocarbon Engineering*, 12(9), 69–73

EQUIPMENT

9

ENERGY EFFICIENCY IN FURNACES AND BOILERS

Bala S. Devakottai

Chevron Phillips Chemical Company, Houston, TX, USA

INTRODUCTION

In any manufacturing plant, energy consumption is a major portion of operating cost. Energy consumption worldwide is expected to increase 1.4% per year over the next 25 years, from 200 to 307 quadrillion Btu, as shown in Figure 9.1 [1]. Energy is required for running process equipment, for generating steam, for lighting, and for heating and air conditioning. The petrochemicals and refinery industries are energy intensive, consuming over 25% of total energy.

Furnaces and boilers are an integral part of most sites in the process industries. Furnaces are commonly used to process hydrocarbon feeds, and they often also include convection banks that produce steam. Boilers, in contrast, are used to produce only steam. In many process plants, fired duty in furnaces and boilers is the largest single energy input. Individual furnaces and boilers are energy intensive, and therefore it is important to design and operate these units at the best possible efficiency. While many boilers operate under near-constant conditions, furnace operation is often very dynamic.

A major product group in the heart of petrochemical industry is ethylene and other olefins, which are used in the manufacture of plastics and many other derivatives. More than 60% of the energy consumption in a typical olefins plant is in the furnaces. Olefins

Energy Management and Efficiency for the Process Industries, First Edition. Edited by Alan P. Rossiter and Beth P. Jones.

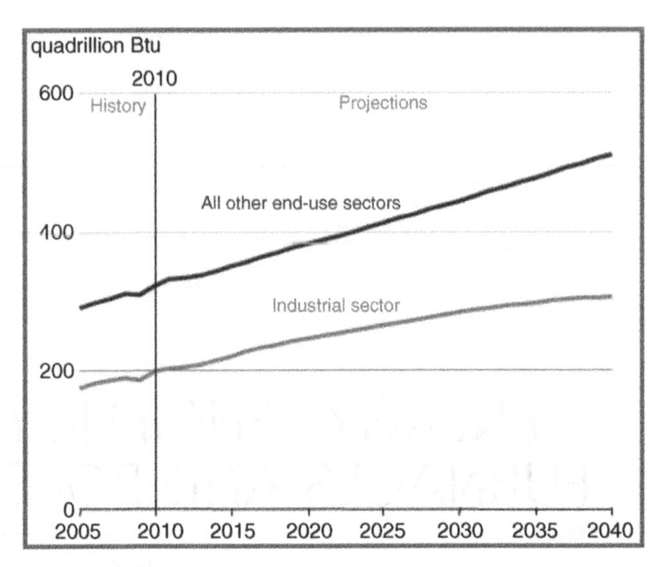

Figure 9.1. Energy demand projection. ([*Source:* U.S. Energy Information Administration [1]].)

furnaces are among the largest items of fired equipment within the process industries, and they typically run at higher firebox and process temperatures than other types of process furnaces and boilers.

In this chapter, we will focus on olefins furnaces to illustrate the factors that affect the performance of fired equipment. We will also explore principles of energy-efficient design, retrofit, maintenance, and operation that apply across all types of furnaces and boilers, as well as some specific issues applicable to individual equipment types.

DRIVERS FOR ENERGY REDUCTION

Furnaces and boilers are designed to maximize the thermal efficiency of the combustion and to maximize the energy recovery given the physical capabilities of their components and the material, product, and fuel economics of the time they are designed. While the furnaces and boilers built in recent times are highly energy efficient, many plants operate with assets that were built in the 1970s, which are typically not as efficient. Fuel and furnace material costs and the capabilities of burners, furnace tubes, insulation, and leak prevention have all changed since then. Some of the drivers for modification of older furnaces and boilers are to

- reduce major operating expenses,
- reduce CO_2 and NO_x emissions,
- comply with regulations, and
- maintain the economic viability of assets.

THERMAL EFFICIENCY

API 560 [2] gives a simple definition of thermal efficiency as follows:

$$\text{Thermal efficiency} = \text{total heat absorbed}/\text{total heat input}$$
$$\text{Total heat absorbed} = \text{total heat input} - \text{total heat loss}$$

Heat loss can occur from the radiant section walls, convection section walls, and from the stack gas (Figure 9.2). Stack losses are often exacerbated by allowing more air into the firebox than is needed for safe and effective fuel combustion.

Since the purpose of boilers is to make steam, fuel-to-steam efficiency is sometimes also referred to as boiler efficiency. Utility boiler efficiency is measured using ASME Power test code 4, or ASME PTC 4, an industry standard prescribed by the American Society of Mechanical Engineers.

When comparing efficiencies, one has to be careful in defining whether efficiency is calculated using the lower or higher heating value of fuel. In the United States, boiler efficiencies are usually calculated using higher heating value, whereas in Europe they are usually calculated using lower heating value, which yields a higher numerical value for the efficiency. The higher heating value is the thermodynamic heat of combustion or the enthalpy difference between all combustion products, including condensed water, and the fuel, at a standard temperature (commonly 77 °F). The lower heating value subtracts the heat of vaporization needed to condense the water vapor combustion product—basically assuming that energy cannot be recovered from condensing water in the stack gas.

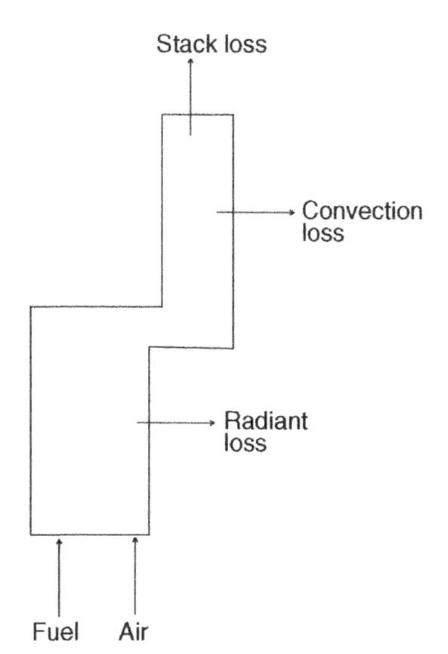

Figure 9.2. Components affecting thermal efficiency.

Stack Loss: Stack Temperature

By modern standards, furnaces built in the 1960s and early 1970s typically have low efficiencies, commonly with stack temperatures between 400 and 550 °F. Figure 9.3 shows thermal efficiency (lower heating value—LHV basis and higher heating value—HHV basis) as a function of stack temperature, at typical operating conditions of 2% O_2 at the top of fire box and 1.5% heat loss through the walls, with two different fuel mixes—90 mol% methane/10 mol% hydrogen and 20 mol% methane/80 mol% hydrogen. As shown in the graph, the overall thermal efficiency of a furnace does not change much over a wide range of fuel gas compositions.

The absorption of heat from stack gases is limited by the heat exchange area (discussed in the next section) and metallurgy limits of the furnace components, which are usually related to the condensation temperature of the acid components of the combustion products—the acid gas dew point. Many modern furnaces have a net (LHV) efficiency above 94%, with a stack temperature of about 230 °F (110 °C) for furnaces that burn clean fuel gas, the water dew point being 212 °F. If there is any sulfur in the fuel, it increases the acid dew point of the flue gas. Even a small amount of sulfur increases the dew point significantly, as shown in Figure 9.4.

Every licensor of olefins or other furnaces will design for the lowest possible stack temperature to maximize efficiency, within the constraints of plant-specific feed and utility temperatures and fuel gas composition. Economics and available stream temperatures determine the approach between the stack and process inlet temperatures. For

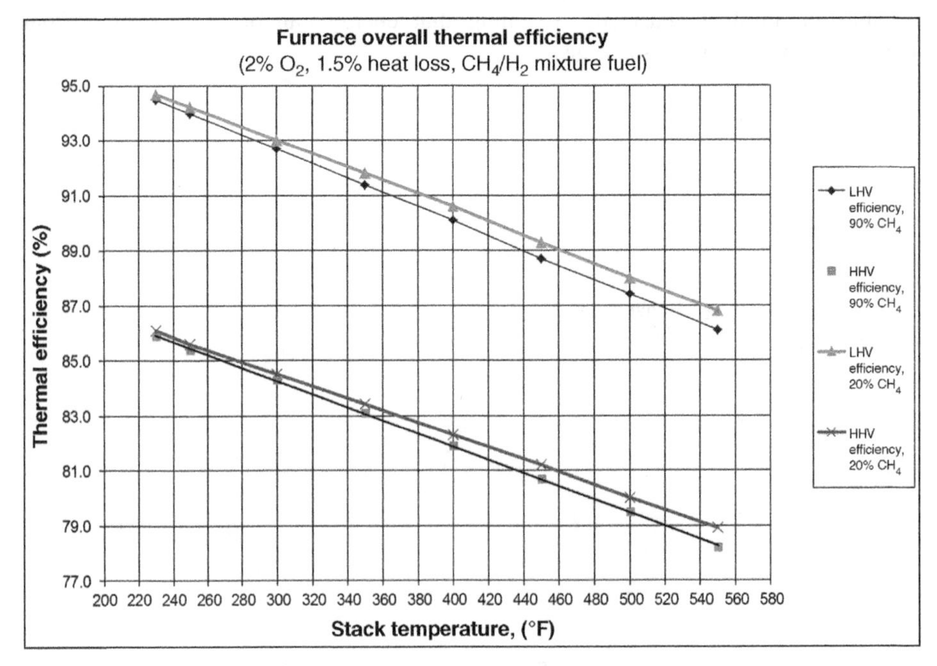

Figure 9.3. Thermal efficiency versus stack temperature.

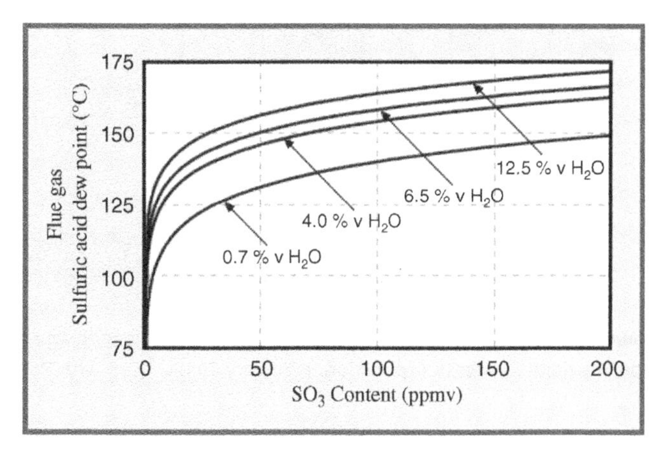

Figure 9.4. Sulfuric acid dew point as a function of SO_3 content. (Courtesy ChemEngineering, chemengineering.wikispaces.com. Used with Permission.)

example, if the feed comes into the convection section at 230 °F at the top, the minimum economical stack temperature, a tradeoff between the cost of additional heat recovery surface area and the benefit of the recovered heat, is typically 260–270 °F.

Boiler stack temperatures are usually in the range of 320–500 °F, with a typical best efficiency of about 85%. Stack loss is the biggest limit on efficiency and is typically in excess of 10% in boilers. Many older boilers have stack losses as high as 20–30%.

Other Components of Thermal Efficiency

In this section, we will discuss how different components of olefins furnace design have contributed to increasing thermal efficiency over the years. Again, the olefins furnace is an extreme example of process furnace design, and most improvements in olefins furnace technology can be applied to boilers and other process furnaces.

Insulation and Sealing. The firebox is the heart of a furnace. In olefins furnaces, this is where the cracking reaction takes place to thermally break down hydrocarbon feeds into valuable products. The cracking process is highly endothermic, and large amounts of heat must be input to the process fluids to crack the hydrocarbons to lighter components. The reaction takes place inside the radiant coil, where radiant heat is provided by combusting fuel in burners.

In the 1970s and 1980s, tube alloy materials were less resilient and their corresponding maximum firebox temperatures were lower than those for modern furnaces. Olefins furnace fireboxes were once operated at 2000 °F or less. The furnaces typically had a refractory thickness of 9 in. to maintain the outer skin temperatures under the API recommended 180 °F. Insulation bricks were designed for 2300 °F.

Over the years, both furnace capacity and radiant heat intensity have increased. Most modern furnaces operate with firebox temperatures in excess of 2200 °F and could have

Figure 9.5. Lower wall of firebox with firebrick with ceramic fiber in the upper section. (Courtesy Thorpe Specialty Services Corporation, Houston, Texas. Used with Permission.)

fuel-firing rates in excess of 400 MBtu/h (400 million Btu/h) in each furnace. Since firebox temperature has increased over the years, it is even more important to minimize heat loss from the firebox with sufficient insulation to reduce the external skin temperature. A combination of high temperature brick and ceramic fiber insulation with 13 in. thickness, designed for 2600 °F, is not uncommon in modern furnaces. The choice of modern refractory is discussed in detail by Thorpe [3]. Firebox floors are typically lined with 2600 °F castable insulation material. Typically, the bottom 10–12 ft of the firebox is lined with brick due to the possibility of erosion from burners and the possibility of refractory damage due to proximity to the entry door. The middle and upper portion of the firebox, radiant roof, and transition into convection section are lined with ceramic fiber, as shown in Figure 9.5.

An olefins firebox has multiple levels of peep doors—48 or more in each furnace—to view and monitor the firebox performance. These doors require holes in the internal refractory and are routinely opened and closed. It is important that the peep doors have proper refractory, consistent with firebox refractory, to minimize heat loss and to avoid cold air infiltrating into the hot firebox, both of which reduce furnace efficiency.

Radiant tubes hang vertically, with multiple tubes entering and leaving the radiant roof. The roof insulation, with multiple radiant tubes penetrating the roof, is equally critical. Proper sealing of roof openings is critical to avoid excess air leakage into the firebox, which decreases efficiency. Engineered insulation boots covering the openings will minimize air infiltration, as shown in Figure 9.6.

Heat is recovered from the hot firebox flue gases in the convection section. The convection section is critical to maximizing heat recovery and achieving optimum thermal efficiency for the furnace. The bottom section of the convection section sees flue gas temperatures in the range of 1900–2000 °F. This section is typically insulated with ceramic fiber modules. As the flue gas goes further up toward the stack and heat is recovered, its temperature drops. The upper parts of convection section are typically lined with lightweight castable refractory.

Olefins convection sections generally contain 30–40 rows of horizontal tubes, with multiple parallel passes entering and leaving the convection box, creating over

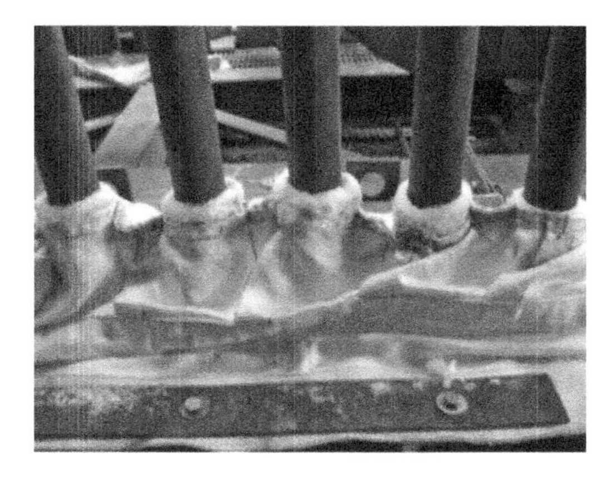

Figure 9.6. HOTSEAL™ - Tube penetration seals. (Courtesy Thorpe Specialty Services Corporation, Houston, Texas. Used with Permission.)

200 penetrations. It is important to seal these penetrations, as shown in Figure 9.7, to prevent cold air from leaking into the convection section and reducing thermal efficiency.

Like furnaces, boilers are insulated to minimize radiant and convection losses and maximize boiler efficiency. Typical modern boilers have a skin heat loss of 1.5% at design conditions. If the boiler operates at lower throughput, the percentage losses will increase. In order to maximize efficiency, it is better to operate fewer boilers at higher load than to operate multiple boilers at turndown.

Furnace Capacity. In the 1970s and 1980s, it was common to see an 800 million lb/year capacity ethylene plant with 10–12 furnaces. Over the years, plant capacities have grown and so have furnace capacities. Today, an ethylene plant capacity in excess of 3 billion lb/year is very common. Correspondingly, the furnace capacities have grown with current capacities exceeding 400–450 million lb/year (200+ kta) of ethylene per furnace. The radiant coils of each furnace could be in two medium-length radiant boxes with a common convection section or in one long radiant box. Today, it is common to see a 3.0 billion lb/year (1500 kta) ethylene plant with only six to seven operating furnaces. Figure 9.8 shows how furnace capacity has grown over time [4].

Figure 9.7. Convection tube penetration and sealing. (Courtesy Thorpe Specialty Services Corporation, Houston, Texas. Used with Permission.)

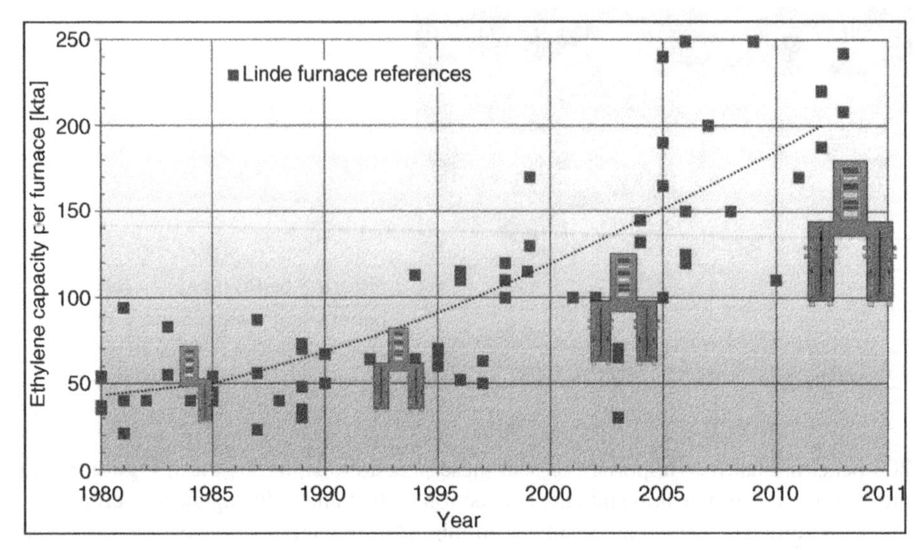

Figure 9.8. Furnace capacity growth over time. (Courtesy of Linde Engineering. Used with Permission.)

In the 1970s and 1980s, this capacity would have required over 24 furnaces. The energy efficiency benefit of reducing the number of furnaces is in easier maintenance of different components of the furnaces, including minimizing air leaks and maintaining refractory, which reduces heat loss.

Boilers with a wide range of capacities are used within the process industries. In most oil refineries and large petrochemical plants, the individual boilers have capacities between 100,000 lb/h and 1 million lb/h. However in other sectors, such as specialty chemicals and food and beverages, boiler sizes down to 10,000 lb/h and less are common. Typical utility boilers, in contrast, produce in excess of 3 million lb/h of steam.

Radiant Coil Design. Many older furnace designs have radiant coil guide support pins penetrating the floor, as shown in Figure 9.9. It is important to seal and insulate these openings properly to prevent cold air leaks that increase fuel consumption and reduce furnace efficiency. In modern furnaces, the guide support pins either have been eliminated or the pins go into a trough on the floor, with no floor penetration, to reduce cold air infiltration and energy loss.

Burners. Before the 1990s, it was common to install multiple rows of wall burners in olefins furnaces, with a typical firing rate of 1 MBtu/h per burner. Wall burners are premix burners, with air coming in with the fuel. It was difficult to control excess air in a furnace with over 144 burners at multiple levels, as shown in Figure 9.10a . Since the mid-1990s, the wall burners have mostly been replaced by floor burners, with each floor burner firing in excess of 10 MBtu/h capacity. For a big furnace, each firebox typically has 24–32 floor burners, as shown in Figure 9.10b, leading to much easier control of

Figure 9.9. Radiant floor penetration and sealing. (Courtesy Thorpe Specialty Services Corporation; Houston, Texas. Used with Permission.)

excess air. In addition, many modern furnaces are fitted with a jack-shaft arrangement to control excess air through the floor burners. This makes it easier to control the excess air over a wide range of operating conditions, including decoke and turn down.

Even though most other process furnaces and boilers operate at lower firebox temperatures and have fewer burners, the same principles apply for maximizing efficiency: using refractory and insulation to minimize heat losses, maintaining good control of excess oxygen, and limiting air infiltration to the furnace.

Boilers can fire a wide range of fuels, including natural gas, fuel oil, and solid fuel. Burners that burn fuel oil need steam as an atomizing medium. The additional steam in

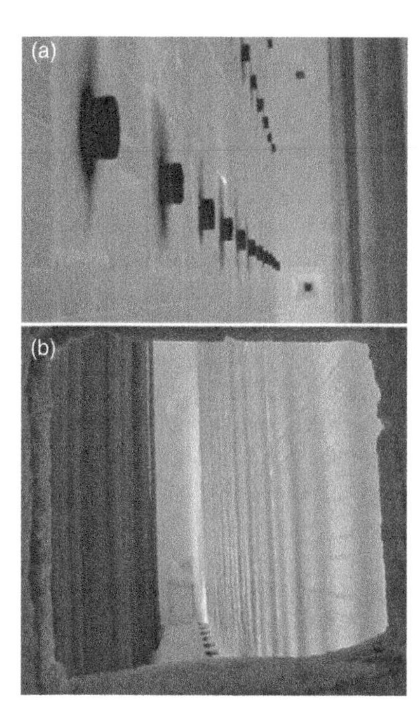

Figure 9.10. Radiant wall burners (a) and radiant floor burners (b). [*Source:* Zeeco Inc., Burner Brochure. Used with permission.]

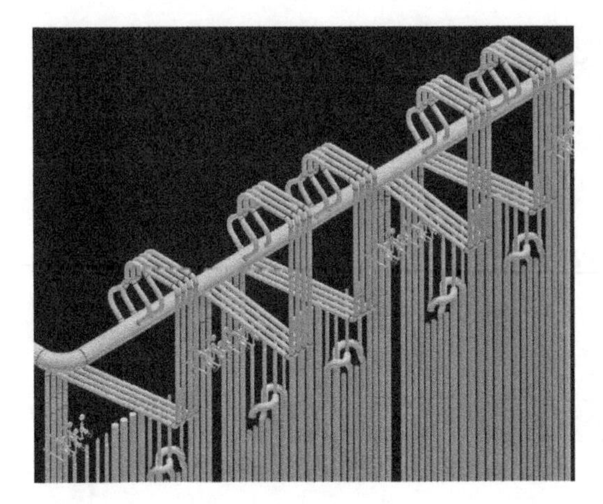

Figure 9.11. Typical modern furnace with multiple radiant tubes. (Courtesy of Linde Engineering. Used with Permission.)

combination with the higher excess air required to burn fuel oil increases the fuel requirement for oil burning. Many older burners have low fuel efficiency and high NO_x production. Over the years, due to regulation, the burners in many boilers have been replaced with low-NO_x burners with high combustion efficiency.

Uniform Flow to Radiant Coil. A typical modern cracking furnace could have in excess of 150 radiant tubes, as shown in a computational flow dynamics (CFD) model of a firebox in Figure 9.11. If the feedstock flow is not uniform in each of the radiant tubes, coke will form unevenly in different tubes, leading to increased firing. In order to ensure uniform feed flow to each radiant tube for maximum furnace efficiency, critical sonic flow venturis are typically installed at the inlet of each radiant tube.

Convection Section. In a process heater or boiler, which operates at lower temperatures, the convection section will be located directly above the radiant section, but the convection section in an olefins furnace must be shielded from high firebox temperatures. The olefins convection section is offset with respect to the firebox so that there is no direct radiation to the bottom rows of tubes (shock tubes), as shown in Figure 9.2. Direct radiation could lead to localized overheating, reduced heat transfer, and premature failure of tubes, all of which lead to reduced efficiency.

The olefins convection tubes are arranged in a triangular pattern between rows of tubes, or in a triangular pitch, to maximize heat transfer to the tubes. In a triangular pattern, end tubes in alternating rows have a larger gap between the tube and the end wall. These larger gaps are filled with refractory corbels to keep the same distance between end tubes and end walls in all rows of tubes, as shown in Figure 9.12. The corbels near the end tubes in each row break up flow and minimize flue gas channeling, thus maximizing efficiency.

The heat recovery in the convection section of an olefins furnace can be divided into two services—process tubes with mixed hydrocarbon and steam flowing into the radiant section, and waste heat recovery tubes with boiler feed water and high-pressure steam. In order to minimize fuel gas usage, the process gas is preheated in the convection section to

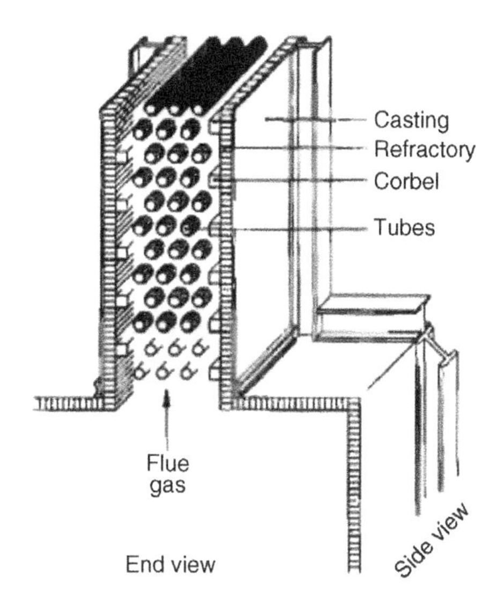

Figure 9.12. Convection tubes showing corbels. (Presented at the Industrial Energy Technology Conference in 1981 [5]. Used with permission.)

the maximum extent possible for a given feed before entering the radiant section. The remaining flue gas heat is recovered partly by preheating boiler feed water before entry to a steam drum and partly by superheating the high-pressure saturated steam generated in the steam drum.

Boilers usually use convection section tubes, or economizers, to recover heat from stack gases and preheat boiler feed water. Extended tube surfaces are often used to increase heat transfer area. Many boilers also recover more energy and gain efficiency by including a steam superheat section, where the high-pressure steam is superheated above saturated temperature.

Blowdown Losses. Some of the hot water from within a boiler—typically between 2 and 10% of the feed rate—is drawn off (blown down) to remove contaminants that would otherwise build up and form scale deposits. Fuel is required to heat the feed water to the drum operating temperature and pressure. The blowdown flow must be made up with colder feed water, and heating this makeup water requires additional fuel. This can lead to reductions of 1% or more in boiler efficiency. Ways to reduce blowdown losses are discussed in Chapter 17.

As noted earlier, olefins furnaces also generate significant amounts of steam. Consequently, they too have blowdown streams—typically 2–3% of boiler feed water flow. However, as the duty is divided between process heating and steam generation, the impact of blowdown losses on overall energy efficiency in olefins furnaces is generally less significant than it is for boilers. Many furnaces in other services (e.g., hydrotreater furnaces in refineries) also generate limited amounts of byproduct steam, and they also require blowdowns.

Heat Integration. In the 1970s and 1980s, as shown in Figure 9.13a, it was common in olefins furnaces to route the saturated steam from the steam drum (typically at

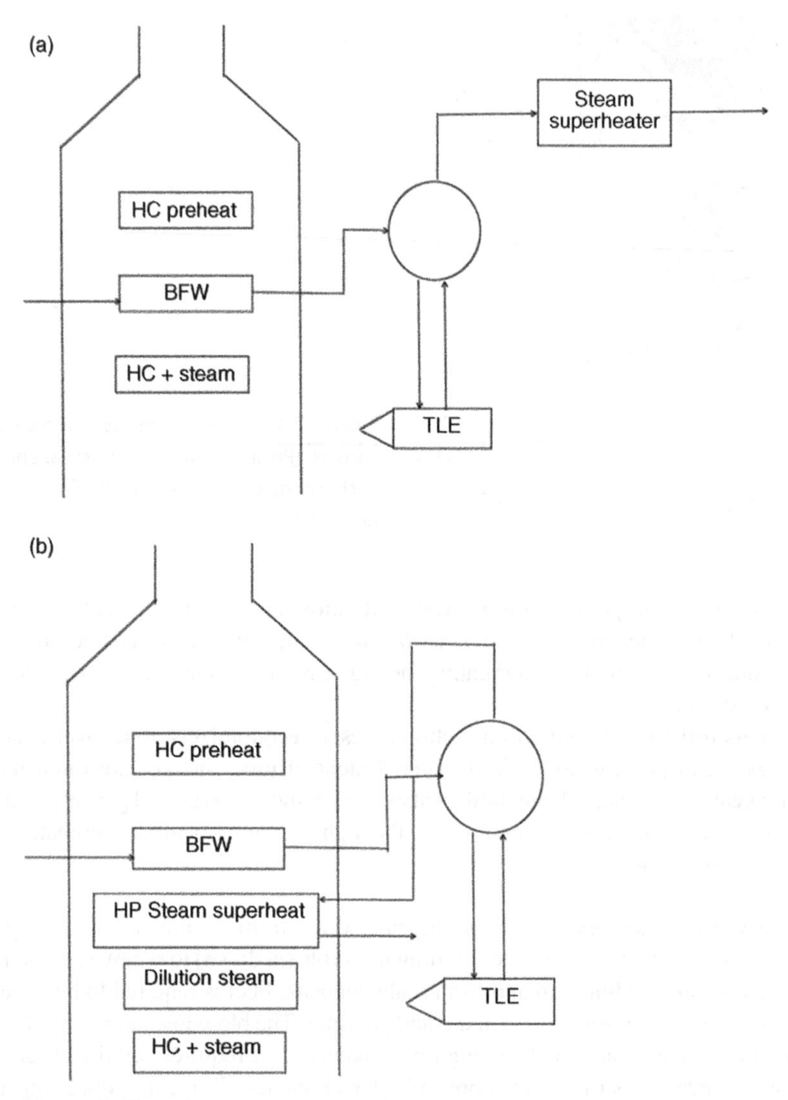

Figure 9.13. Heat integration in olefins furnaces—1970s (a) and modern (b).

1500 psig and 612 °F) to an external steam superheater to superheat the steam to the temperature required for steam turbines (typically 900 °F). The steam superheater consumed additional fuel to superheat steam, often operating at a low thermal efficiency.

Since the late 1980s, the high-pressure steam has been superheated within the olefins convection section, as shown in Figure 9.13b. This increases furnace efficiency and simultaneously eliminates the need for an external steam superheater [6].

While air preheat was used to improve olefins furnace efficiency in the 1980s, it is not commonly used in modern furnaces. The increase in inlet air temperature increases

NO_x formation in the burners and the environmental debit negates any benefit in efficiency improvement.

Quench Exchangers. The olefins cracked gas passes from the radiant tubes to the quench exchangers, where the gas is rapidly cooled from about 1500 to <900 °F. With gas crackers (ethane, propane, and butane feed), the quench exchanger process outlet temperature can be as low as 400–450 °F, further improving overall efficiency.

Older furnaces typically have shell-and-tube heat exchangers, which are often called transfer line exchangers (TLE)s, in the quench service. TLEs are prone to fouling and plugging, which reduce heat recovery and require offline mechanical cleaning. Quench exchanger design has evolved to address these concerns. Most modern furnaces' primary quench exchangers are double pipe exchangers, which have a lesser tendency to foul and are easier to clean online during decoke. The diameter of TLE process tubes is considerably bigger today, too, to minimize plugging and fouling.

Induced Draft Fan and Stack. Until the 1970s, many olefins furnaces were built with a natural draft stack to move the flue gas. In order to minimize stack height, the convection section was built with reduced tube surface for lower flue gas resistance, which led to reduced heat recovery. A convection section height of 15–20 ft was not uncommon.

Since the late 1970s, olefins furnaces have been built with an induced draft fan to move the flue gases, providing the means for higher efficiency heat recovery. A modern convection section could be as tall as 40 ft and a best-in-class furnace will have a thermal efficiency of 94% or higher.

The choice of natural or induced draft in boilers and process furnaces is a tradeoff between equipment costs and heat recovery benefit. Large utility boilers typically use a balanced draft system to maintain a slightly negative draft in the boilers with a combination of a forced draft fan and an induced draft fan. An air damper is used to control the airflow and the draft.

Instrumentation and Control Philosophy

In order to calculate overall thermal efficiency, it is important to have proper instrumentation around furnaces and boilers. The first important measurement is the total heat input, or total fuel gas flow[1]. The fuel gas composition is seldom constant in an olefins furnace. The composition of fuel gas is typically measured online with a gas chromatograph, the output from which is used to correct the fuel gas mass flow for molecular weight and also for calculating the heating value of the fuel gas. If the fuel gas molecular weight ranges widely, for example from 80% hydrogen/20% methane to propane or butane as a backup fuel, it is common to use a Wobbe calorimeter and a Wobbe index [7] for fuel composition (Figure 9.14).

[1] This discussion assumes that the furnace uses a site-specific fuel gas, which is common in refineries and petrochemical plants. Many different fuels can be used for furnaces and boilers in other facilities—for example, natural gas, fuel oil, coal, or biomass. In all cases, the basic requirement of measuring heat input applies, but the details of the measurement can vary significantly depending on the fuel type.

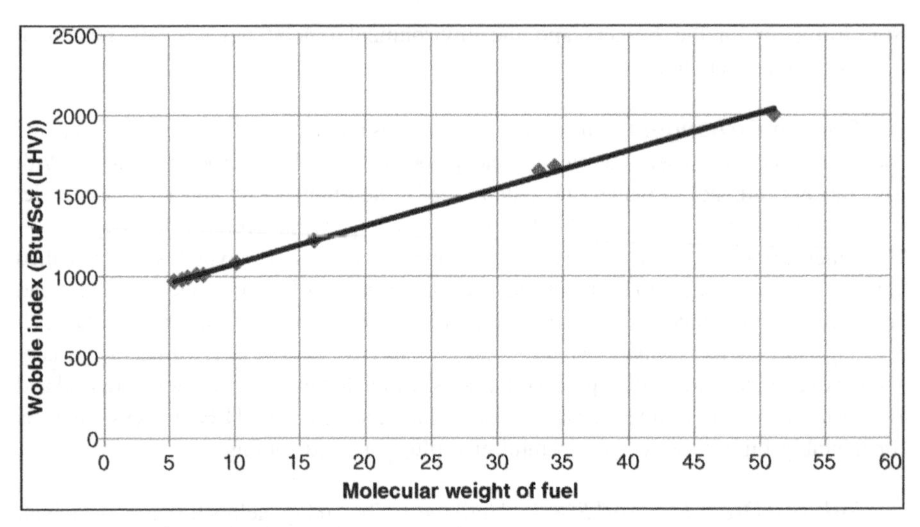

Figure 9.14. Wobbe index of fuel gas versus molecular weight.

Another important measurement for furnace efficiency calculations is excess oxygen in the firebox. Oxygen is often measured at the top of the firebox, and some plants install redundant oxygen analyzers. Oxygen in the stack can also be measured online or manually at a regular frequency with a hand held portable meter to estimate air leakage in the convection section.

One way to minimize air leakage is to operate the arch draft as high as possible, typically at 0.1 in. of water. The draft measurement is based on the difference between firebox pressure and atmospheric pressure. Draft in the furnace is controlled by using a stack damper installed at the inlet to the induced draft fan. Stack temperature measurement is another critical parameter in the calculation of furnace efficiency.

In an olefins furnace, steam is added to the feedstock to decrease the partial pressure for the cracking reaction and to reduce coking. For a given operating condition in a furnace, fuel consumption is impacted by feed flow and dilution steam flow, both of which must be measured properly. The higher the dilution steam flow, the higher the firing rate for a given feed flow. Efficiencies are often compared between furnaces in a plant by calculating specific fuel consumption, or Btu/lb of feed.

Furnace control philosophy is equally important in maximizing furnace efficiency. The desired cracking severity, as measured by radiant coil outlet temperature, controls the fuel gas flow to the furnace. Draft at the arch is controlled automatically by opening the stack damper. Oxygen, or excess air, is controlled by opening burner registers, either manually or by using a jackshaft.

For boilers, the two major control loops are feed water control and combustion control. Steam demand can change suddenly, and in order to maintain a constant steam drum level, feed water flow rate must be adjusted quickly. Similar to olefins furnace feed control, a robust three element feed water control will help with quick response to changes in steam drum level. Combustion control involves controlling both the fuel and

TABLE 9.1. Typical Flue Gas Oxygen Content Control Parameters

	Automatic Control		Positioning Control	
Fuel	Minimum[a]	Maximum[a]	Minimum[a]	Maximum[a]
Natural gas	1.5	3	3	7
No. 2 fuel oil	2	3	3	7
No. 6 fuel oil	2.5	3.5	3.5	8
Pulverized coal	2.5	4	4	7
Stoker coal	3.5	5	5	8
Stoker biomass	4	8	5	8

[a]Flue gas O_2 content, % full gas sample.
Source: U.S. Environmental Protection Agency [9].

the air to maintain the correct amount of excess air in a boiler at a given steam load. Boilers are high-energy devices and can become unsafe without proper combustion controls. Boiler burner management systems are designed per the NFPA 85 code, which provides standards for operating safety and avoiding boiler explosions.

Excess Oxygen Control in Boilers. Excess oxygen (or excess air) is also a key factor in the efficiency of all types of furnaces and boilers. Ideally, only the stoichiometric amount of oxygen in air should be supplied to a boiler or furnace, but in order to ensure essentially complete combustion, it is always necessary to provide some excess. This increases the flue gas flow rate, and as the flue gas is hotter than ambient temperature, it also increases the amount of heat lost up the stack. Flue gas oxygen should, therefore, be controlled to minimize excess oxygen subject to the requirement that the fuel supplied to the furnace or boiler is, for practical purposes, fully combusted [8].

In order to control excess oxygen, it is necessary to measure it. This is accomplished using an analyzer—either a portable device used periodically, or a permanent one installed within the boiler or furnace. As a minimum, the analyzer measures the excess oxygen content. Some analyzers also measure combustibles or carbon monoxide (to ensure substantially complete combustion), or NO_x and other chemical species required for environmental compliance.

A control mechanism is also required. Control is generally achieved in boilers by using a combustion airflow control device such as a fan outlet damper, inlet guide vane (IGV) control, or variable speed control (see Chapter 15).

Table 9.1 shows typical excess oxygen control limits for different types of boilers, based on full gas sample (wet basis) measurements[2] leaving the combustion zone.

Table 9.1 also shows two different types of control. Positioning control is typically used in boilers without continuous flue gas oxygen measurement. Combustion airflow is

[2] Flue gas consists primarily of the products of combustion (mostly carbon dioxide and water vapor), together with unreacted oxygen and nitrogen from the air. Wet basis measurements include all of the components of the flue gas, including the water vapor, in the denominator when computing the percentage of excess oxygen in flue gas. In contrast, the water is condensed and removed in dry basis measurements.

adjusted in a preset manner in response to changes in the fuel flow setting. The relationship between the airflow and fuel flow settings can be adjusted; however, changes can be implemented when periodic excess oxygen measurements are made. In contrast, when continuous measurements are available the more efficient automatic control can be implemented. Here the flue gas oxygen content is monitored continuously, and the combustion airflow is automatically adjusted to stay within the desired oxygen limits.

Maintenance of Furnace Components for Peak Efficiency

Olefins furnace operation is very dynamic, with a wide range of operating conditions and modes—start-up, decoke, feed-in, feed-out, normal operation, changing feedstocks during a run, and shut down. These various operations lead to different components of the furnace being exposed to cyclic temperatures from ambient to over 2000 °F, with possible quick temperature swings of over 250 °F. In this section, we will discuss how the different furnace components should be serviced to maintain the best possible efficiency.

Refractory Maintenance for Furnaces and Boilers. Refractory vendors recommend a maximum heating and cooling rate of about 200 °F/h to avoid excessive shrinkage of ceramic fiber. The ceramic fiber blocks are held together by compression and repeated temperature cycles cause gaps between modules. If these gaps are not filled back, heat will seep through the gaps and heat up the furnace casing, leading to heat loss and other maintenance issues. Over the years, hot spots in carbon steel casings, and in some cases buckling of casings, have been reported by many plants. Buckling will also lead to uneven surface temperatures and corresponding uneven heat distribution in the firebox, causing a loss of efficiency.

In order to extend the longevity of furnace refractory and prevent deterioration and increased heat loss, it is important to inspect the refractory and make minor repairs as required during every outage and furnace entry. Furnaces are designed typically for a 1.5% heat loss from the refractory. It is not uncommon to see heat loss increase to as much as 3% over the years due to lack of refractory maintenance, resulting in a corresponding 1–2% drop in furnace efficiency (Figure 9.15).

One of the major problematic areas in a boiler is refractory failure. Often, refractory in a boiler is subjected to rough treatment. Seldom do boilers burn clean fuel. It is important to choose refractory appropriate to the fuel being burned to avoid refractory failures and loss of efficiency. Equally important is the curing and dry out of refractory after repair.

Operating Above Design Firing Rates. Another reason for increased heat loss compared with design could be operating the furnace at a higher firing rate than design. The refractory thickness is designed based on inside surface temperature, thermal conductivity of the refractory material, ambient temperature, and wind conditions. Many plants operate the furnaces at higher than original design throughput, gradually pushing the firing rate above the design and causing the firebox temperature to increase. If the refractory thickness or material is not changed correspondingly, this will lead to higher skin temperatures and higher heat loss.

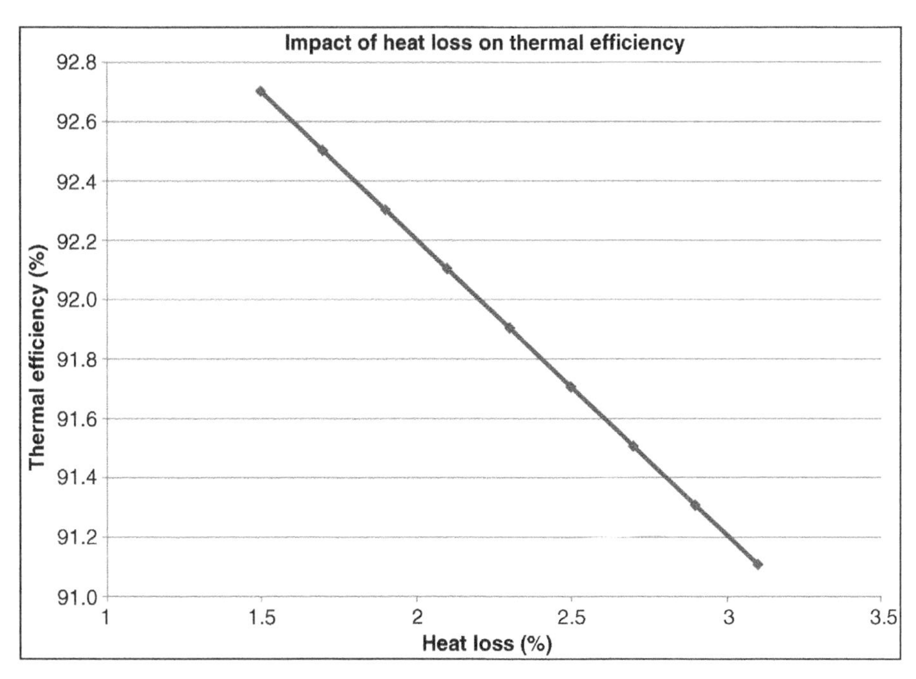

Figure 9.15. Impact of heat loss through refractory on thermal efficiency.

Burner Maintenance. Significant effort goes into burner design and testing to ensure that the burners can provide a uniform heat distribution. But during real life operations, burners often see fuels that may not be in the range of the initial design. Fuel gas could be contaminated from methane used during regeneration of catalysts, from upset conditions (introducing ethylene, acetylene, etc.), or from waste streams being mixed into the fuel gas drum. These contaminants often lead to burner tip plugging and inefficient combustion, and a resulting loss of efficiency. Burner tip maintenance is one of the most easily neglected parts of maintaining furnace efficiency. It takes diligent effort to inspect all the burners routinely while in service and to clean them online as required. Many plants employ a dedicated maintenance crew for this effort. The crew makes daily burner inspections, tags dirty burners, and cleans the tips as required. The added benefit is that clean burners help keep NO_x emissions low.

Tramp Air Infiltration. While the furnaces are designed for minimum excess air to maximize efficiency, they are often operated differently in the field. The oxygen required for combustion is controlled with draft and burner register positions to the desired oxygen and draft levels, typically between 1.5 and 2.0% O_2 and a pressure of −0.1 in. of water at the arch. Since the furnace firebox operates under a negative pressure, any opening in the furnace leads to unwanted, cold tramp air entering the furnace. This lowers efficiency, as shown in Figure 9.16. The fuel consumption could increase 1–2% from tramp air incursion.

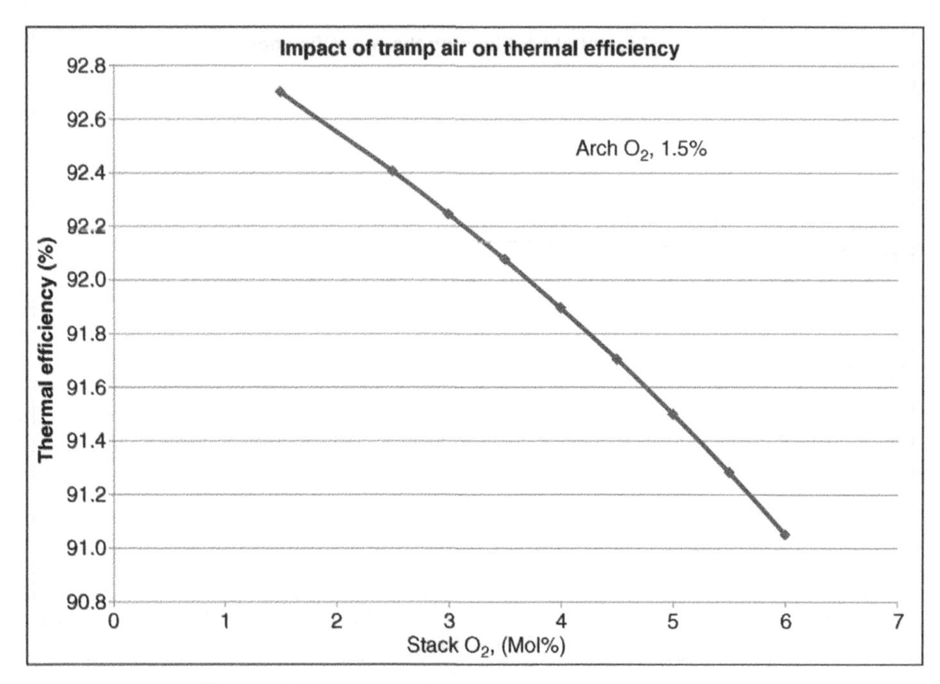

Figure 9.16. Impact of tramp air on thermal efficiency.

There are several sources for tramp air leaks. Many furnaces have radiant support guide pins through the floor and tube supports through the roof of the firebox. During normal operation the tubes expand and contract, moving the insulation around these penetrations. Many times, during a coil retube or other maintenance, this insulation is removed. If these gaps are not sealed back properly, tramp air will increase leading to a loss of efficiency.

As discussed earlier, modern convection sections could have over 40 rows of tubes, with six to eight pass arrangements, which lead to over 240 tube penetrations through the convection wall. These tubes expand and contract horizontally during normal operation. It is important to seal these penetrations and inspect the seals frequently. These seals are typically overlooked during routine furnace maintenance.

Another major source of tramp air is peep doors in the firebox. For furnace monitoring, multiple peep doors—typically 48 or more per furnace—are provided along the length of the firebox and in the end wall at multiple levels. These peep doors are used every day and need to be sealed completely. Attention must be paid to the peep door design, insulation, and sealing to ensure that the doors do not act as a source of tramp air infiltration.

In boilers, too, unwanted air can infiltrate into the combustion chamber from cover leaks, observation ports, gaskets, and other openings that are not sealed properly. Since fuel must be supplied to heat up the unwanted air that is not used for combustion, this in turn will lead to lower boiler efficiency.

Quench Exchanger Cleaning. Olefins furnace effluent temperature leaving the radiant coil is in the range of 1475–1575 °F. Heat is recovered from this stream in quench exchangers by producing high-pressure steam. Gas crackers do not produce much fuel oil, which can condense and foul at low temperatures, and hence their quench exchangers (TLEs) are typically designed for a final outlet as low as 400 °F. As the furnace run progresses, the outlet temperatures remain fairly steady.

Liquid crackers produce a large amount of fuel oil, which will condense and foul the quench exchangers. The heavier the liquid feedstock, the higher will be the fouling rate of the quench exchangers. Liquid feedstock with a heavy tail (high boiling point portion), such as condensate, will also cause increased fouling of TLEs. For example, light naphtha will tend to foul exchangers less than heavy gas oil. A liquid cracker is typically designed for a clean quench outlet temperature of 650–800 °F and an end-of-run condition of 1200 °F. While an online decoke will partially clean the TLEs, outlet temperatures will never be restored to clean condition without an off-line mechanical cleaning. The higher the quench outlet temperature, the lower will be the overall energy efficiency of the plant. The loss in heat recovery steam must be made up by other means such as external boilers. Depending on steam demand and furnace availability, some plants clean the TLEs every three furnace decoke cycles, while others may choose to clean once a year. In order to maintain energy efficiency, quench exchanger cleaning frequency needs to be taken into account in furnace availability calculations.

Convection Cleaning. Extended surfaces are often used to increase heat recovery in furnace and boiler convection sections (see Chapter 10). Olefins convection section tubes have fins with tight spacing in order to maximize efficiency. Over the years, dust and debris collect over convection tube fins [10], as shown in Figure 9.17a, and efficiency tends to drift downwards until the convection section is cleaned [Figure 9.17b].

The convection tubes need to be cleaned on some frequency to restore efficiency. Some furnaces have seen a stack temperature reduction of 50 °F or more when the convection section is cleaned after 5 years of operation, while other plants have not seen much of a drop. The cleaning frequency can be adjusted based on individual site experience. It is also a good practice to make sure that the furnace floor is cleaned of any debris after maintenance, since debris could be sucked onto the tube surface by the induced draft fan.

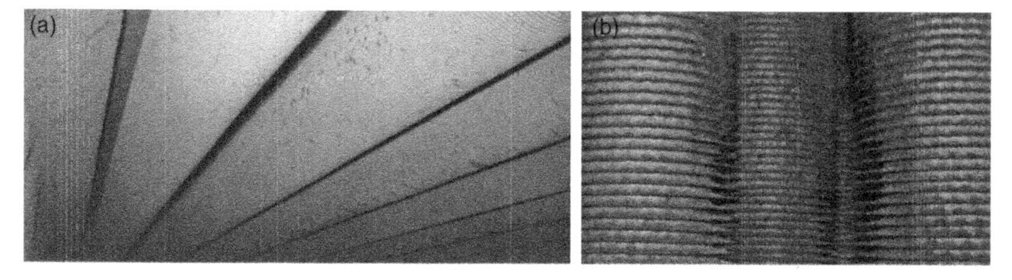

Figure 9.17. Fouled convection section (a) and cleaned convection section (b). (Courtesy of FINFOAM, a Division of Thompson Industrial Services, LLC. Used with Permission.)

Furnace Instrumentation Maintenance. Several critical components for furnace efficiency calculations (as described in the previous section) need constant attention. Furnace fuel gas composition analyzers must be maintained properly to get an accurate fuel flow measurement and heating value. Over the years, plants often make changes to the fuel gas system, such as the addition of new branches or localized branches to a few furnaces. These changes should be reflected properly in the fuel composition measurement. Firebox oxygen analyzers must be calibrated on a regular basis to maintain peak efficiency at all times. If excess oxygen is too low, this will lead to incomplete combustion and spikes of carbon monoxide and possible after burning in the convection section. If excess oxygen is higher than required, this will lead to higher fuel requirement and a drop in efficiency.

Stack temperature is often easy to measure and requires the least maintenance. Draft measurement, to ensure minimal air leakage and fuel requirements, is trickier. Draft measurement involves measuring a steady atmospheric pressure. Reliable and steady draft measurement requires protection from wind fluctuations.

Feed composition changes should be reflected in the feed flow measurement devices. Debris coming in with the feed, especially in liquid feed, can mislead feed flow measurement devices. In many plants, dilution steam pressure changes over time and may not be reflected in the dilution steam measurement orifice calculations, which could lead to improper dilution steam flow and the corresponding increased fuel requirement.

In order to control fuel firing to the furnaces at a minimum, it is important to service radiant coil outlet thermocouples during every furnace maintenance shutdown. The thermowells and thermocouples exposed directly to process flow are subject to erosion from velocity and coke particles. In many furnaces, the thermowells are rotated during every outage. Many furnaces use skin-type thermocouples for coil outlet measurement. These thermocouples cannot be replaced online and must be diligently repaired during furnace outages.

Furnace dampers are another critical component for optimizing furnace efficiency. Dampers and registers on burners and stacks must be operable in order to control oxygen and draft. Regular preventive maintenance is important.

ENERGY MANAGEMENT STRATEGY

Furnaces and boilers can be designed for maximum efficiency but it takes a tremendous effort to operate and sustain efficiency over a long period of time. The following are some of the sound strategies followed by best-in-class plants:

- management commitment,
- key performance metrics monitoring,
- cross functional discipline commitment, and
- audits against baseline.

Operation is very dynamic, and especially in furnaces there are multiple thermal cycles. It takes people from various disciplines—operations, maintenance, instrumentation, controls, and inspection, to work as a team, to continuously maintain and operate furnaces and boilers at the greatest possible efficiency. The American Boiler Manufacturers Association (ABMA), a nonprofit trade association of manufacturing and design companies of boilers, provides ample guidance to maintain and operate boilers efficiently.

REFERENCES

1. International Energy Outlook 2013, Report No. DOE/FIA-0484, *U.S. Energy Information Administration*. July 2013, Available at: http://www.eia.gov/forecasts/ieo/pdf/0484%282013%29.pdf.
2. API Standard 560 (2007) *Fired Heaters for General Refinery Services*, 4th Edition, American Petroleum Institute.
3. Thorpe, J.T. Refractory life cycle considerations in ethylene cracking heaters, *Proceedings of the 20th Ethylene Producers' Conference*. AIChE, New Orleans, Louisiana, April 6–10, 2008.
4. Feigl, J. Energy improvements of cracking furnaces of the 1960s and 1970s, *Proceedings of the nineteenth Ethylene Producers' Conference*. AIChE, Houston, Texas, April 22–27, 2007.
5. Sento, H. Restoration of refinery heaters using the technique of prefabricated ceramic fiber lined panels, *Third Industrial Energy Technology Conference, Houston, Texas*, April 26–29, 1981.
6. Zhang, G. and Evans, B. (2012) Progress of modern pyrolysis furnace technology. *Advances in Materials Physics and Chemistry*, 2, pp. 169–172.
7. Hobre Instruments BV, Purmerend, The Netherlands, *General Information: Wobbe index and calorimeters*, accessed July 17, 2014. Available at: http://www.hobre.com/images/stories/pdf/wobbe-index-general-information.pdf.
8. Harrell, G. Boiler tune-up guide for natural gas and light fuel oil operation, *U.S. Environmental Protection Agency*, accessed July 11, 2014. Available at: http://www.epa.gov/ttn/atw/boiler/imptools/boiler_tune-up_guide-v1.pdf.
9. U.S. Environmental Protection Agency, *Boiler tune-up guide*, accessed July 5, 2014. Available at: http://www.epa.gov/ttn/atw/boiler/imptools/tune-up_guide.pdf.
10. Ghetti, J. Furnace convection section cleaning systems – engineered solutions for cracking furnaces, *Proceedings of the 24th Ethylene Producers' Conference*, AIChE, Houston, Texas, April 1–5, 2012.

10

ENHANCED HEAT TRANSFER AND ENERGY EFFICIENCY

Thomas Lestina

Heat Transfer Research Inc., College Station, TX, USA

The overall efficiency of an industrial process is largely dependent upon the design of the heat exchanger system. The traditional design practice involves specifying process capacities and temperatures plus any mechanical, maintenance, operation, and size constraints. The heat exchanger is then designed as the best fit to these specifications. It is not uncommon that competing requirements for ease of maintenance and low installed cost limit the overall process efficiency. This is illustrated in Figure 10.1 where a process is required to heat a cold feedstock using hot effluent. The temperature profile on the left has a nominal temperature difference between the hot and cold fluid streams, ΔT_{nom}. The temperature profile on the right provides for higher process efficiency and has a temperature difference half the nominal.

For common heat exchanger types, temperature profiles as shown in Figure 10.1 can be used to determine the number of shells in series needed to meet the process requirements [1]. Each step shown (dotted line) illustrates the temperature change for one shell in the series of exchangers. These steps are selected to ensure there is no temperature cross in an exchanger. (This practice is typical for stacked shells consisting of TEMA [2] E-shells with an even number of tube passes.) For the nominal process design, two shells-in-series are required, and for the high-efficiency process, four shells-in-series are needed. Unless there is a change in the design practices for the high-

Energy Management and Efficiency for the Process Industries, First Edition. Edited by Alan P. Rossiter and Beth P. Jones.

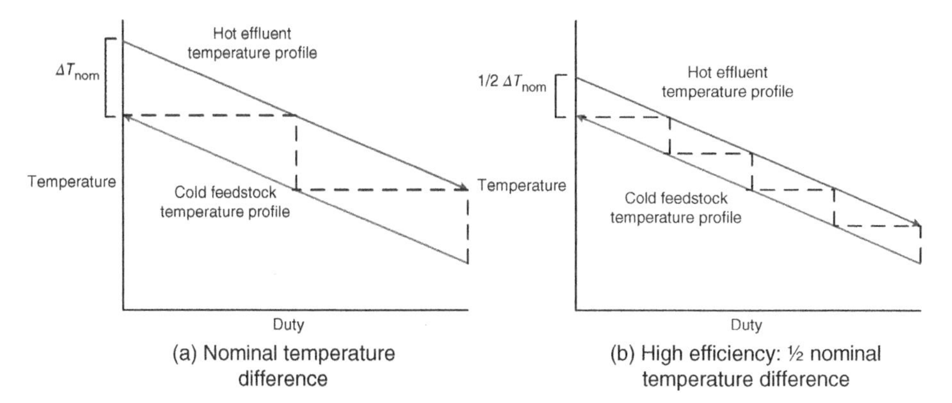

Figure 10.1. Feed-effluent temperature profile for (a) nominal and (b) high-efficiency process design.

efficiency process, twice the heat exchanger area is needed for high efficiency. This can be readily seen from the classical rating equation for process heat exchangers:

$$Q = UA\Delta T_{\mathrm{m}},$$

where Q is the exchanger duty, U is the overall heat transfer coefficient, A is the heat transfer area, and ΔT_m is the mean temperature difference. If ΔT_{m} is halved, then A needs to double to attain the same duty (assuming the heat transfer coefficients are unchanged). Clearly, improved design methods are needed for high-efficiency processes.

For new facilities, the largest gains in process efficiency can be attained if the process designer works with the heat exchanger designer in a preliminary design phase. With this approach, the process temperatures and exchanger selection can be optimized to avoid pinch points, to promote designs which mitigate fouling, to determine the most economical pressure drop for the exchanger, and to utilize enhancement technologies to the fullest benefit. To attain the full benefit of enhancement technologies, designers should consider the following four steps:

1. Optimize the temperature profile. The temperature profile is the shape of the hot and cold fluid process temperature curves throughout the exchanger. Common profiles are shown in Figure 10.2. To optimize thermal efficiency, flow streams should be countercurrent with a uniform temperature difference between the hot and cold streams (Figure 10.2a). In these ideal cases, the mean temperature difference can be as low as 1 °C, which provides for high-efficiency processes. For shell and tube exchangers in the process industries, nonideal profiles are more common. For process condensers with superheated vapor and some subcooling (Figure 10.2b), there are typically large variations in the temperature difference, marked by sharp discontinuities at the dew point and the bubble point. For applications where the hot and cold stream heat capacities differ (Figure 10.2c), the temperature difference can also vary greatly. Fluid streams with greatly different Prandtl numbers (such as

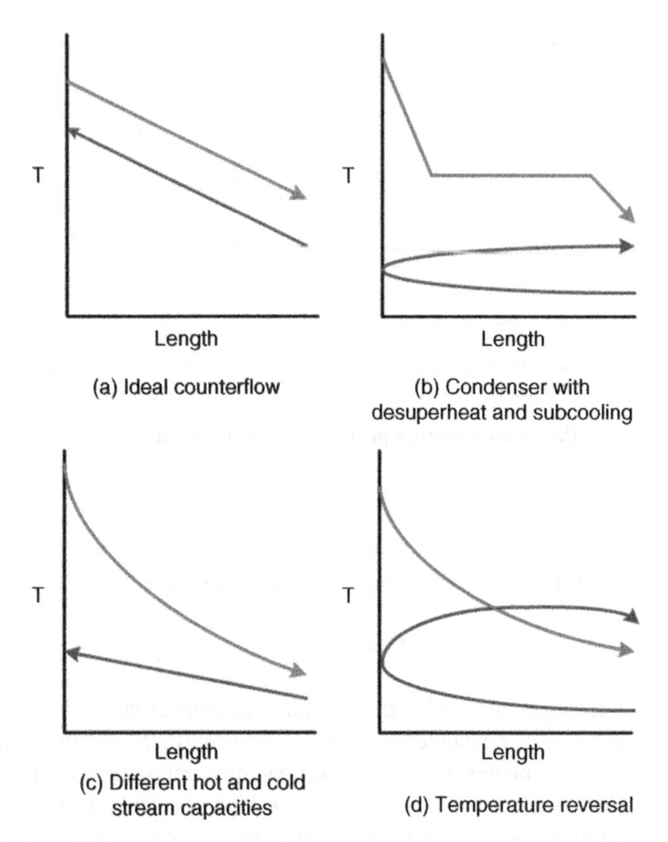

Figure 10.2. Typical temperature profiles for process heat exchangers.

gas–water or oil–water) often have these profiles. The most troublesome temperature profile is one where the cold stream is cooled and the hot stream is heated over a portion of the heat transfer surface (Figure 10.2d). This profile can occur in shells with an even number of tube passes, especially where the original design is based on large fouling factors. This profile is all too common in the process industries and will negate the benefit of applied enhanced heat transfer technology.

2. Mitigate fouling. In this context, fouling is not only the adherence of undesirable material to the heat transfer surface, but also the buildup of heavy, nonboiling components near hot boiling surfaces plus accumulation of noncondensable vapors near cold condensing surfaces. Fouling reduces energy efficiency inherently. Poor heat exchanger design can promote fouling and good heat exchanger design can mitigate fouling. In general, enhancement technologies are not effective when fouled. More information about design practices to mitigate fouling can be found in the literature [3–5].
3. Use small flow passages. Many heat exchangers in the process industries use ¾-in. or 1-in. diameter tubes. With a smaller diameter tube and the same velocity,

pressure drop will increase, which is undesirable. However, designers can distribute the flow with a reduced velocity, and adequate heat transfer can be achieved with acceptable pressure drop. As a result, exchangers with smaller hydraulic diameters generally have higher heat transfer capability for a given volume and allowable pressure drop. Smaller flow passages facilitate higher surface area densities where more heat transfer area is available for a given volume. Lower exchanger volumes result in shorter flow lengths, which help to limit pressure drop. In addition to increasing the compactness of the exchanger, tubes less than ¾-in. in diameter are also less susceptible to stratification in condensing and boiling applications, which provides more efficient wet-wall heat transfer. Tube diameters of less than ¼-in. are not very common in the process industries. Compact plate-fin exchangers are used to achieve smaller flow passages. Plate-fin exchangers have corrugated fins sandwiched between plates and brazed into a core. Plate-fin exchangers can be fabricated with hydraulic diameters as low as 1 mm. Smaller passages can be broadly classified as microchannels. Despite a great deal of recent research into microchannel heat transfer, microchannels are still used in relatively few applications in the process industries.

4. Use enhancement technology where possible. Commercial enhancement technology can increase the heat transfer coefficient, which allows for efficiently sized exchangers at smaller mean temperature differences.

Successful implementation of heat transfer enhancement requires a multidisciplinary study of the process, heat transfer, and mechanical design of the exchangers. A solid understanding of the key enhancement technologies is an essential first step to successful design. The study and development of a wide variety of enhancement technologies has been robust for more than 50 years. Bergles is credited with classifying the various enhancement techniques as summarized in Reference 6:

- Passive techniques do not require external power and include surface treatment, additives to the flow stream, extended surfaces, and inserts for flow passages.
- Active techniques require external power and include stirring or scraping, vibration of surface or fluid, applying electrical fields, and injection and suction.

Passive techniques are the most common in the process industries and Webb [7] provides a thorough, easy-to-read overview of the various techniques and the underlying research that quantify their benefit. Thome [8] provides insight into various types of enhancement techniques and quantifies their benefit. A brief survey of the technologies in use in the process industries is provided below.

SURFACE TREATMENT AND MODIFICATION

Heat transfer is a surface phenomenon and it is not surprising that texture and surface properties can enhance heat transfer. Surface enhancement technologies can be further classified as single phase, boiling and condensing applications.

Single-Phase Surface Enhancements

Roughening the heat transfer surface increases the heat transfer coefficient. For that matter, any change that disrupts the boundary layer of a plain smooth surface tends to increase heat transfer. Engineers and researchers have observed increased heat transfer following startup of new or recently cleaned heat exchange surfaces. This improved heat transfer is attributed to roughening due to initial adhesion of fouling product. Roughened surfaces also have a higher pressure drop, which is the trade-off a thermal designer must consider when evaluating rough surfaces. While some researchers have correlated roughness with heat transfer and pressure drop, these idealized correlations are not adequate to estimate the capability of commercially roughened heat transfer surfaces. Due to the lack of a generalized heat transfer theory for roughness, multipliers to plain surface are applied on a case-by-case basis for a particular enhancement.

Dimpled tube is one of the more common roughened surface technologies used in the process industries. The dimples protrude through the smooth tube boundary layer and improve the turbulent mixing throughout the tube cross section. Dimpled tubes are used in food and chemical applications where the mitigation of fouling is critical for success of operation. Figure 10.3 shows a commercial dimpled tube with a conjugate roughened texture. This tube has been tested to provide a heat transfer enhancement of approximately 50% and a pressure drop increase of 60% under turbulent conditions (Reynolds Number [Re] > 2000).

Boiling Surface Enhancements.

Surface texture can dramatically increase the heat transfer coefficient for nucleate boiling. Enhancements providing as much as 10 times the coefficients of plain surfaces are possible with commercial technology. For plain tubes, a substantial amount of wall superheat (difference between the wall temperature and the saturation temperature) is needed for bubble formation and growth. The resulting boiling heat flux is a function of this superheat in addition to the reduced pressure (reduced pressure is the operating pressure divided by the critical pressure) and tube roughness. Enhanced surfaces promote

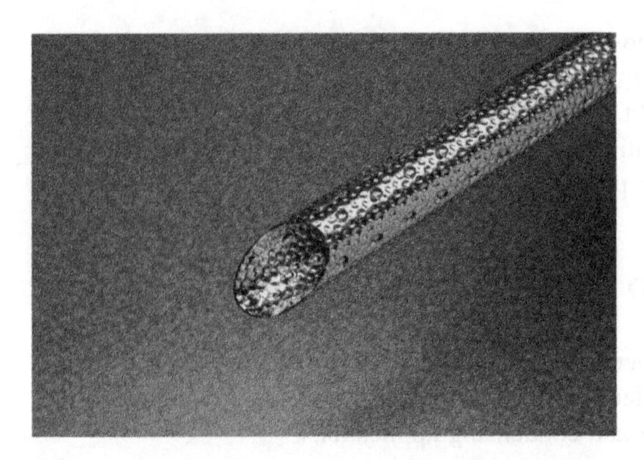

Figure 10.3. Vipertex dimpled tube with conjugate roughened surface finish. (Courtesy of Rigidized Metals Corporation.)

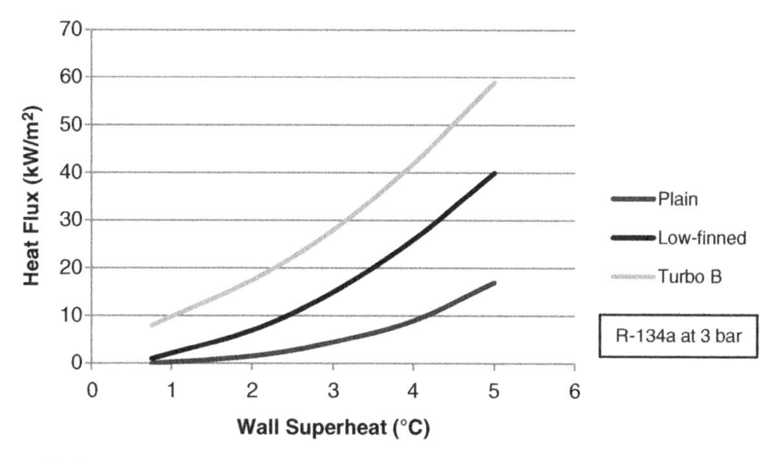

Figure 10.4. Pool boiling of R-134a on plain, low-finned and Wolverine Turbo B tubes.

the formation and growth of bubbles. To this end, porous coatings are one of the early enhanced surface treatments used to increase boiling heat transfer coefficients. UOP's High Flux™ tube design is one of the more common of this type of enhanced boiling surface treatment, and it is often selected for column reboiler applications where the wall superheat is inadequate to support boiling on a plain surface.

More recently, techniques to form channels or pores on tube surfaces have been shown to not only to promote bubble formation but also to create a pumping action where liquid is drawn into a cavity of a departing bubble. These tubes are sometimes formed from low-finned tubes by deforming the fins to form a densely packed surface of boiling pores and re-entrant channels. Figure 10.4 shows the thermal performance of one of these enhanced tubes (Wolverine Turbo B) compared with low-finned and plain tubes [8,9].

Condensing Surface Enhancements.

The condensing coefficient on plain heat exchanger surfaces tends to be high compared with the coefficient on the cold fluid side of the exchanger. Traditionally, the incentive to develop surfaces that enhance condensation has been low. Recently, research activity has increased into developing surfaces that reduce or eliminate the formation of a condensate film. If the condensate film can be effectively eliminated, dropwise condensation results, and the heat transfer coefficient approaches infinity. The challenge is to remove the condensate from the heat transfer surface and maintain the properties of the heat transfer surface that facilitate drop formation. Surface treatments that facilitate dropwise condensation have not yet been adopted in industrial applications, due to the limited durability of the surface treatment. Recent advances with surface nanostructures [10] offer renewed hope that dropwise condensation can be achieved commercially.

EXTENDED SURFACES

Extended surfaces provide additional heat transfer area to one side of the wall separating the hot and cold fluid streams. They are suitable for fluid streams that are thermally limiting and therefore are common for gases and viscous liquids such are oils. Fins are the most common extended surface and they are available outside tubes, inside tubes, and in between plates for a plate fin exchanger. Figure 10.5 shows the three most common fins outside tubes:

- The integral low finned tube (Figure 10.5a) is formed from an extrusion process where a plain tube is compressed to form fins 1–3 mm in height. The plain tube diameter is greater than the finned tube diameter so that the tube can be pulled through baffles in shell & tube heat exchangers. The low-fin outside area is about 3–5 times greater than the inside area.
- High fins are fabricated by attaching fins to a plain tube surface (Figure 10.5b). Fins can be attached by welding (for high-temperature applications), but more commonly they are attached by compression between the fin root and the tube outside diameter. The high-fin outside area is approximately 20–25 times the inside tube area.
- Plate fins with a tube pattern drilled out are sometimes referred to as continuous fins (Figure 10.5c). A bundle is formed by sliding the plates over a bundle of tubes and expanding the tubes into the plate. Area ratios of 10–25 are typical. It is common to enhance the plate surface by adding texture. Figure 10.6 shows a dimple pattern on a continuous plate fin.

Extended surfaces are not 100% efficient. The temperature of the fin varies from the base to the tip, and therefore the heat flux varies accordingly. The ratio of the actual heat transferred to/from the fin to the maximum possible heat transferred (if the entire fin was at the root temperature) is called the fin efficiency. For fins with uniform cross section, the fin efficiency can be calculated based on the simple equation originally attributed to Gardner [11]:

$$\eta_f = \frac{\tanh(mL)}{mL}$$

where m = fin parameter $\equiv \sqrt{\frac{2h}{k_f t_f}}$

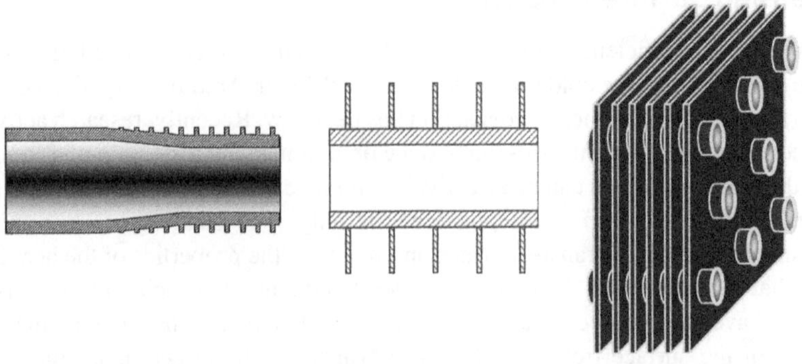

Figure 10.5. Common types of fins outside tubes. (Courtesy of Heat Transfer Research, Inc.)

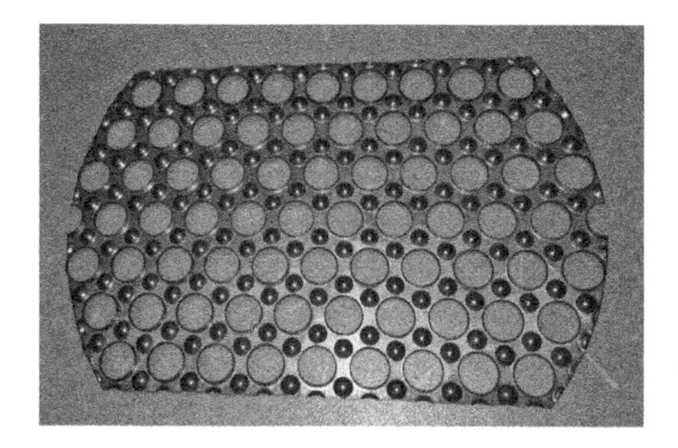

Figure 10.6. Dimple pattern to enhance continuous fin. (Courtesy of Heat Transfer Research, Inc.)

and L is the fin height, h is the heat transfer coefficient, k_f is the fin thermal conductivity, and t_f is the fin thickness. Thin fins with large height have a low efficiency, whereas wide fins with small height have high efficiencies. Fins with low efficiency add excess weight and volume without much benefit. Fin efficiencies greater than 80% are typical and fin efficiencies less than 60% are uncommon.

INSERTS

Inserts are used to optimize the flow distribution and maximize the heat transfer for a given pressure drop. For shell-side flow with TEMA exchangers, segmental baffles divert flow across tubes to increase heat transfer by crossflow. However, the flow is not efficiently distributed and there is substantial bypass flow, plus regions of stagnation and recirculation. Non-TEMA baffles, shown in Figure 10.7, improve the flow distribution, and they can lower pressure drop and mitigate fouling. These non-TEMA baffles are ideal for retrofit applications, as capacity and heat transfer in an existing shell can be increased without exceeding pressure drop limits. These improvements are attained by increasing the longitudinal velocity component and distributing the velocity more uniformly.

Tube-side inserts should be considered when the tube-side heat transfer is thermally limiting. Compared with a plain, empty tube, a tube with inserts has a higher pressure drop. However, a skilled designer can develop a small, efficient design with inserts that does not increase the pressure drop when compared with empty tubes. This reduction in pressure drop can be accomplished by judicious reduction in tube length and reduction in the number of tube passes. Shilling [12] has characterized the augmentation mechanisms for tube inserts into static mixing, boundary layer interruption, swirl flow, and displaced flow.

- Static mixers, Figure 10.8a. Static mixers exchange fluid particles in the center of the tube with fluid at the tube wall. They are used in deep laminar flow where fluid particles do not migrate in a radial direction in an empty tube. Static mixers can increase the heat transfer as much as six times compared with empty tube flow.

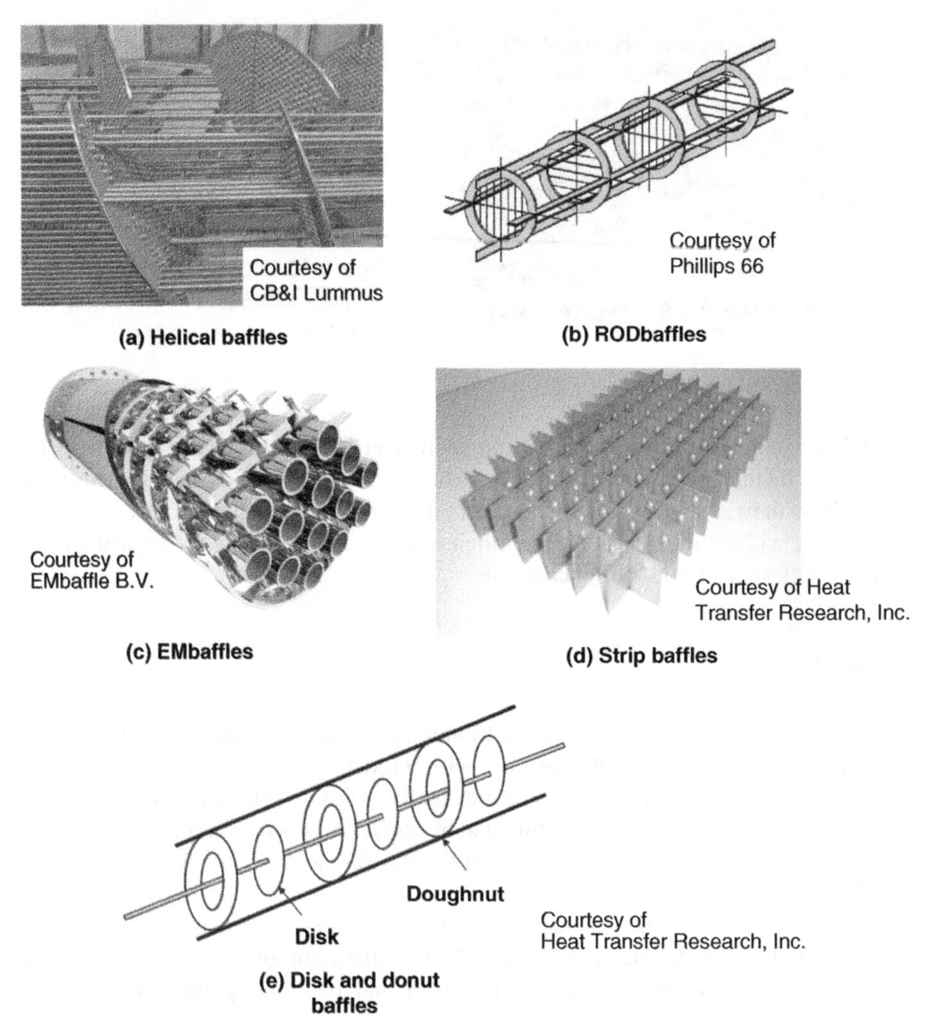

Figure 10.7. Non-TEMA baffles to enhance shell-side flow.

- Boundary layer interrupters. These devices "trip" the laminar boundary layer at periodic intervals so that it becomes thin and thus increases heat transfer. Wire mesh inserts and a wire coil (Figure 10.8b) are the most common of these types of inserts. Heat transfer increases by as much as five times compared with an empty tube.
- Swirl flow inserts. These inserts establish a flow rotation such that the velocity at the wall is higher than with an empty tube. Twisted tape (Figure 10.8c) is the most common of these inserts, and heat transfer can be increased as much as five times compared with an empty tube.
- Displaced flow inserts. These inserts block flow in the center of the tube and increase the velocity at the tube wall. A simple rod in a tube can increase heat

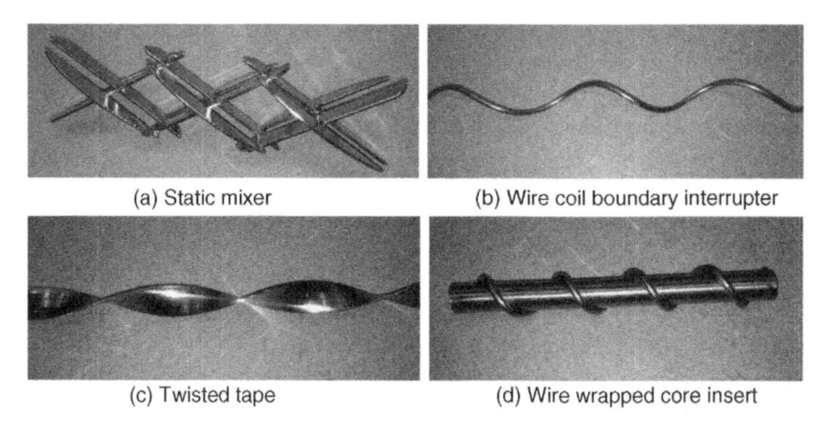

(a) Static mixer (b) Wire coil boundary interrupter

(c) Twisted tape (d) Wire wrapped core insert

Figure 10.8. Tube insert devices. (Courtesy of Heat Transfer Research, Inc.)

transfer to as much as 2.5 times that of the empty tube. A wire-wrapped core insert (see Figure 10.8d) combines displaced flow with swirl flow. Increases in heat transfer as much as 10 times that of the empty tube is possible.

Each of these types of inserts has an optimal flow regime and regimes where they provide less than optimal enhancement. As a rule of thumb, static mixers are often selected where Reynolds number (Re) <10, boundary layer interrupters are selected with $1 < Re < 2000$, swirl flow inserts are selected where $200 < Re < 10000$, and displaced flow inserts are used where $Re > 2000$.

APPLICATIONS IN NEW PLANT DESIGN AND REVAMPS

All of the enhanced heat transfer strategies discussed in this chapter, including surface treatment, extended surfaces and inserts, can be applied not only in new plant designs but also in revamps, and they can be particularly helpful where space constraints demand compact solutions. However, some of the technologies are particularly helpful in one or other of these situations. For example, in process plants the most common revamp approach is to retain the shell and replace the bundle. Many original designs are poor in their existing service, and a bundle replacement can optimize both the shell-side and tube-side performance. All of the enhancement technologies can be exploited in these situations, but in our experience the non-TEMA baffles are probably the most common and effective form of bundle replacement. In contrast, integrating process design with heat exchanger design is very difficult on an existing plant, and in general this can only be applied to new process designs.

CLOSING THOUGHTS

In summary, there is a large body of commercial enhancement technology that can increase heat transfer area and overall heat transfer coefficient. These enhancement technologies can be used to reduce the energy required for a process provided that the design considers practical matters such as material compatibility, fouling, transients and other upset conditions, and maintenance. Acceptance of enhanced technologies has been slow in some industries and additional successful installations will facilitate general acceptance in the future.

REFERENCES

1. Bell, K.J. and Mueller, A.C. (2001) *Wolverine Engineering Data Book II.* Wolverine Tube Inc. Available at http://www.wlv.com/heat-transfer-databook/.
2. Tubular Exchanger Manufacturers Association (TEMA) (2007) *Standards of the Tubular Exchanger Manufacturers Association*, 9th edition, Tubular Exchanger Manufacturers Association, New York.
3. Wilson, D.I. and Polley, G.T. (2001) Mitigation of refinery preheat train fouling by nested optimisation, *Advances in Refinery Fouling Mitigation,* Session #46, AIChE Spring Meeting, Houston, TX, April 23–26, pp. 287–294.
4. Gilmour, C.H. (1965) No fooling—no fouling. *Chemical Engineering Progress*, 61(7), 49–54.
5. Nesta, J. and Bennett, C.A. (2004) Reduce fouling in shell-and-tube heat exchangers. *Hydrocarbon Processing*, 83(7), 77–82.
6. Jensen, M.K., Bergles, A.E. and Shome, B. (1997) The literature on enhancement of convective heat and mass transfer. *Enhanced Heat Transfer*, 4(1), 1–6.
7. Webb, R.L. and Kim, H.N. (2005) *Principles of Enhanced Heat Transfer*, 2nd edition, Taylor & Francis, New York.
8. Thome, J. *Wolverine Engineering Data Book III.* Wolverine Tube Inc., 2004–2008. Available at http://www.wlv.com/heat-transfer-databook/.
9. Palm, B.E. (1995) Pool boiling of R22 and R134a on enhanced surfaces, *Proc. 19th International Congress of Refrigeration*, The Hague, 4a, pp. 465–471, August 20–25, 1995.
10. Miljkovic, N., Enright, P., Nam, Y., Lopez, K., Dou, N., Sack, J. and Wang, E.N. (2013) Jumping droplet enhanced condensation on scalable superhydrophobic nanostructured surfaces. *Nano Letters*, 13(1), 179–187.
11. Gardner, K.A. (1945) Efficiency of extended surfaces. *Trans. of ASME*, 67(8), 621–631.
12. Shilling, R.L. (2012) Selecting tube inserts for shell-and-tube heat exchangers. *Chemical Engineering Progress*, 108(9), 19–25.

11

HEAT EXCHANGER CLEANING METHODS

Joe L. Davis

PSC Industrial Outsourcing, LP, Houston, TX, USA

Refineries and chemical plants depend on clean heat exchangers to maintain efficient plant operation. But over time, exchangers can build up fouling on both the tube side and shell side of the bundle. In the case of crude unit exchangers, heavy asphaltenes from the crude can accumulate on the walls of the exchanger tubes. Fouling is common in both crude preheat trains and tower overhead condensers. Calcium and other minerals in the plant cooling water system can plate out on overhead exchanger tubes. Designers endeavor to size exchanger tubes such that the velocities through the tubes are maximized in order to extend the time period before fouling begins to impede flow (Chapter 10). But eventually, fouling is an unavoidable reality in the use of heat exchangers.

As fouling increases, heat exchange is compromised and efficiency is diminished, resulting in reduced throughput as well as increased energy requirements. For example, as crude preheat exchangers foul, the crude temperature entering the furnace is reduced and additional fuel is required to fire the furnace. In the case of tower overhead exchangers, fouling can result in higher tower pressures, which impedes throughput and/or fractionation quality. Chapter 12 describes a technique to determine optimum heat exchanger cleaning cycle lengths in complex preheat trains, based on overall process energy impacts. This chapter focuses on the methods used for cleaning.

Energy Management and Efficiency for the Process Industries, First Edition. Edited by Alan P. Rossiter and Beth P. Jones.

There are two primary methods used for cleaning shell and tube heat exchangers:

1. Hydroblasting
2. Chemical cleaning

HYDROBLASTING

In hydroblasting applications, water is pumped to pressures ranging from 10,000 to 40,000 psi in specialized pumps, and then conveyed through high-pressure hoses to engineered lance assemblies that spray the high-velocity water directly on the exchanger tube surfaces. This action is very effective at removing foulants from the tubes.

During unit turnarounds, exchanger bundles are typically removed from the shells and sent to a centralized wash pad, where both tube-side and shell-side cleaning takes place. Figure 11.1 shows equipment that is commonly used on the cleaning pad.

In some instances, removing heat exchanger bundles from their shells and sending them to the wash pad is not practical. In these cases, specialized equipment is used to "clean in place" (CIP). The CIP equipment must be highly portable and able to be positioned in areas with limited space. Figure 11.2 shows examples of CIP applications.

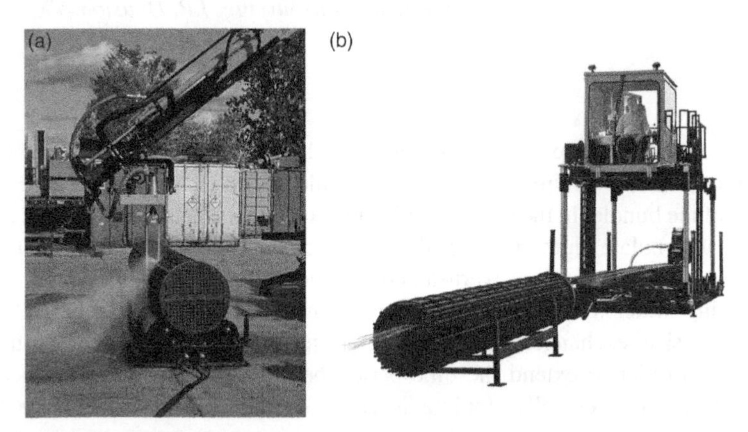

Figure 11.1. Shell-side (a) and tube-side (b) hydroblasting equipment. (*Source:* Photos courtesy of NLB Corporation.)

Figure 11.2. Equipment for clean in place (CIP) applications. (*Source:* Photos courtesy of StoneAge Inc.

CHEMICAL CLEANING

Chemical cleaning methods fall into two broad categories: *reactive cleaning* and *decontamination*. Reactive chemistry is used to remove inorganic, non-carbon foulants such as calcium, carbonates, oxides, phosphates, and sulfates. Decontamination chemistry is applied where hydrocarbon foulants are present.

In both cases of reactive cleaning or decontamination, chemical cleaning utilizes a combination of three cornerstones: chemistry, flow, and temperature. Each of these cornerstones can determine the degree of success in removing the deposits and is the primary reason chemical cleaning often requires more planning and experience than hydroblasting.

Reactive chemical cleaning is most often used to improve energy efficiency, especially on cooling water exchangers and utility boilers. Depending on the type of foulant, mild to strong acids are circulated throughout the system with inhibitors to remove the foulants while minimizing impact on the base materials. Care must be taken not to overexpose the tubes to the chemicals, as metal degradation can occur. After cleaning, passivation is usually performed utilizing various methods such as an application of sodium nitrite for carbon steel or an alkaline solution such as sodium carbonate, commonly referred to as soda ash, for stainless steel.

When fouling consists of hydrocarbon deposits, decontamination chemistry is utilized to remove the deposits. These chemistries use various methods ranging from pH adjustment to surfactant packages targeted at specific hydrocarbon types.

Flow and temperature are controlled through various means as the chemicals are applied. The desired flow can be achieved through circulation, filling and soaking, foaming, or even vapor-phase application. Temperature is often controlled through direct injection of steam or the use of an additional heat exchanger.

COMPARING THE OPTIONS

The following are some of the benefits of chemical cleaning over hydroblasting:

- *Cost:* Performed without costly, time-consuming unit dismantling and reassembly, thus reducing labor requirements and system downtime.
- *Effectiveness:* Vastly improves system cleanliness. Deposits are often totally removed, thereby increasing efficiency of flow and heat transfer.
- *Thoroughness:* Complete deposit removal prevents “seeding” of new deposits and extends unit run time.
- *Speed:* Appreciably faster than most methods, especially on complex systems.
- *Access:* Allows cleaning to take place in otherwise inaccessible areas.
- *Online options:* Online cleaning allows continued operation without having to perform a shutdown.

The following are some of the disadvantages of chemical cleaning compared with hydroblasting:

- *Upfront time:* Increased planning and scheduling requirements.
- *Waste:* Waste must be neutralized and or disposed of by appropriate means.
- *Unexpected effects:* In the case of reactive chemistry, chemical cleaning removes all deposits, including those that may be assisting in maintaining tube integrity where tube walls have thinned.

12

MONITORING OF HEAT EXCHANGER FOULING AND CLEANING ANALYSIS

Bruce L. Pretty[1] and Celestina (Tina) Akinradewo[2]

[1]*KBC Advanced Technologies, Inc., Houston, TX, USA*
[2]*KBC Process Technology, Northwich, UK*

Heat exchanger fouling has long been recognized as a significant cause of energy inefficiency across many process industry sectors. In the U.S. refining industry, the economic impact has been estimated to be of the order of 0.2 quadrillion Btu/year [1], or $1 billion/year at current U.S. natural gas prices. Almost 50% of this cost is attributable to fouling in crude preheat trains alone. However, the true cost of fouling often extends beyond the direct lost energy recovery opportunity. The cost of reduced production can and often does dwarf the additional energy consumption costs.

Some of the most important fouling costs include the following:

- Additional preheat furnace firing from low crude furnace inlet temperatures.
- Fouling-related pressure drop.
- Artificial cooling medium limits. Fouling effects here are twofold: the direct impact of reduced cooler duty and indirect effects due to the increased load on coolers as heat recovery exchanger duties are reduced.
- *Throughput constraints*: Low crude inlet temperatures can also push furnaces to firing or environmental emission limits, and increased pressure drop can restrict

Energy Management and Efficiency for the Process Industries, First Edition. Edited by Alan P. Rossiter and Beth P. Jones.

feed rate as the charge pump capacity limit or preheat train design pressure limits are reached. Reduced air or cooling water exchanger duty capacity can also limit production where product rundown temperatures are critical.

- Reduced yield from lost pumparound cooling capacity.
- Lost throughput and reduced equipment reliability from unplanned unit shutdowns or slowdowns as equipment hydraulic or firing constraints are reached.
- Labor and material costs for cleaning fouled equipment.
- Cost of antifoulants or other fouling prevention programs and the associated disposal and environmental costs.
- Additional capital costs for excess exchanger surface area, online cleaning equipment, and piping.

Most recently in the U.S. oil industry, processing light shale oils with variable composition and physical properties has resulted in significant increases in severe equipment fouling and has spurred a renewed interest in both understanding and predicting the potential for fouling, especially in heat exchangers, fired heaters, and reactors.

In the refining industry, crude feed types can change significantly, both in the short term and in the longer term, depending on opportunity crude availability and prevailing economics. As a result, actual fluid properties and operating conditions may be quite different from the original equipment design basis. Furthermore, the characterization of most crude oil feeds is a complex process given the nature of the oil itself and the complex fractionation and reaction processes the oil components undergo between raw crude input and final product to tankage. Crude oil fouling characteristics, both in general terms and more specifically with respect to parameters that help predict fouling, are neither well developed nor available for everyday engineering practice.

Much good work has been undertaken to characterize fouling tendencies and develop fouling mitigation methods (see Chapter 10), especially for crude oil systems. Predictive fouling methods have been proposed and evaluated but as yet the quality and quantity of available plant operating data limit the ability of researchers to fully validate predictive methods against real operating data.

In the absence of established fouling mitigation practices, industry relies on methods such as the following:

1. Heat exchanger fouling monitoring and cleaning optimization.
2. Use of antifoulants.
3. Modification to shell-side flow characteristics using rod baffle, EMBaffle®, or helical baffle exchangers.
4. Use of tube-side devices such as SpirElf®, Koch Twisted Tubes®, Turbotal®, and HiTran® tube inserts.
5. Real-time side stream fouling monitors.
6. Surface treatments.

Mitigation methods 1 and 2 are most common for existing equipment.

Chapter 11 discusses heat exchanger cleaning methods. This chapter focuses on fouling monitoring and heat exchanger cleaning applications based on the authors' experience in developing and deploying software-based solutions primarily in the oil refining industry. With nearly half of all refinery fouling costs attributable to the crude atmospheric tower preheat train exchangers [2], and the inherent complexity of these preheat trains, this chapter will focus on crude units in particular to present a comprehensive consideration of what it takes to successfully monitor fouling and optimize plant operation in a practical and achievable way. Chapter 26 includes a description of a relatively simple crude unit, including a diagram of the preheat train (Figure 26.3).

FOULING MONITORING IN COMPLEX SYSTEMS

Crude unit engineers have long used simple spreadsheets to track trends in exchanger overall heat transfer coefficient (U) values to infer the extent of fouling in individual exchangers. This practice can be effective in identifying services that might need attention for cleaning, especially if the simple U value analysis is backed up with pressure drop information to confirm that the performance loss is fouling related. This sort of analysis can provide reliable trigger points for a maintenance program of periodic cleaning of large sections of the preheat train.

However, there are several problems with this simplistic approach:

1. It provides no diagnostic information to engineering, operations, or management staff on the reasons for the fouling buildup.
2. It does not account for fouling changes that may occur from changes in fluid flow rates or physical characteristics.
3. It provides no information on the economics associated with addressing the fouling in a systematic manner.

The decision to clean a fouled heat exchanger is usually based on the energy benefit of a clean exchanger compared with the cost of the cleaning. Direct cleaning costs can be substantial—anywhere from $60,000 to $80,000 per exchanger bundle for labor, bundle pulling equipment, cleaning chemicals, and waste disposal. If exchangers are cleaned while the unit is online, increased fuel firing costs will be incurred while the exchanger is isolated for cleaning, and there is always the potential for lost production if furnace firing or emissions limits are reached. These economic losses must be included in any decision-making process for exchanger cleaning. Exchangers also continue to foul after cleaning, and different exchangers foul at different rates. An estimated refouling rate or a rate based on a particular exchanger's history should be included in the economic evaluation of each cleaning.

The impact of cleaning an exchanger must be determined beyond the simple assessment of the immediate increase in its duty. For example, in crude unit preheat trains, interactions between exchangers dampen the duty change in any one exchanger such that the reduction in furnace duty is invariably less than the single exchanger's duty increase. Cleaning exchangers located at the cold end of the preheat train will often result in only a negligible reduction in furnace firing.

The determination of expected furnace duty savings can be further complicated by crude preflash drums and prefractionation towers, especially if the tower or drum overhead vapors are integrated back into the cold end of the preheat train. Furthermore, reconciliation of temperature and flow data is a nontrivial task. Rarely will raw data show consistent shell-side and tube-side heat balances across most exchangers. Heat balancing exchanger duties on a stand-alone basis may throw out the heat balances on other exchangers, often severely, and it is very difficult to see and understand these interactions with isolated exchanger analyses.

For all of these reasons, the process of identifying the right exchangers to clean in complex heat exchanger networks requires dedicated tools that accurately predict exchanger fouling levels and provide the right economic data to guide cleaning decisions.

ELEMENTS OF A STATE-OF-THE-ART FOULING MONITORING PROGRAM

There are three primary components to establishing a state-of-the-art fouling mitigation program at a manufacturing facility:

1. The technical component to determine fouling levels and properly analyze the economic and production impacts of exchanger cleaning.
2. The reporting component to visualize individual exchanger and system performance over time and provide effective decision-making support for exchanger cleaning actions.
3. The organizational component to provide the right organizational drivers and allow allocation of appropriate levels of financial and manpower resources and to integrate fouling monitoring into sustainable best-practice workflows.

TECHNICAL CONSIDERATIONS

There is nothing more damaging to the credibility of a cleaning program than to open an exchanger for cleaning only to observe minimal or no fouling, or to see no economic benefit realized after the exchanger is cleaned. To successfully implement an exchanger fouling monitoring program requires a systematic fouling assessment process that carefully addresses all of the following elements:

1. Instrumentation and data access.
2. Data quality and data reconciliation.
3. Network thermodynamic performance analysis.
4. Cleaning cycle economic analysis.

Instrumentation and Data Access

Most refineries have sophisticated plant data historians that record detailed plant operating conditions with high resolution. Electronically accessible temperature and flow data are crucial for effective exchanger performance monitoring. Manual temperature data entry for monitoring most exchangers is feasible in principle. However, a great deal of effort is required to collect field temperature readings from local temperature indicators or "temperature guns." Manual monitoring is very difficult to maintain at any meaningful frequency level. The most frequent surveys of this type that the authors have seen were completed every 3 months, and then only when a team from a corporate engineering support group provided the staff.

Ideally, data should be available for each exchanger's shell-side and tube-side outlet temperature and each fluid's temperature as it first enters the heat exchanger network. Inlet temperatures for all other exchangers can be set equal to the outlet of an upstream connected exchanger without appreciable loss in accuracy. The exchanger outlet temperatures must be adjusted for any bypasses. Exchangers can also be modeled by grouping identical exchangers and identifying common configurations (two series/one parallel, two series/two parallel, etc.). Note, however, that in such groupings some resolution is lost as to the exact location of any fouling.

Exchanger grouping for monitoring may also be determined by the availability of bypass and isolation valves to allow online exchanger isolation and removal from service. In general, though, monitoring resolution should be determined by available temperature instrumentation.

Crude feed, product rundown, and tower pumparound flow rates are generally well measured. Ideally, flow rates should be corrected for flowing density versus design for each meter and an overall mass balance check should be performed across the unit to ensure flow meter consistency.

Flow splits are usually less well measured unless the splits are actively manipulated by operators. Temperatures around an associated mixer can be used to estimate an unmetered flow split. This is especially useful around a controlled bypass. Failing this, valve position in conjunction with valve flow characteristics can sometimes be used to estimate flow distribution between branches. However, this is complicated by the unequal pressure resistance on different branches, particularly when fouling is involved. Finally, in the absence of any of these, bypasses should be assumed closed and other flow splits to have an equal distribution.

Alternatively, data reconciliation can be used to find best-fit flow split data, though data reconciliation then becomes a more difficult nonlinear optimization exercise. This can also introduce more uncertainty as the model now has extra optimization variables.

Rigorous calculation of film heat transfer coefficients requires thermal and physical data about each stream. Distillation data and liquid specific gravities are sufficient to synthesize the required liquid properties of density, heat capacity, thermal conductivity, and viscosity at flowing conditions in each heat exchanger. The best approach is to use rigorous physical property simulation to calculate these properties from composition or crude assay information. However, standard shortcut methods can also be used in spreadsheets.

If available, electronically sourced laboratory data should be used to characterize crude and oil product properties. It is important to include all components when characterizing the whole crude oil stream. Different distillation types and light ends analysis can be combined in the process simulation and the result checked against the crude unit material balance.

Data Quality and Data Reconciliation

Data quality and data reconciliation methods can have a huge impact on fouling analysis. Table 12.1 shows a portion of the raw data heat balances for a large heat exchanger network. The uncorrected plant data for a crude preheat train are taken directly from a plant data historian. Exchanger tube-side heat duties (Q_t) and shell-side heat duties (Q_s) are calculated independently and compared for consistency. Shell- and tube-side heat imbalances as shown in this example are not uncommon. It is rare that raw plant data give reliable (±5%) heat balance closures for all exchangers, especially with larger preheat train systems that may have up to 60 exchangers excluding product and pumparound coolers.

Tables 12.2 and 12.3 represent manual processes to balance the exchanger heat duties using reasonable logic, but different strategies. In Table 12.2, the crude-side temperatures for the first six exchangers are varied to balance the exchanger duties. Note the final duties (Q_t and Q_s) and fouling factors (R_f). In Table 12.3, the crude-side outlet temperature for the first exchanger is varied, but product-side (shell-side) temperatures are varied for the remaining five exchangers. Note how different the exchanger duties and fouling factors can be!

With the right tools it may be possible to manually adjust exchanger outlet temperatures to come up with acceptable heat balances across the preheat train system; however, this is not a trivial undertaking due to the complexity of interactions between exchangers. The final solution is also subjective as it involves a set of sequential decisions as to which exchanger to start with and whether to adjust the tube- or shell-side temperature in order to achieve a heat balance. There are many possible heat balance solutions. The question is: Which is the most representative solution?

Mathematical data reconciliation can effectively answer this question. It is a vital component of any successful fouling monitoring application. Reconciliation is a least-squares optimization process where the objective function minimizes the deviation between reconciled, heat balanced data and the raw plant data. Temperature, fluid flow, and flow split data can be included in the data reconciliation model. The relative weight the reconciliation model places on different data types is governed by assigning trust factors or typical errors to each plant measurement. Occasionally, though not often, it may be necessary to include local exchanger bypasses in the reconciliation problem. Given that local bypass flows are never measured, bypasses must be estimated and thus very low trust factors (or high typical errors) should be assigned to these variables.

A measure of "acceptability" should be established for any optimized solution, to determine whether to accept or reject the current data set. If reconciled plant data deviations on a particular plant measurement are consistently higher than expected instrumentation errors, the flow meter or temperature indicator should be checked.

TABLE 12.1. Exchanger Tube- and Shell-Side Duty Comparison From Raw Data

	Tube						Shell									
Exchanger	Fluid	$T_{t,in}$	$T_{t,out}$ Balance	$T_{t,out}$	Split	Bypass	Fluid	$T_{s,in}$	$T_{s,out}$ Balance	$T_{s,out}$	Split	Bypass	Q_t	Q_s	Q_s/Q_t	R_f
E-1	A3SS	325	195	*107*	100	0	New crude	70	97	87	100	0	23.62	14.09	0.60	0.0016
E-2	New crude	87	102	111	50	0	A3SS	220	118	157	100	0	10.80	6.70	0.62	−0.0011
E-3	New crude	111	129	116	50	0	A5SS	255	228	163	75	0	2.25	7.78	3.46	0.0376
E-4	New crude	87	118	117	50	0	A4SS	208	144	142	100	0	13.59	13.99	1.03	0.0178
E-5	New crude	117	150	157	100	0	VTPA	330	237	255	100	0	36.27	29.46	0.81	0.0032
E-6	New crude	157	174	173	100	0	A4SS	278	214	208	100	0	14.11	15.38	1.09	0.0019
E-7	New crude	173	188	190	100	0	A2PA	349	308	312	100	0	15.20	13.83	0.91	0.0004

TABLE 12.2. Reconciliation Results When Only Exchanger Crude Outlet Temperatures Are Changed

	Tube						Shell									
Exchanger	Fluid	$T_{t,in}$	$T_{t,out}$ Balance	$T_{t,out}$	Split	Bypass	Fluid	$T_{s,in}$	$T_{s,out}$ Balance	$T_{s,out}$	Split	Bypass	Q_t	Q_s	Q_s/Q_t	R_f
E-1	A3SS	325	110	*107*	100	0	New crude	70	97	97	100	0	23.62	23.27	0.98	−0.0035
E-2	New crude	97	112	111	50	0	A3SS	220	162	157	100	0	6.21	6.70	1.08	0.0172
E-3	New crude	111	128.6	129	50	0	A5SS	255	163	163	75	0	7.78	7.78	1.00	0.0063
E-4	New crude	97	128.6	129	50	0	A4SS	208	142	142	100	0	13.98	13.99	1.00	0.0149
E-5	New crude	129	161	161	100	0	VTPA	330	255	255	100	0	29.49	29.46	1.00	0.0058
E-6	New crude	161	178	178	100	0	A4SS	278	208	208	100	0	15.39	15.38	1.00	−0.0002
E-7	New crude	178	194	190	100	0	A2PA	349	322	312	100	0	10.19	13.83	1.36	0.0026

TABLE 12.3. Reconciliation Results When Exchanger Product Outlet Temperatures Are Changed on All but E-1

	Tube						Shell									
Exchanger	Fluid	$T_{t,in}$	$T_{t,out}$ Balance	$T_{t,out}$	Split	Bypass	Fluid	$T_{s,in}$	$T_{s,out}$ Balance	$T_{s,out}$	Split	Bypass	Q_t	Q_s	Q_s/Q_t	R_f
E-1	A3SS	325	106	107	100	0	New crude	70	97	97	100	0	23.62	23.68	1.00	−0.0037
E-2	New crude	97	111	111	50	0	A3SS	220	164	163	100	0	6.00	6.02	1.00	0.0252
E-3	New crude	111	116	116	50	0	A5SS	255	229	228	75	0	2.25	2.31	1.03	0.2250
E-4	New crude	97	117	117	50	0	A4SS	214	173	173	100	0	8.80	8.82	1.00	0.0653
E-5	New crude	117	157	*157*	100	0	VTPA	330	237	237	100	0	36.27	36.00	0.99	−0.0005
E-6	New crude	157	173	173	100	0	A4SS	278	214	214	100	0	14.11	14.13	1.00	0.0042
E-7	New crude	173	188	190	100	0	A2PA	349	308	312	100	0	15.20	13.83	0.91	0.0004

Most often, only temperature manipulation is required to achieve consistent reconciliation of the preheat train heat balance with acceptable levels of deviation versus raw plant data inputs. In this situation, linear optimization/reconciliation methods that manipulate only exchanger duties can be used. These methods are very quick and very robust. If flows or flow splits are included in the reconciliation problem formulation, the problem becomes nonlinear. This significantly increases the time required for reconciliation and can affect the robustness of the solution. A robust reconciliation model should allow both linear and nonlinear options depending on plant data quality.

All reconciliation methods must allow constraints to be introduced into the system. Obvious constraints are nonnegative exchanger duties, minimum exchanger approach temperatures to assure heat transfer feasibility (e.g., no temperature crosses, minimum approach temperatures, or minimum F_t correction factors), and nonnegative flows and flow splits. Optional constraints such as nonnegative fouling factors in the back-calculated exchanger performance provide additional confidence in the final reconciled solution.

Network Thermodynamic Performance Analysis

There are three basic analytical requirements of any effective rigorous fouling monitoring application:

1. It must be able to accurately determine the clean film heat transfer coefficients from first principles calculations for each exchanger in the network for current operating and projected conditions. The basic heat exchanger performance equation is expressed as follows:

$$Q = UA\Delta T_{\mathrm{lm}}F_{\mathrm{t}},$$

where

$$\frac{1}{U} = R_{\mathrm{f,shell}} + R_{\mathrm{f,tube}}\frac{D_{\mathrm{o}}}{D_{\mathrm{i}}} + \left(\frac{D_{\mathrm{o}}}{2k_{\mathrm{tube}}}\right)\ln\left(\frac{D_{\mathrm{o}}}{D_{\mathrm{i}}}\right) + \frac{1}{h_{\mathrm{f,shell}}} + \frac{1}{h_{\mathrm{f,tube}}}\frac{D_{\mathrm{o}}}{D_{\mathrm{i}}},$$

Q is the exchanger heat duty, U is the overall exchanger heat transfer coefficient, A is the exchanger surface area, ΔT_{lm} is the log mean temperature difference, F_{t} is the log mean temperature difference correction factor, $R_{\mathrm{f,shell}}$ is the shell-side fouling factor, $R_{\mathrm{f,tube}}$ is the tube-side fouling factor, $h_{\mathrm{f,shell}}$ is the shell-side film heat transfer coefficient, $h_{\mathrm{f,tube}}$ is the tube-side film heat transfer coefficient, D_{o} is the tube outside diameter, D_{i} is the tube inside diameter, and k_{tube} is the tube metal thermal conductivity.

This is the only reliable method for determining the actual operating thermal fouling resistance and the impact of changing the fouling resistance through cleaning. Clean pressure drop prediction is also valuable as a guide to support observed thermal fouling, to ascertain on which side of the exchanger fouling is occurring, and to estimate how much pressure drop reduction might be expected from cleaning the exchanger.

2. It must be able to evaluate not only the performance of the cleaned exchanger but also the impact of the changed duty on all other exchangers in the heat exchanger network.
3. It must be able to calculate the total cycle costs for the exchanger cleaning process including the following:
   a. Determining the loss of heat recovery in the heat exchanger network when the exchanger is taken off-line.
   b. The fixed costs of cleaning the exchanger, such as labor, cleaning materials, and equipment hire. This information is input by the user.
   c. The change in heat recovery benefits over time as the cleaned exchanger or exchangers refoul at a higher initial rate than the exchangers that were not cleaned.

Exchanger Thermal Performance Prediction: Single-Phase Systems. Accurate thermal rating of existing exchangers requires mechanical construction details for the tube bundle and shell baffle configuration. Most of the required data are generally available from manufacturer's data sheets or exchanger drawings. Clearances for bundle to shell, tube to baffle, and baffle to shell may be less consistently available. In these instances, typical TEMA clearances can be used. Physical property data can be generated from laboratory data as described earlier.

Tube-side film heat transfer coefficient and clean tube pressure drop calculations are relatively straightforward for single-phase systems. Shell-side thermal and pressure drop calculations are much more complex. In general, the Bell–Delaware method [3–5] is sufficiently accurate for most exchanger configurations and exchanger types encountered in refining operations. It has been found to be very quick, reliable, and sufficiently accurate for regular fouling monitoring applications. More advanced exchanger rating and design software such as that provided by Heat Transfer Research Institute (HTRI) and HTFS can provide more advanced prediction methods, but these applications are not well suited to performing the very large number of exchanger simulations required in a networked exchanger configuration, where a large number of iterative calculations are required.

Exchanger Thermal Performance Prediction: Two-Phase Systems. Two-phase heat transfer in crude preheat systems primarily involves tower overhead condensing against crude and crude vaporization. Vaporization in crude preheat trains is not common, but is appearing more frequently where refineries have shifted to lighter shale oil crude feeds. Crude vaporization can cause significant crude train pressure drop and heater pass flow balancing issues.

Reliable methods for first principles rating calculations for condensing and vaporizing services are not publicly available, particularly for shell-side vaporization. In general, these services require detailed computational exchanger rating methods only available through providers such as HTRI or HTFS. Although individual exchanger rating calculations for these services may take less than a minute each, a thorough cleaning cycle analysis for a complex heat exchanger network will require hundreds, sometimes thousands, of iterative exchanger calculations. This makes it impractical to include integrated, detailed simulations of these exchangers in fouling monitoring and cleaning optimization applications.

An acceptable approach for two-phase services is to use the advanced exchanger rating and design applications to run parametric studies for the two-phase exchanger services and to develop correlated relationships for the two-phase film heat transfer coefficients, and pressure drop if required, that are a function of key operating variables such as flow rate and fractional vaporization.

For vaporizing services in particular, the main inaccuracy in any predictive methodology is determining the extent of crude vaporization. The sophisticated thermodynamic fluid property calculations required to generate the pressure- and temperature-dependent heat and vaporization curves are available in commercial simulation software such as Petro-SIM™ from KBC, Hysys™ from Aspen Technology, UniSim™ from Honeywell, or Pro-II™ from Invensys.

Integrated Exchanger Network Analysis. The determination of operating fouling factors for individual exchangers is called rating. This is only the first step of an overall fouling monitoring and cleaning analysis application. Determining the overall heat recovery benefits from cleaning interconnected heat exchangers requires heat exchanger network simulations capable of treating exchangers as interdependent, interacting as temperatures change throughout the network due to changes in exchanger fouling factors and due to the temperature dependence of exchanger film heat transfer coefficients.

Rigorous network simulation calculations have been developed in spreadsheet-based solutions. This platform can be particularly effective in solving highly interdependent systems such as heat exchanger networks. However, developing these rigorous calculations is a complex exercise. When combined with the need for plant data reconciliation functionality, extensive spreadsheet macro programming is required to develop an application that is reliable, accurate, and integrated into plant data systems. Commercially, KBC's Persimmon™ application is a long-established spreadsheet-based application that has both rating and simulation capabilities specifically focused on cleaning benefit analysis and automated linear and nonlinear data reconciliation capabilities.

For simulation calculations, it is recommended that exchanger performance evaluations use the effectiveness-NTU approach rather than the more widely recognized $UA\Delta T_{lm}F_t$ approach. Effectiveness equations are shown below for shell and tube heat exchangers. The more conventional $UA\Delta T_{lm}F_t$ approach requires an initial guess for one of the exchanger outlet temperatures and some form of convergence routine to solve for the correct outlet temperatures. As can be seen in these effectiveness-NTU equations, for an exchanger of known UA, both shell and tube outlet temperatures can be calculated explicitly knowing only the inlet temperatures. This makes the iterative solution process quicker and convergence much more stable. The "effectiveness-NTU" performance equations for an exchanger are

$$Q = E(MC_p)_{\min}(T_{\mathrm{h,in}} - T_{\mathrm{c,in}}),$$

$$E = \frac{\Delta T_{\mathrm{actual}}}{T_{\mathrm{h,in}} - T_{\mathrm{c,in}}}.$$

If the cold stream $M \cdot C_p$ product is lower than the hot stream $M \cdot C_p$ product, then

$$\Delta T_{\text{actual}} = T_{\text{c,out}} - T_{\text{c,in}},$$

If the hot stream $M \cdot C_p$ product is lower than the cold stream $M \cdot C_p$ product then:

$$\Delta T_{\text{actual}} = T_{\text{h,in}} - T_{\text{h,out}},$$

$$E = f(\text{NTU}_{\min}, C_{p,\text{hot}}, C_{p,\text{cold}}, \text{exchanger configuration}),$$

$$\text{NTU}_{\min} = UA/(M \cdot C_p)_{\min},$$

where Q is the exchanger heat duty, E is the exchanger effectiveness, $(M \cdot C_p)_{\min}$ is the minimum of hot fluid and cold fluid mass flow times heat capacity, M is the fluid mass flow, C_p is the fluid heat capacity, $T_{\text{h,in}}$ is the hot fluid inlet temperature, $T_{\text{h,out}}$ is the hot fluid outlet temperature, $T_{\text{c,in}}$ is the cold fluid inlet temperature, $T_{\text{c,out}}$ is the cold fluid outlet temperature, U is the overall exchanger heat transfer coefficient, and A is the exchanger surface area.

Commercial refinery hydrocarbon flow sheet simulation programs are also well suited to evaluating complex heat exchanger network systems. With the current computing capabilities of standard desktop computers, complex networks, even those including rigorous column simulations, can solve quickly. Their inherent advantages as commercial grade software applications with sophisticated thermodynamic and physical property prediction capabilities make these platforms the state-of-the-art choice for refinery heat exchanger network analysis. If required, the analysis can be extended to include fractionation operations, including preflash drums and towers, prefractionators, atmospheric fractionation towers, and vacuum towers. Applications such as Petro-SIM™, Hysys™, UniSim™, and Pro-II™ all have powerful exchanger rating and network simulation capabilities. KBC's Petro-SIM™ includes its HX Monitor utility extension, which offers a full fouling and cleaning cycle economic assessment capability integrated into the simulation flow sheet.

Cleaning Cycle Economic Analysis

Many sites take an "opportunistic" approach to heat exchanger cleaning. Opportunistic cleaning is governed by planned or unplanned unit shutdowns or driven by a fouling problem that has become so severe that unit capacity or yields are being affected. This is not unlike waiting until you are critically ill before going to the emergency room as a health management strategy. It is clearly nonoptimal.

An optimized cleaning strategy requires predictive analytical capabilities where the analysis can simulate the economic impact of the key events associated with any exchanger cleaning process. The key components are the following:

1. The labor, materials, equipment hire, chemical, and disposal costs required to pull an exchanger out of service, clean it, and bring it back online.
2. The economic impact of the lost network heat duty while the exchanger is out of service.
3. The economic benefit, over time, of the increased heat duty of the network with the exchanger in a clean condition, relative to the current network with the exchanger in a fouled condition.

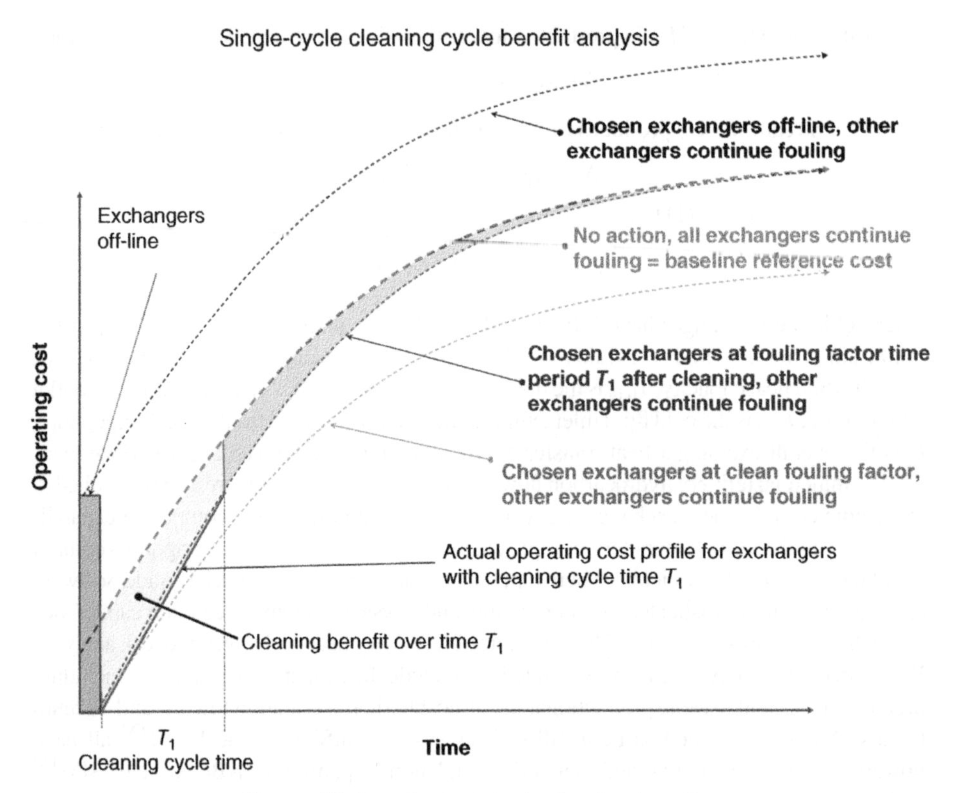

Figure 12.1. Exchanger single cleaning benefit.

The analysis must recognize that economic benefits from cleaning will diminish over time due to refouling of cleaned exchangers and the continued fouling, usually at a slower rate, of exchangers that are not cleaned. The cycle benefit for a single cleaning event is shown graphically in Figure 12.1.

The model must include the dates of previous exchanger cleaning events and the expected exchanger fouling profiles. Extensive research has been conducted in this area, and fouling models have been proposed and tested, but to date no first principles predictive fouling models are readily available. Fouling profiles must therefore be developed from historical fouling rate data. In practice, these data are generated by running the rating portion of the heat exchanger fouling monitoring application for a range of historical dates, typically over the past 2 or more years. This exercise can also be used to identify exchanger cleaning events, which can be used to test the model's predicted system response against actual system responses.

Cleaning Cycle Optimization. The trade-off between the cost of cleaning and the benefit to be gained from the cleaning can be systematically evaluated. The application should be capable of performing the required individual exchanger and network thermal analyses and should contain a representative model for the events included in a cleaning, namely: fixed costs, off-line impact, and clean condition and refouling profiles. Performing sequential or stand-alone analyses for every exchanger, or a subset of the whole network of exchangers, will identify those exchangers with the most favorable economic return based on current operating characteristics.

However, once one exchanger is selected for cleaning, the cleaned exchanger will affect the benefits "left on the table" for the remaining fouled exchangers. Benefits for the second, third, and all subsequent exchanger cleanings will be lower than each exchanger's stand-alone benefits. This is the first level of complication in trying to develop an optimum heat exchanger cleaning program. The second question to be answered in the optimization process is the frequency of cleaning.

The problem of identifying the mathematically optimal selection is truly a daunting exercise, requiring thousands of scenarios of potential exchanger combinations to be evaluated. Methods such as simulated annealing [6] have been used to generate optimal or near-optimal cleaning schedules. However, these programs still exist primarily in the academic arena and are not available for practical application in day-to-day plant engineering environments. Many different methods for generating optimized cleaning schedules have been proposed. We describe two practical approaches below.

The first approach is a series of steps of best economic choices over time:

1. Identify how many exchangers you can clean now or in the near future with realistically available resources.
2. Use a sequential process to identify the most economical exchanger to clean using simulation techniques to generate the cleaning cost benefits and cleaning cost penalty profile shown in Figure 12.1. Ensure that the operating cost benefits versus cost penalty profile meets a target minimum economic benefit and payback. The process must include operating cost penalties for periods of exchanger downtime.
3. Repeat the analysis steps with the chosen exchanger's clean fouling factor, and repeat the analysis to determine the next best exchanger to clean. Again, ensure that the target minimum economic benefit and payback are met.
4. Repeat this process until either the resource limit for the maximum number of exchanger cleanings is met or there are no further fouled exchangers that meet your economic payback criteria for cleaning.

The exchangers selected for cleaning form the basic plan for the near future. The benefits are generally projected out over a certain time period, typically representing the frequency at which cleaning actions might be considered by the site, for example, every 3 months.

Using forward-looking sets of fouling factors, the above process can be repeated at future points in time to determine the next set of exchangers to consider for cleaning. The schedule can be revisited at regular intervals to check on the consistency of the plan.

The second method of cleaning cycle optimization is the "time slice" methodology. This is represented graphically in Figures 12.2 and 12.3. The method will determine the

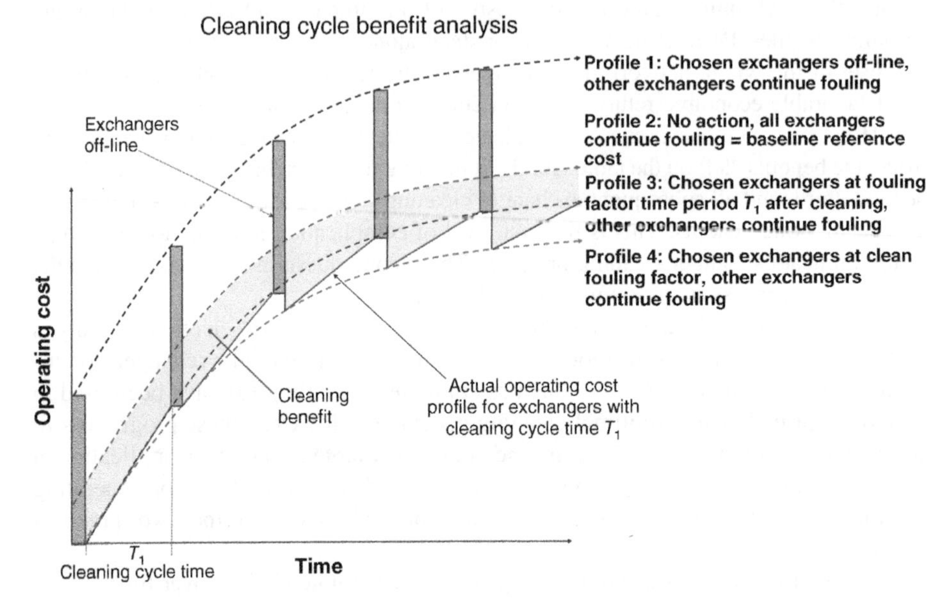

Figure 12.2. Exchanger cleaning operating cost benefits: time slice method.

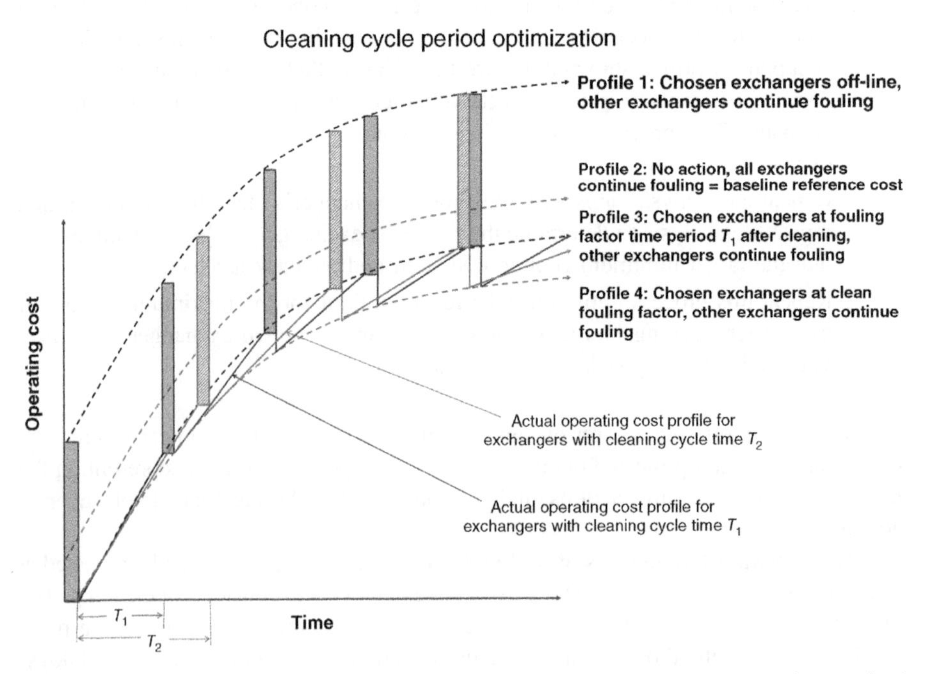

Figure 12.3. Optimizing exchanger cleanings to minimize operating cost: time slice method.

optimum cleaning cycle for any selected set of candidate exchangers. The operational cost profile for a particular set of exchangers within the network chosen for cleaning is also shown. The exchanger combinations investigated are typically sets of the best exchanger candidates defined from the individual exchanger analysis approach outlined above.

It is possible, for any given selection of exchangers, to develop a curve of total cost penalty and total cleaning benefit versus cleaning frequency. Four network performance profiles must be determined with a network simulation:

1. The network cost profile with the exchangers selected for cleaning off-line at future points in time: profile 1 in Figure 12.2.
2. The reference operating cost case where no exchangers are cleaned and the whole network keeps fouling over time: profile 2 in Figure 12.2.
3. The network cost profile with the exchangers selected for cleaning operating with their projected fouling factors at a time period T_1 (the cleaning cycle time) after cleaning: profile 3 in Figure 12.2.
4. The network cost profile with the exchangers selected for cleaning operating with clean fouling factors: profile 4 in Figure 12.2.

Each of these profiles can be defined by running only a few simulation cases per curve, using the time-dependent fouling factor projections for all the relevant exchangers. These are used to curve-fit mathematical cost expressions (typically polynomial) for each operating curve.

With these profiles, the actual operating cost profile (as indicated in Figure 12.2) for any selected cleaning cycle time T_1 can be quickly determined from the four known curve profiles. Generally, the program's total operating cost is determined for a specified crude run length or projection period in the future. Adding in the fixed costs per cleaning event (labor, chemicals and materials, equipment hire, etc.) gives the total operating cost for the required future projection period.

Repeating the mathematical process of determining the actual operating cost profile and total period operating costs for a series of different cleaning cycle times, as illustrated in Figure 12.3, generates a total operating cost for each cleaning cycle time. No further network simulation runs are needed. The various operating cost profiles are generated using mathematical expressions. These are summed to provide total costs and differentials are generated to identify the optimum cycle time. The end result, as shown in Figure 12.4, is a determination of the operating fouling cost, off-line operating cost, and overall labor and material costs as a function of cleaning cycle time. The optimum cleaning period is identified as the overall minimum in the summation curve for all three cost components.

In the absence of a viable method for identifying a true mathematical optimum, the analysis methods described above can be deployed in a practical and meaningful manner. The objective is to enable plant engineers to have tools and processes at their disposal that can generate sufficiently accurate information for the site to take action.

The most important aspects of any application used for generating optimized cleaning schedules are that the optimization can be performed automatically with

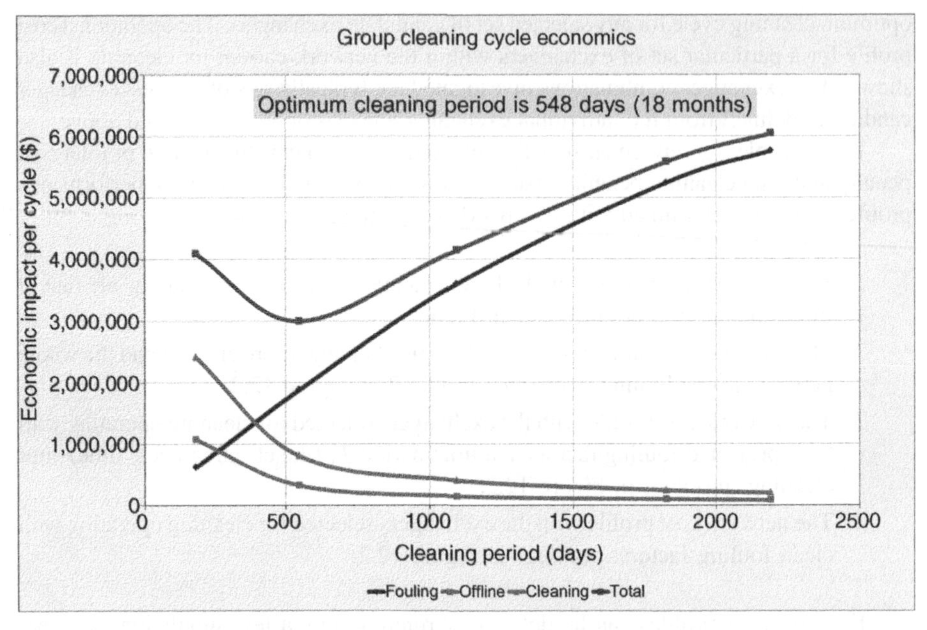

Figure 12.4. Cost curves for optimizing cleaning periods.

relatively easy-to-generate user inputs, and it can be completed in a manageable time frame by the engineers charged with developing the exchanger cleaning schedules.

Commercial programs that have built-in heat exchanger fouling analysis and cleaning cycle optimization capabilities include the following:

- Petro-SIM™ HX Monitor from KBC (supersedes Persimmon).
- EXPRESS™ from IHS ESDU.
- MONITOR™ from Nalco.

HEAT EXCHANGER CLEANING COST BENEFITS OTHER THAN ENERGY SAVINGS

The discussion throughout this chapter has been focused on evaluating the energy benefits arising from cleaning exchangers. Emissions benefits such as reduced NO_x, SO_x, and CO_2 emissions can be directly attributed to reduced energy consumption from crude unit fired heaters. A secondary benefit from improved heat recovery, by heat balance, is reduced cooling loads, and these have both cost and environmental benefits. More significant economic benefits arise from cleaning when firing or cooling loads constrain unit capacities. Relieving these constraints allows unit feed rate to be increased.

The calculation methodologies described above for determining the economic incentive for cleaning exchangers can be modified for determining throughput benefits. The procedures are more complex and time-consuming, but they can be implemented in most of the commercial flow sheet simulation applications. The HX Monitor cleaning cycle analysis feature in Petro-SIM, for example, can be adapted to consider throughput benefits by making use of the application's "Adjust" feature.

REPORTING CONSIDERATIONS

Plant engineering, operations, and supervisory personnel are busy people who need clear, timely information. Effective reporting and visualization of the analysis results is therefore a key component of the success of any program.

Components of an effective reporting application include the following:

- Easily accessible reports of key performance information such as fouling factors, fouling influence variables, exchanger heat duties, furnace firing impacts, cleaning economic benefits, and other unit Key Performance Indicators (KPIs) affected by both fouling and exchanger cleaning.
- Ability to access more detailed calculation results when needed, and archiving of the evaluation process results.
- Archiving the performance parameters and economic results into an enterprise-level database structured for easy data retrieval into Microsoft Excel, SharePoint, or other Web-based reporting and performance dashboard applications. A rich graphical environment for outputs with well-structured information presentation will make the application easier to use and explore. Keeping the archived results in industry standard database formats also allows for integration with other decision-making and planning applications that will be in use throughout the facility.

ORGANIZATIONAL CONSIDERATIONS

An easy-to-use and accurate fouling monitoring and cleaning cycle optimization software system is essential for a successful heat exchanger cleaning program. However, it is not the only requirement. The application owners must also be committed, and senior management must dedicate financial and resource support to ensure the success of the program.

Upfront resources are needed to build the application and to train the appropriate plant engineers on its use and maintenance requirements. Ongoing resources are required to enable the software to be run, maintained, and analyzed. Both the application itself and user training must be updated for inevitable changes in plant configuration and plant personnel. These updates must be embedded into routine plant work processes or the programs can easily fall into disuse. Furthermore, the KPIs from the fouling monitoring and the training program must be embedded into plant workflow processes, and the correct KPIs must be reported to each part of the organization to ensure that the appropriate people review and act on the information.

For the program to be sustainable, its benefits must be seen as significant to the plant. This will require actions that affect the way the plant is operated, and this will add work for the maintenance and planning staff. It is therefore vital that as a part of the program rollout plan, operations and supervisory staff should be consulted to map out the needs and the accountabilities for each of the functional roles with a stake in the program effectiveness and success.

THE FUTURE OF FOULING MONITORING

The most significant development for effective fouling monitoring and cleaning cycle optimization in the foreseeable future is the potential to incorporate predictive fouling based on shear threshold models into an overall fouling mitigation strategy. These methods are based on deposit and removal rate models that involve shear stress and other fluid properties. With reliable predictive models, optimization of operating conditions to avoid or minimize fouling could be combined with exchanger cleaning strategies to further reduce overall fouling related operational costs. The theory behind these first principles models has matured over the past decade, but as yet there is insufficient industrial operating data available to generalize them.

Methods do exist to tune a fouling model for any given fluid to plant data to create a predictive model, and today's computing capabilities are sufficient for the task. Such methods have been used to tune reactor models with actual plant operating data. However, the authors are not aware of any commercial application of these methods for predictive fouling.

Software simulation capabilities also continue to increase in speed and capability. It is now feasible to perform the many scenario calculations required for cleaning cycle optimization for networked systems of heat exchangers in manageable day-to-day working time periods. Recent developments such as workflow management and time-based scenarios integrated into the flow sheet simulation environment enable more detailed, complex, and realistic future operating scenarios to be constructed whereby variables such as crude oil type and throughput can be specified and evaluated.

CLOSING THOUGHTS

The economics of cleaning fouled heat exchangers in complex heat exchanger networks such as crude unit preheat trains are complicated, and at many facilities little effort is made to schedule cleanings based on economic criteria. However, commercially available software can be used to develop near-optimal cleaning programs, even for these complex systems. Adopting this approach to heat exchanger cleaning can lead to significant energy savings. Future advances will most likely be based on predictive fouling models and improvements in simulation software.

ACKNOWLEDGMENT

The authors wish to thank Ant Waters of KBC Process Technology for providing input and reviewing material.

REFERENCES

1. Van Nostrand, W.L., Leach, S.H., and Haluska, J.L. (1981) Economic penalties associated with the fouling of refinery heat transfer equipment, in *Fouling of Heat Transfer Equipment* (eds E.F.C. Somerscales and J.G. Knudsen), Hemisphere, New York, pp. 619–643.
2. Panchal, C.B. Fouling mitigation in the petroleum industry: where do we go from here? *Proceedings of Engineering Foundation Conference on Fouling Mitigation in Industrial Heat Exchangers*, June 18–23, 1995, San Luis Obispo, CA.
3. Perry, R.H. and Green, D.W. (1997) *Perry's Chemical Engineers Handbook*, 7th edition, McGraw-Hill Professional, New York.
4. Hewitt, G.F. (1992) *Handbook of Heat Exchanger Design*, Begell House, Danbury, CT.
5. Saunders, E.A.D. (1988) *Heat Exchangers: Selection, Design and Construction*, Longman, London.
6. Rodriguez, C. and Smith, R. (2007) Optimization of operating conditions for mitigating fouling in heat exchanger network. *Chemical Engineering Research and Design*, 85(A6), 839–851.

13

SUCCESSFUL IMPLEMENTATION OF A SUSTAINABLE STEAM TRAP MANAGEMENT PROGRAM*

Jonathan P. Walter and James R. Risko

TLV Corporation, Charlotte, NC, USA

Steam trap management is an essential component of any comprehensive energy management program in a facility that uses steam. The health of the steam trap population can also significantly affect safety, reliability, and product quality, yet these impacts are sometimes not fully understood. As a result, companies may neglect steam traps for long periods of time, which can be a costly mistake.

The key component of the steam trap management program is the survey. However, effective management involves much more than the survey itself. The survey identifies problems and possible improvements; the program corrects the problems and executes and sustains the improvements.

An effective steam trap management program focuses on three areas:

1. Preimplementation strategic planning.
2. Onsite program implementation.
3. Ongoing program oversight.

* Adapted with permission from *Chemical Engineering Progress* (CEP), January 2014 [1].

Energy Management and Efficiency for the Process Industries, First Edition. Edited by Alan P. Rossiter and Beth P. Jones.

This chapter discusses each of these elements. It also explains why a company might struggle with implementing a program to manage steam traps, as well as ways to justify such a program.

WHAT ARE STEAM TRAPS?

A steam trap is a device used to discharge condensate and noncondensable gases with little to negligible loss of live steam when functioning to manufacturer's specifications (Figure 13.1). These devices are essential to the proper operation of steam systems and the recovery of condensate, and they are critical in ensuring the energy efficiency of the steam system. However, even a properly functioning steam trap can consume a small amount of steam during normal operation. This is referred to as functional steam loss [2].

There are many different types of steam traps, based on several different physical principles. They can be broadly classified into three main groups: thermodynamic, thermostatic, and mechanical. Each design has its own strengths and weaknesses, so it is important to select an appropriate type of steam trap for any given application [3].

Steam traps can fail through two general modes: leakage, in which the trap continues to perform its job of removing condensate, but leaks steam; and drainage (i.e., cold traps, low-temperature traps), in which the flow of condensate is blocked, preventing the removal or draining of condensate from the system. Most leakage-failed traps are easy to spot if discharging to atmosphere due to their visible steam plumes, and their cost impacts are easy to quantify in terms of steam loss. Steam lost through a leaking steam trap can be categorized as failure steam loss.

Drainage-failed traps are often less obvious to the casual observer, and their impacts are less widely recognized and may be more difficult to quantify. When traps are in a drainage-failed condition, condensate builds up in steam lines, and this can lead to many serious consequences, including flare outages, steam leaks, rupture of steam lines, and damage to equipment, notably steam turbines. Incidents of this type can lead to expensive plant outages, with attendant loss of production, and they can even result in injuries to personnel due to flying pipe shrapnel [4].

Figure 13.1. This normally functioning steam trap discharges condensate and noncondensable gases, with negligible loss of live steam. (Courtesy TLV Corporation. All rights reserved.)

Steam trap failure rates can be high, and if left unattended for a few years a population of steam traps can easily have between 20 and 40%—or even more—in a failed condition. On a large site, this can represent millions of dollars in failure and functional steam losses, as well as significant additional costs in production losses from drainage failures. For these reasons, the steam trap population should be maintained in good condition, and a steam trap management program is essential to achieve this goal. Yet, even though end user sites recognize the valuable cost reduction opportunity, some do not implement or take full advantage of a sustained long-term steam trap management program.

WHY ARE TRAP MANAGEMENT PROGRAMS DIFFICULT TO IMPLEMENT?

Sites may struggle to implement a steam trap management program for several reasons, including

- insufficient resources;
- lack of engagement or focus by the site;
- inadequate understanding of both the potential benefits and the challenges of steam trap maintenance;
- insufficient knowledge of how to implement a trap management program and resolve steam trap problems;
- focus on other activities that are either easier to accomplish or perceived to have a higher priority;
- concerns about creating necessary management of change (MOC) documents for new steam trap technology;
- reluctance to spend a limited maintenance budget on steam system improvements due to higher perceived return on investment for other projects;
- difficulty in estimating the financial gains needed to justify costs of starting or continuing the program.

MOTIVATIONS

Before a steam trap management program can be implemented, the facility's management and associated personnel must be sufficiently motivated. Several factors illustrate the benefits of a typical program and can serve as motivators:

- *Leaking steam costs:* Leaking steam traps can be expensive due to steam energy loss [5] and other overhead costs related to the leaking traps [6], such as running a standby boiler or water treatment facilities.
- *Production impacts:* Failed steam traps, especially cold traps [4,7], can affect production (e.g., turbine trips, freezing process lines, and unscheduled process unit shutdowns due to freezing instrumentation).

- *Maintenance costs:* Costs associated with repairing and ensuring reliability of steam equipment are often significant.
- *Personnel safety:* Steam traps that are not operating properly can cause injuries (e.g., burns from leaking steam and slips on pooled or frozen condensate).
- *Environmental impact:* Leaking steam traps result in higher energy consumption, which increases emissions of greenhouse gases and other pollutants.

The magnitude of these impacts and their associated costs increase dramatically with the length of time the plant has not consistently managed its steam trap population. To illustrate this cost, consider a facility that includes 1000 traps with a failure rate of 20%/year (Table 13.1). As shown in Table 13.2, for every year that the facility waits to implement a steam trap management program, 20% more of the traps fail, resulting in losses that accumulate over time. The unrecoverable losses are the costs that accumulate throughout the year as more traps fail at a cost of $800/year for each trap. (This assumes that the same number of traps fail each month, so that on average the traps are in failed operation for half of the year.) Forward-looking recoverable losses are the losses that will start to accumulate over the next year assuming no more traps fail.

The total cost shown in Table 13.2 is the investment required to undertake a survey of all traps ($12/trap for 1000 traps) and to take corrective maintenance action for failed traps (replacement steam trap cost and maintenance labor). If the plant makes this investment, then the forward-looking recoverable losses can be avoided. These savings can be used to justify the costs associated with implementing a trap management program.

TABLE 13.1. A Plant Has 1000 Steam Traps That Have an Average Failure Rate of 20% per Year

Steam trap population	1000
Failure rate of steam trap population	20%/year
Cost impact of failed traps	$800/trap
Trap survey cost	$12/trap
Steam trap purchase cost	$300/trap
Steam trap installation cost	$100/trap

TABLE 13.2. For the Trap Population Described in Table 13.1, the Cost of Delaying Steam Trap Maintenance Rapidly Accumulates with Time

Survey and Follow-Up Delay (Years)	Trap Population Failed (%)	Total Accumulated Trap Failures	Unrecoverable Accumulated Losses ($)	Forward-Looking Recoverable Losses ($/Year)	Total Cost ($)
1	20	200	80,000	160,000	92,000
2	40	400	320,000	320,000	172,000
3	60	600	720,000	480,000	252,000
4	80	800	1,280,000	640,000	332,000
5	100	1000	2,000,000	800,000	412,000

GETTING WIDESPREAD INVOLVEMENT

For the successful implementation of a steam trap management program, all levels and groups within the organization should be involved in developing the program, especially

- corporate-level management, such as the company's board of directors (including directors who drive initiatives related to energy and environmental issues as well as plant reliability initiatives), who can support and fund the program, as well as hold other employees accountable for the program's implementation;
- senior site managers, such as plant managers and business unit managers, who are responsible for what happens at the plant level (rather than across all plants within the company);
- maintenance managers and technicians, who have the time and budget to undertake the hands-on work, and for whom steam trap management is a priority;
- operations managers, operations supervisors, and operators, to support testing and steam trap commissioning.

Finally, the program needs a champion to manage its implementation. This person is typically an energy manager with support from supervisors.

PLANNING AND PREPARATION

Once the plant has decided to implement a steam trap management program, the next step is planning and preparation. At this point, people may be tempted to rush into the steam trap survey (Figure 13.2) in the hope of quickly identifying, and then replacing, failed

Figure 13.2. The steam trap survey is the key component of a steam trap management program. However, a plant should not jump into performing the survey without proper planning and preparation. (Courtesy TLV Corporation. All rights reserved.)

traps. However, it is prudent to spend additional time in the planning phase. The planning activities that typically have the largest impact on the success of the program are

- selecting and correctly installing the best steam traps for the site's conditions;
- identifying the most accurate diagnosis technology and testing resources;
- defining the scope of the trap management program.

TRAP SELECTION AND INSTALLATION

Many plants spend considerable effort testing steam traps and then repairing or replacing them, usually based on a replace-in-kind strategy. This approach will reestablish the plant's steam system to its original design, with no improvement over that design, which may be many years old. If the original design used inappropriate or suboptimal trap technology, the full benefits of a trap management program may not be realized.

Therefore, before starting the survey (Figure 13.3), the plant should evaluate current steam trap practices and determine how they might be improved. Using this approach, higher performance traps can be selected.

An effective method for assessing the existing steam trap technology involves a life cycle cost model, which accounts for four basic cost components related to the four phases of the trap's life cycle:

- purchase and installation costs of a new trap;
- operational costs of a correctly operating trap related to functional steam loss (FSL), which can be estimated based on international standards [8,9];

Figure 13.3. Before starting the steam trap management program, the plant should evaluate the current steam trap practices and determine whether the optimal technologies are being used. (Courtesy TLV Corporation. All rights reserved.)

- operational costs associated with a failed trap (e.g., leaking or cold trap) [7];
- repair or replacement costs.

Life cycle costs should be evaluated for key applications (i.e., those that have the largest impact on plant performance and energy efficiency), such as drip applications (including high-pressure drips) and tracing applications (including copper tracing). Each application has its own challenges and therefore requires a different type of trap. For example, high-pressure drip duty often deals with superheated steam, which typically causes traps to wear out quickly, whereas copper tracing traps may be susceptible to blockages caused by dissolved copper that precipitates and forms deposits within the trap internals when condensate flashes.

Trap selections should be documented to create a plant standard. This documentation should also include plant-specific installation guidelines and piping drawings to ensure that the steam traps are installed correctly. By selecting and documenting the optimal trap models in advance, corrective actions can be taken promptly.

The plant standards and installation drawings should be updated on a regular basis to take advantage of new trap technology as it becomes available. This helps with maintenance and also with outsourced projects to ensure traps are supplied and installed to the plant's best-practice requirements.

If the trap assessment identifies a better trap model, an MOC procedure should be initiated and purchasing and inventory processes should be updated to reflect the new equipment. Thought needs to be given to the disposition of any existing inventory of old models and to preventing the automatic reordering of old traps that are no longer preferred.

Training on the trap standards and installation guidelines is invaluable, but may not be effective in the long term if the standards can only be reviewed from a company computer or library. Trap selection and installation requirements need to remain visible to, and easily accessed by, the maintenance technicians undertaking installation or repairs in the field. This can be done by summarizing the content of the standards on a wall chart or in a plastic pocket-sized flip book. All of this material should be prepared before the survey starts to ensure that any new and improved best practices for trap selection and installation supersede past practices.

DIAGNOSIS ACCURACY

Correctly diagnosing the operational status of the steam trap has a significant impact on the profitability of the trap management program (Figure 13.4). Four potential scenarios can occur:

1. *Correct diagnosis of trap condition:* A correct diagnosis does not add extra costs to the program.
2. *Incorrect diagnosis of a good trap as either leaking or blocked:* The site may then needlessly spend money (typically $400/trap) purchasing and installing a replacement trap. This situation not only results in unnecessary expenditures

Figure 13.4. The accuracy of the diagnosis method used during the trap survey is very important. (Courtesy TLV Corporation. All rights reserved.)

but could also take resources away from other, more valuable, improvement projects.

3. *Incorrect diagnosis of a leaking trap as good:* Steam leakage from such a trap that is left in place can translate to an average annual energy loss of up to $800 per trap.
4. *Incorrect diagnosis of a blocked trap as good:* A blocked trap that is missed and left in place could have a potential impact of $800 per trap [4].

To understand the impact of misdiagnosing steam trap health, assume an average financial penalty of $600 per misdiagnosis. If, on average, three condition diagnosis errors are made for every 100 tests, then for a 1000-trap population, the hidden misdiagnosis cost is $18,000. For a facility with 6000 traps, the corresponding cost would be $108,000. The misdiagnosis cost can be allocated as a testing penalty of $18 per trap for each trap in the population ($18,000 divided by 1000 traps). The significant costs associated with misdiagnosis may influence the choice of testing strategy, and persuade you to choose a more expensive, yet more accurate, method. This example also highlights the importance of correctly diagnosing a trap before undertaking costly maintenance action.

Factors to consider when evaluating the accuracy of a testing methodology include the following:

- *Technology type:* Typically, a combination of ultrasonic and contact temperature measurement is the most accurate method of steam trap testing.
- *Objectivity:* Objective methods that diagnose a trap's operating condition based on empirical reference data or reference standards [2] specific to each trap model are more accurate than subjective methods (such as visual observation). The

more specific the reference data to each particular model, the more accurate the diagnosis.

- *Outside validation:* The diagnosis method should be validated by a recognized third-party verification agency to ensure that the diagnosis methodology and results obtained are accurate according to sufficient, in-depth, empirical confirmation testing.
- *Survey speed:* A typical survey can accurately evaluate between 50 and 150 traps per day, depending on the type of facility and accessibility of the traps. By estimating an average time to find a trap, test it, and record the condition data, and comparing that with the time stated in the survey proposal, a plant can determine whether the testing prices quoted in contractors' bids are reasonable.

Certification, experience, and training of the people undertaking the trap survey are also important, although there are no industry standards defining certification and training requirements.

SURVEY SCOPE

It is important to determine the scope of the survey, replacement models, installation improvements, and maintenance actions before starting the survey. The survey scope should include

- safety training requirements;
- plant areas to be surveyed and the numbers of traps to be tested;
- steam trap diagnostic technology to be used;
- the tagging with equipment numbers needed to identify steam traps;
- instructions for marking failed traps in the field (e.g., red paper tags and orange spray paint);
- data to be collected on the steam trap location (e.g., pipe connections, sizes, isolation valve locations, and bypass valve details);
- accessibility and provisions to access traps (e.g., personnel lifts and scaffolds);
- special requirements (e.g., testing in confined spaces or hazardous areas);
- instructions for updating the plant's steam trap database;
- suggested replacement models for failed traps;
- additional testing steps if traps are identified as cold (i.e., not in service);
- additional testing for locations susceptible to vibration;
- content of the survey report and format (e.g., an Excel spreadsheet);
- survey report presentation requirements;
- survey progress report requirements.

In defining the survey scope, pay special attention to onsite data collection, special requirements for testing steam traps, and trap replacement philosophy.

Onsite Data Collection

During the survey, the testing team typically populates a database with such information as trap identification number, line pressure, trap model, connection type, and simple application notes such as drip or tracer. This application information should be expanded into as many different classes as the surveyor can reliably identify—for example, stainless steel tracing, copper tracing, sulfur line tracing, instrument tracing, turbines, and flare lines—because each of these applications may require a different type of trap. More detailed application classifications can also be valuable in identifying trends and root causes of failure, which is important for improving future trap selections and installations.

Another database field is priority. This may be as simple as differentiating among critical, important, and normal application significance, although additional classifications may be beneficial. Typically, site personnel will need to populate this field or provide onsite support to contract surveyors. The priority together with the type of application and other survey results can be used to prioritize maintenance responses beyond simply fixing the largest leaks.

For example, a cold trap on critical instrument tracing or a critical turbine may warrant immediate attention to avoid an erroneous alarm and subsequent plant shutdown.

Special Requirements for Testing Steam Traps

It is crucial to determine the root cause of any cold trap in order to develop an appropriate maintenance response. A trap may be cold for several reasons—for instance, it might have failed closed or been blocked; it might have been valved out because it was leaking, on an abandoned line, or on a line that is temporarily out of service; it might have been mistakenly diagnosed as low-temperature trap based on an incorrect pressure assumption; or an upstream strainer might be blocked. While this work may accrue additional costs, it typically has a good return on investment. For example, the simple act of blowing down an upstream strainer and retesting the trap often eliminates the need to replace the trap, and may even prevent shutdowns due to catastrophic turbine damage or a turbine trip.

Traps on high-pressure (e.g., >1000 psi) steam lines can be difficult to diagnose with ultrasonic technology, so alternative methods, such as thermal imaging, may be necessary. Ultrasonic testing can also be affected by ultrasound propagating from local sources, such as flow through a nearby control valve or a leak through an adjacent bypass valve, as well as by vibration from turbines or rotating equipment. In these cases, additional testing may be required. This should be defined before the survey begins, so that testers understand the requirements and are technically able to carry them out.

Trap Replacement Philosophy

Maintenance programs to diagnose and repair unhealthy steam traps often focus on leaking traps, while ignoring or placing a lower priority on cold traps. Cold traps are often perceived as less critical than leaking traps. However, the impact of cold traps is much more serious. The philosophy and budget for replacing cold traps should be considered during the planning stage.

A survey of a large facility might identify a significant number of traps that have small leaks. The company may not have the resources to repair all of these leaking traps, and the costs associated with a small leak may not justify the expense of replacement. Consequently, the plant's philosophy may be to address traps only when the steam leak exceeds a specified quantity. If the trap is leaking even a small amount of steam on a continuous basis, the leak will get worse and will likely justify replacement during the next survey.

PROGRAM IMPLEMENTATION

If the trap management program has been planned well, execution should go smoothly. The areas that are most critical to the success of the implementation phase are testing, maintenance response, and oversight.

The first decision that needs to be made at the outset of implementation is to identify the individuals who will perform hands-on trap testing and follow-up maintenance actions. This often involves determining whether onsite personnel will have sufficient time and training to properly diagnose trap operation, or whether it would be better to contract an external specialist to perform the survey. It may also be a challenge to keep general maintenance technicians focused on trap testing and repair/replacement due to other maintenance or process priorities.

In deciding who will perform testing and maintenance, consider the following credentials of the person or team:

- safety record and safety training;
- experience using testing equipment that makes automatic and accurate trap condition judgments based on empirical reference data;
- experience and training, particularly regarding the correct identification of steam trap models, principles of operation, and installation practices;
- availability to undertake the work without being diverted to other issues;
- references, if an external testing service is being considered.

Most facilities either use an external specialist for testing and a dedicated in-house contractor for trap replacements or form an in-house team (possibly including an embedded contractor) to perform the testing and to make any necessary trap repairs or replacements.

Before testing starts, a training session on trap selection and installation should be conducted for all technicians. This is essential, because the benefits of the program could be lost if a trap is not correctly diagnosed during testing, the proper replacement trap is not selected, or the replacement trap is not installed correctly. In addition, the trap installation drawings should be reviewed, modified as needed, and approved in advance of any installation. It is beneficial for the steam trap vendor to review and validate several of the initial replacements to ensure that they are properly installed according to the manufacturer's recommendations. This allows for early identification of deviations, minimizes ineffective expense allocation, and enables quick retraining of maintenance personnel before a problem becomes significant or difficult to correct.

Once the resources have been identified, other testing logistics, such as nondisclosure agreements, work contracts, site access, permits, and licensing, should be addressed according to the company's standard procedures.

TESTING

Three aspects of testing are commonly overlooked and warrant special consideration:

1. *Locating and identifying the trap to be tested:* The first survey in a plant or production unit may require operations support to locate all of the traps. During the initial survey, specific trap location information should be recorded so that future surveys can more easily locate the traps. This could be done by entering notes that describe trap locations into a field in the steam trap database, or by marking trap locations on a map of the facility. The cost of plotting trap locations on a map without software specifically designed for this purpose can be significant, so the additional cost to collect and document trap location data should be weighed against the expected future economic savings.
2. *Trap access:* Some traps might require a ladder, scaffolding, special access permits, or lifts; some might be covered with insulation or screening that prevents testing. Provisions for testing these difficult-to-access traps should be included in the request for proposal given to contractors before they bid on the job.
3. *Process operations support:* Support from process operators can be extremely valuable. Because they are familiar with the unit, experienced operators can help to ensure that all traps are located. They can provide accurate data on trap operating conditions, as well as whether the trap is genuinely in service by locating and determining the flow status of isolation valves. They can also blow down the strainers of cold traps and help facilitate the testing of traps that are valved out but should be in service.

Once a trap has been located, identified, accessed, and confirmed to be in operation, data logging, testing, and tagging are straightforward. However, some people find it difficult to classify and judge traps that are not being tested or are not in service. Consequently, guidelines should be developed for classifying traps whose operating condition is unknown. Traps that are not operating at the time of testing—for example, those on winterization lines, lines that are used only occasionally, or abandoned lines, should be classified as not in service. Traps that were not tested because they could not be accessed should be classified as not checked or inaccessible.

At large plants with several thousand or more steam traps, the testing results should be handed over to the maintenance team at least once a week while the survey is underway so that failures can be addressed as soon as possible. Such quick communication minimizes the negative effects of failed traps. It also avoids overwhelming the maintenance department with a long action list.

MAINTENANCE RESPONSE

The maintenance planners who prioritize personnel time and ensure that materials are ordered and ready for scheduled jobs should get involved early in the program planning phase. Once the survey starts, the planners will manage the purchase, inventory, and allotment of steam traps and repair parts, place work orders, arrange scaffold access, and facilitate maintenance work.

All maintenance actions (i.e., trap repair or replacement) should be documented by recording the date, action, new trap model, and comments. The data should be logged in the database that contains the diagnosis of the failed trap's operating condition. This will enable the analysis of accurate and historical results to improve the trap management program. This final step of connecting maintenance records to the steam trap database is often overlooked, but it is necessary for the ongoing improvement of the trap management program.

OVERSIGHT

Successful programs are often driven by the involvement of at least two enthusiastic individuals, one in a management oversight role and the other in a supervisory and/or execution role. It also helps if higher levels of management in both operations and engineering support the trap management initiative. Companies that have been most successful also usually have senior executives watching over program implementation.

Accurate and timely reporting of activities is essential. Since the benefits of the trap management program are not attained until a trap failure is corrected, it is more important to track maintenance response actions than to track survey findings. The program status report should include data on

- number of leaking traps replaced, the amount of steam loss prevented, and the monetary savings associated with preventing that steam loss;
- number of leaking traps waiting for maintenance and the amount of steam being lost;
- number of cold traps replaced;
- number of cold traps still in place, especially those classified as critical or on turbines or instrument tracing;
- classifications of traps of unknown status;
- cost of the survey and repairs (including parts and labor).

Milestone replacement targets should be established, reports reviewed monthly or quarterly, and the site held accountable for repairing failed traps according to the corresponding schedule. The infrastructure, content, and procedures to create and circulate the performance reports should be carefully planned to ensure that the reports can be generated quickly and easily. For example, software is available that is designed for the effective management of a steam trap population and that provides detailed trap condition analysis, as well as reports on failure rate and economics.

It can also be very helpful to involve a knowledgeable steam trap vendor representative in the program on an ongoing basis to provide expertise and to help make steam system and trap management program improvements.

SUSTAINING AND IMPROVING THE PROGRAM

After the survey is completed, the failure data should be analyzed to identify common failure modes, and the effectiveness of the maintenance response should be reviewed to identify areas for improvement.

Once a company has survey data for 3–5 years, it should analyze the historical data to identify common failure modes, trends in failure rates, trends in the number of traps classified as not in service and untested, the number of failures being carried forward to the next survey without being addressed, failure trends by applications or types of traps, and the locations where traps have frequently failed after maintenance action has been taken. This analysis may indicate an application, piping, or trap selection problem. The root cause of frequent failure should be identified and addressed. It is not uncommon for a significant percentage of a site's failures to be repeat failures in certain locations.

The Failure State (% failed) typically decreases rapidly soon after a program is initiated and then levels off. At this point, the company must determine whether the steady-state failure rate can be further reduced by improving the management program or if the program is already as good as it can be. The historical analysis may provide clues to help with this determination. For example, it is possible for annual failure rates to decrease (which looks good) while the number of not-in-service and not-tested traps increases, which may indicate that problems are being hidden.

Finally, the annual survey analysis should be documented, and the trap testing leader, site champion, and trap vendor representatives should hold a review meeting to discuss the analysis and lessons learned that year.

FINAL THOUGHTS

If the trap management program is run well, the annual savings will gradually decrease as new traps and better technology are deployed. However, there are still many other opportunities to improve the performance of the steam system [10], including initiatives to improve specific steam-using applications such as reboilers, heat exchangers, turbines, and air heaters, as well as to optimize the entire steam and condensate system balance.

REFERENCES

1. Walter, J.P. (2014) Implement a sustainable steam-trap management program. *Chemical Engineering Progress*, 110(1), 43–49.
2. Risko, J. R. (2011) Understanding steam traps. *Chemical Engineering Progress*, 107(2), 21–26.

3. The Engineer's Toolbox, Steam Trap Selection Guide. Available at http://www.engineeringtoolbox.com/steam-traps-d_282.html
4. Risko, J.R. (2013) Beware of the dangers of cold traps. *Chemical Engineering Progress*, 109(2), 50–53.
5. Risko, J.R. (2006) Handle steam more intelligently. *Chemical Engineering*, 113(12), 44–49.
6. TLV Corporation, (2010) *Steam trap losses: what it costs you.* TLV Corporation, Kakogawa, Japan. Available at http://www.tlv.com/global/US/steam-theory/cost-of-steam-trap-losses.html.
7. Risko, J.R. (2011) Use available data to lower system cost. *Presented at the Industrial Energy Technology Conference*, New Orleans, LA, May 18, 2011. Available at www.tlv.com/global/US/articles/use-available-data-to-lower-system-cost.html.
8. International Organization for Standardization (1998) *Automatic steam traps—Determination of steam loss—Test methods.* ISO 7841, ISO, Geneva, Switzerland.
9. American Society of Mechanical Engineers (2005) *Steam traps.* PTC39, ASME, New York.
10. Risko, J.R. (2008) Optimize the entire steam system. *Chemical Engineering Progress*, 104(6), 32.

14

MANAGING STEAM LEAKS

Alan P. Rossiter

Rossiter & Associates, Bellaire, TX, USA

Several chapters in this book deal with various aspects of steam systems, and no discussion of steam system management would be complete without some consideration of methods to minimize and manage against steam leaks. In many refineries and chemical plants, leaks are extremely prevalent, and steam plumes can be seen all over the facility. Other plants have a "zero tolerance" policy toward steam leaks, and the difference in the appearance and noise level of the facility can be striking. Addressing steam leaks can often yield improvements worth hundreds of thousands—even millions—of dollars per year in steam losses, as well as significant improvements in reliability and safety.

In this chapter, we will demonstrate the scope of the steam leak problem and the magnitude of the potential savings through a case study. We will also discuss estimation of steam leaks, causes of steam leaks, and an approach to managing steam leaks.

CASE STUDY

Rohm & Haas reported an illustrative case study in 2003, from their Deer Park, TX, chemical plant [1]. At the time steam production across the facility varied between around 1,000,000 and 2,000,000 lb/h, depending on unit rates. However, the bulk of the

Energy Management and Efficiency for the Process Industries, First Edition. Edited by Alan P. Rossiter and Beth P. Jones.

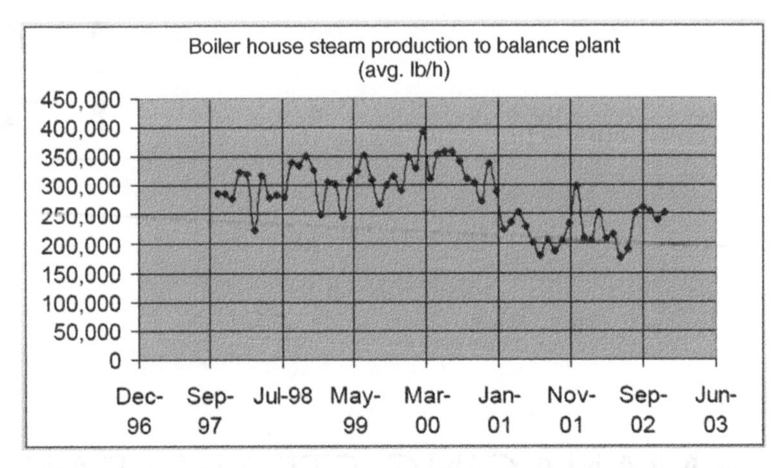

Figure 14.1. Rohm & Haas Deer Park boiler house steam production [1].

steam came from waste heat boilers; the amount of steam production from the boiler house required to balance the site was routinely below 400,000 lb/h (see Figure 14.1). There were three main steam headers at the site, at 600, 150, and 75 psig, and over 2,000 steam traps. The plant had been in existence for a considerable time, and some of the piping was more than 50 years old.

The site initiated an audit of steam traps and steam leaks in March 1999. The audit identified steam losses of 90,000 lb/h, which triggered a $500,000 capital project and a program of repairs in partnership with local service providers.

Follow-up audits were carried out in September 2000 and July 2002, by which time the losses had been reduced to 44,000 and 28,000 lb/h, respectively (Figure 14.2). The program continued to evolve, with the adoption of software tools to aid in auditing and recordkeeping.

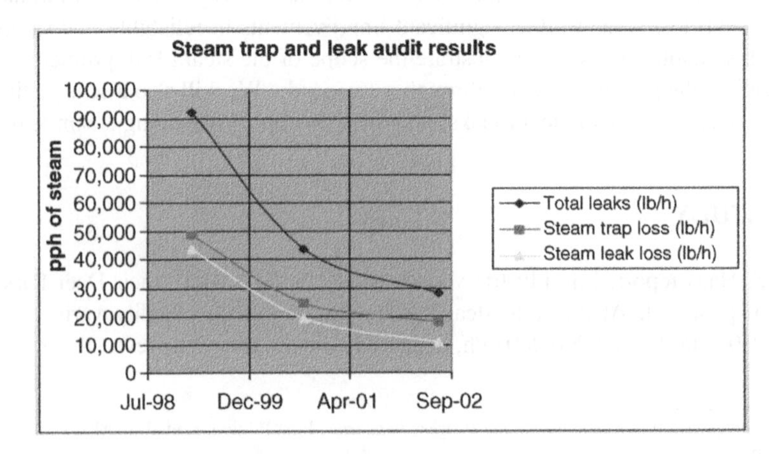

Figure 14.2. Steam trap and leak audit results [1].

Steam trap management programs are discussed in detail in Chapter 13, so in the present discussion we will focus on steam leaks. As can be seen from Figure 14.2, the losses from steam leaks were almost as large as those from steam traps, with over 40,000 lb/h of leakage at the start of the program. To put this in perspective, at the start of the program steam leaks alone represented over 10% of the steam produced in the boiler house. Assuming a typical cost of $5.00/klb, the loss due to leaks was around $1,750,000 in a full year of continuous operation. However, in just 3 years of the repair program the steam leak losses were reduced by roughly 75%.

Even though the steam loss due to leaks was greatly reduced, survey results indicated that new leaks appeared at a rate of more than one every 3 days. This rate did not change appreciably over the 3-year period for which information was provided. This point is significant, and is discussed further below. However, most of the leaks were fairly small: only about 10% were estimated at more than 100 lb/h. Higher pressure systems tended to have the highest failure rates, though higher pressure systems are a smaller portion of the population. These results are consistent with findings for steam systems in general [2].

ESTIMATING STEAM LEAKS

Direct measurement of steam loss through leaks is extremely difficult. However, reasonable estimates can be made using the "plume length" method [3]. The length of the plume created by a steam leak is the distance from the source of the leak to the region where water condenses out of the steam. This point is usually beyond the visible plume. Used in conjunction with steam pressure and ambient temperature, the plume length can be used to estimate the steam loss, in lb/h, as shown in Figure 14.3.

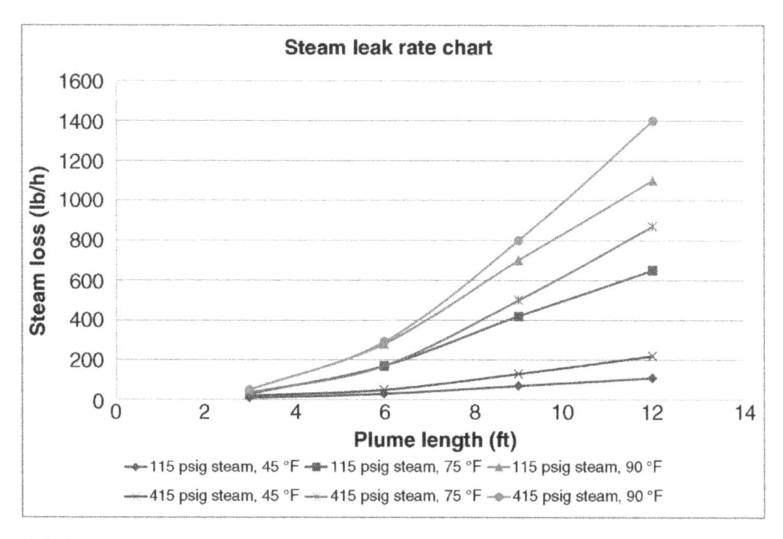

Figure 14.3. Steam leak rates at different header pressures and ambient temperatures, estimated using the plume method. (Based on data from Ref. 3.)

CAUSES OF STEAM LEAKS

Steam leaks are commonly associated with improper piping design, corrosion problems, and joint and valve seal failures [2]. Water hammer is very often a key factor contributing to these failures, and it is discussed further below.

The steam produced by many of the boilers used in chemical manufacturing plants is wet (i.e., less than 100% of the water has been evaporated). While some boilers include devices such as separators to remove this water before it exits the boiler, not all of the water is removed and some remains entrained in the steam supplied to the plant. Steam produced from state-of-the-art boilers can still contain 3–5% water at the boiler exit.

Even if a boiler produces superheated steam, there is still a high probability of liquid water forming within the steam distribution system. Heating the cold pipeline of the steam system at start-up generates condensate that must be drained via drop legs and condensate discharge locations (CDLs): that is, drainage systems consisting of a steam trap and the associated piping, check valves, blowdown valves, isolation valves, strainers, tees, and so on. Once the start-up condensate has been drained and the system reaches superheated conditions, the vertical collection piping of the CDLs, commonly known as "drop legs," become stagnant-flow heat sinks that cool down the superheated steam, generating condensate. Other situations can also change the quality of the steam from superheated to wet: for instance, when desuperheaters go awry and inject too much water into the steam flow.

As the steam travels through the distribution pipeline, various mechanical and thermodynamic influences can cause the entrained water to fall out of the steam (see Figure 14.4). If not removed through steam traps—typically because the trap is blocked or isolated to stop the trap itself from leaking steam if failed—the disentrained

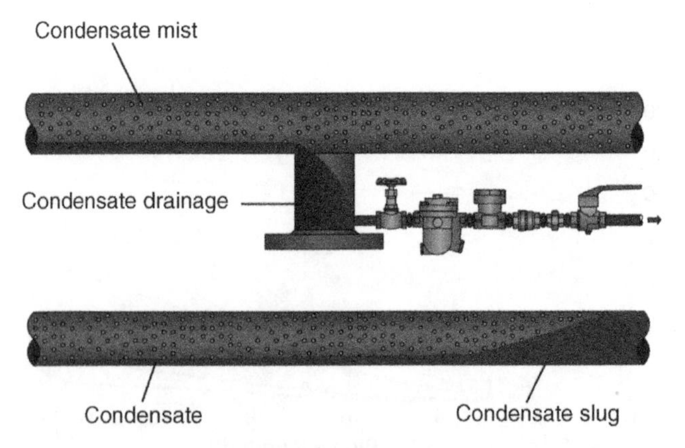

Figure 14.4. Steam produced in a typical boiler contains condensate, which must be removed from the system via a steam trap (top). When not properly removed, the condensate will eventually form a slug of water (bottom) that will be thrust forward by the fast-moving steam. (Reprinted with permission from *Chemical Engineering Progress* (CEP), February 2013 [4]. Copyright 2013, American Institute of Chemical Engineers (AIChE).)

condensate continues downstream and can be propelled forward by the steam, which is flowing at high speeds, typically about 8,800 ft/min (100 mi/h). Slugs of condensate will eventually encounter an elbow, nozzle, valve, flange, and so on and come to an abrupt stop, causing water hammer. This pressure surge can damage equipment, including valve packings, fittings, and flange gaskets, or cause erosion of piping elbows, all of which can lead to steam leaks. In extreme cases, water hammer can also cause personal injury. High-velocity retained condensate can cause severe damage in turbine blades, from pitting to even catastrophic destruction.

The root cause of water hammer is often poor drainage from CDLs due to failed steam traps ("cold traps" or "blocked traps"). When a CDL is blocked, the condensate is retained in the system. However, CDLs are not typically repaired until a catastrophic event occurs. Some facilities seem to take an "out of sight, out of mind" approach to the handling of retained condensate and repair of cold traps. Unlike steam leaks that are visible, retained condensate is effectively invisible as long as it is contained inside the pipeline, and steam can carry significant amounts of destructive condensate throughout the system. However, the "invisible" retained condensate is very often the cause of the "visible" steam leak, particularly when the leak is the result of erosion, a blown flange gasket, or damaged isolation valve packing. In some instances, there are simply not enough CDLs installed in the system, and in those cases it is necessary to add the required number to remove retained condensate.

MANAGING STEAM LEAKS

A successful steam leak management program must address a number of issues:

- *Detection and identification of leaks:* Most moderate to large steam leaks are easily identified by their plumes. However, they are often ignored, especially in plants where leaks are commonplace and are accepted as the norm. In such situations, a culture change is needed, and the operations staff, in particular, must be trained to recognize that steam leaks are not acceptable. Expectations for plant rounds should include that steam leaks will be identified, and that when leaks are found, they are tagged and entered into the maintenance system for repair.

  Smaller leaks are sometimes harder to detect, especially when they occur under thick insulation or in high-noise areas. Ultrasonic detectors [3] can be used to assist in leak detection in these circumstances. This demands a more intentional approach to finding leaks.
- *Estimation and prioritization:* Plant operators can be trained to use the plume length method to estimate steam loss rates. This information, together with the site's steam balance (see Chapter 18) and steam price data for each steam header (see Chapter 17), can be used to estimate the savings that will accrue when any given leak is fixed.

  Steam systems are often complex, and as illustrated in the Rohm & Haas case study earlier in this chapter, much of the steam can come from "waste heat," usually from process sources but in some cases also from power turbine exhausts. The amount of steam produced from these waste heat sources is generally fixed by the process throughput or by the power output of the turbines. The remaining steam required for the site's steam balance (often called "on-purpose steam") typically comes from fired boilers, which are almost always the most expensive source of steam on a site. In most cases, reductions in steam leaks directly reduce on-purpose steam production. This can result in large monetary savings—especially if the reduction in steam demand is sufficient to shut down a low-efficiency boiler.

Additional factors, such as safety or reliability concerns, may also affect prioritization for maintenance.

- *Maintenance resources:* The most successful steam leak programs generally have dedicated resources and budgets allocated specifically to steam leak repairs. Without dedicated resources, it is likely that personnel and maintenance money will quickly be diverted to other activities that tend to get higher management attention.

  Different leaks can require different maintenance approaches. Some can be repaired without taking equipment off-line. Others cannot, and this can lead to delays—especially if the leak point is hard to isolate. In some cases, the best option may be to use specialized leak repair contractors, while in other instances it may be more appropriate to use the facility's own personnel.
- *Addressing root causes:* Many facilities have noted that even though their steam leak programs do lead to a significant reduction in steam losses, new leaks continue to form at an undiminished rate (see the case study earlier in this chapter). In many cases, these leaks are in the same places as the ones that have been repaired. One option is simply to repeat the repairs. A better approach is to explore root causes. Often the underlying issue is poor piping design or inadequate drainage, either because of insufficient or poorly located drop legs or because of drainage-failed steam traps. Correcting the root cause (e.g., rerouting steam lines, adding drop legs, or replacing steam traps) can eliminate the cost and inconvenience of future steam leak repairs.

CLOSING THOUGHTS

Fixing steam leaks is not glamorous, and like all forms of energy management it requires discipline and persistence. However, it is worth the effort. In addition to significant energy savings, it can also yield important safety and reliability benefits, while also making the plant environment quieter and generally more pleasant. No overall energy management program is complete without it!

ACKNOWLEDGMENTS

Numerous people have assisted with this chapter. The author wishes especially to thank Joe Davis of PSC Industrial Outsourcing and Jim Risko of TLV Corporation for providing input and reviewing material.

REFERENCES

1. Dafft, T. (2003) *Plant Steam Trap and Leak Repair Program.* Texas Technology Showcase, Houston, TX, March 17–19, 2003. Available at http://texasiof.ceer.utexas.edu/texasshowcase/pdfs/presentations/a6/tdafftraps.pdf.
2. U.S. Department of Energy, Office of Industrial Technology (2002) *Steam System Scoping Tool.*
3. U.S. Department of Energy, Office of Industrial Technologies (1999) *Energy Tips: Quantify and Eliminate Steam Leaks.* Available at http://www.apmnortheast.com/Department%20of%20energy%20steam_leaks.pdf (accessed March 4, 2014).
4. Risko, J.R. (2013) Beware of the dangers of cold traps. *Chemical Engineering Progress*, 109(2), 50–53.

15

ROTATING EQUIPMENT: CENTRIFUGAL PUMPS AND FANS

Glenn T. Cunningham

Mechanical Engineering Department, Tennessee Tech University, Cookeville, TN, USA

INTRODUCTION

Typical rotating equipment utilized by the process industries includes centrifugal pumps and fans of sizes less than 1 hp up to 2000 hp or larger. While proper maintenance and alignment are necessary for efficient operation and equipment upgrades are sometimes possible, energy efficiency studies usually concentrate on the type of capacity modulation scheme employed, the system configuration, and controls. This chapter focuses primarily on commonly used capacity control schemes for centrifugal pumps and fans, and illustrates how to identify potential energy-saving projects.

CENTRIFUGAL PUMP CAPACITY CONTROL SYSTEMS

Pumping energy assessments often focus on identifying inefficient control schemes that are devised to align the system's delivered flow with the requirements of the process but waste a considerable amount of energy. Common approaches to varying the capacity of a pumping system are:

Energy Management and Efficiency for the Process Industries, First Edition. Edited by Alan P. Rossiter and Beth P. Jones.

- *recirculation*, where fluid leaves the discharge of the pump and flows, via a recirculation line, directly back to the suction tank (very wasteful of pumping energy);
- *throttling valves*, where a control valve on the discharge side of the pump partially closes, forcing the operating point back up the pump curve and reducing the flow delivered (wastes a considerable amount of energy if the valve stays heavily throttled most of the time); and
- *parallel pumping with throttling valves*, where multiple pumps are piped in parallel discharge into a common header. Throttling control valves are often employed with this arrangement to control flow in different lines coming off of the main supply header (often with the maximum number of pumps ever needed operating all of the time, and valves modulating the flow to each branch line).

All of these control schemes may waste a considerable amount of energy when used in systems where the flow requirements vary over time or where the pumps are oversized.

System designers typically oversize pumps in order to ensure that the pump will be large enough to provide the needed flow when the system is actually built and operated. An oversized pump can be "made smaller" by closing the control valve (or balancing valve in a static application) at the pump discharge to reduce the flow to the amount needed. Other methods of reducing a pump's capacity are to trim the impeller if the required flow is always the same, or to install a variable-frequency drive (VFD) to slow the rotational speed of the pump. Impeller trimming is an efficient way to reduce the capacity of an oversized pump if the pump is oversized all of the time and the amount of flow reduction is not too large. If the flow requirements vary widely, then trimming the impeller is probably not the correct solution. The pump manufacturer should always be consulted before trimming an impeller. Installation of a VFD is often the best solution, since it allows the capacity of the pump to be varied over a wide range and saves energy as the pump's speed is reduced.

The main drawback to the installation of a VFD for flow control is the cost of purchasing and installing the drive. Low-voltage drives (below about 1000 V) typically cost about $65–100/hp of motor size to purchase and a considerable amount to install. Medium-voltage drives are more costly and generally take longer to pay back. Recent pricing for a VFD for a 400 hp low-voltage motor was $26,000 and for a medium-voltage (4160 V) 400 hp motor was $140,000. VFD installation must be properly supervised by a qualified electrical engineer to avoid possible problems such as electrical harmonics, bearing currents, and voltage spikes that can drastically reduce motor life.

Trimming an impeller can usually be accomplished for $2000 or less. The difficulty with impeller trimming is that the pump's capacity is permanently reduced. If the original flow capacity is needed later, another new impeller must be purchased and installed.

Centrifugal Pump Recirculation Flows

Recirculation lines are sometimes installed on the main discharge pipe a short distance downstream of the pump. These lines carry flow from the pump discharge back to the tank supplying liquid to the pump suction. Recirculation can be used to provide minimum flow protection on pumps and for flow control, utilizing throttling valves in the recirculation line. Without a recirculation line, if flow control valves on the process fluid users are mostly or fully closed, the fluid within the pump heats rapidly and can damage the pump. Induced pressure pulses originating within pump suction and/or discharge areas by recirculation vortices can also cause damage. Fluids carrying solid particles often require constant circulation to prevent the solids from settling out on the bottom of piping and storage tanks. Recirculation for flow control whereby the pump produces the same flow all the time and excess flow is returned to the source tank is the least efficient form of flow control.

Boiler feedwater pumps are often equipped with recirculation lines to protect the pump from cavitation damage caused by the vaporization of hot water within the pump casing when valves controlling water flow into the boiler close. Hot boiler feedwater already has a high vapor pressure (compared to cool water), and any additional temperature increase from deadheading the pump will usually result in boiling within the pump and pump damage. Recirculation lines and valves are also frequently used to protect light (low atmospheric boiling point) hydrocarbon pumps from cavitation damage.

Centrifugal pumps can be protected more efficiently by automatically controlling the recirculation flow. Automatic control valves or specialized recirculation valves open only when the system flow is low or stopped and recirculation flow is needed. Valves can be controlled by a pressure sensor in the pump discharge line and opened at a high pressure or by a spring-controlled mechanism set to open automatically at a set pressure.

Centrifugal Pump Recirculation Flows: Example System

Two 1250 hp horizontal split-case double suction centrifugal pumps are used to pump spent liquor at a bauxite refinery. Each pump has a 6″ diameter recirculation line installed just downstream of the pump, returning flow to the suction side supply tank. These lines are always open and have no control valves. Plant process engineers estimate the typical flow through the recirculation lines to be 1520 GPM.

The total flow provided by these two pumps for 2000 h of operation is illustrated in Figure 15.1. The discharge head on the pumps is 425 ft and the specific gravity of the spent liquor is 1.284. The pump curve predicts an efficiency of 64% and the motor efficiency is assumed to be 94%. Electric energy at this site is valued at $60.50/MWh. Motor power can be calculated from Equation 15.1 (motor input power from pump head, flow, pump efficiency, and motor efficiency).

$$\text{Motor}_{\text{kW}} = \frac{(\text{head}_{\text{pump}})(Q)(\text{specific gravity})}{(5308)(\eta_{\text{pump}})(\eta_{\text{motor}})} \tag{15.1}$$

where $\text{head}_{\text{pump}}$ = pump head developed (feet) and Q = pump flow rate (GPM).

Motor power used to provide the recirculation flow is 256 kW/pump and two pumps are operating in the same manner. Typical operation is 8000 h/year. Thus, electrical consumption to support the recirculation flow is (for both pumps combined): Power: 514 kW, Consumption: 4096 MWh, Cost: $247,800 annually. If the recirculation flow is stopped and the pump capacity not altered, the operating point will shift to lower flows at

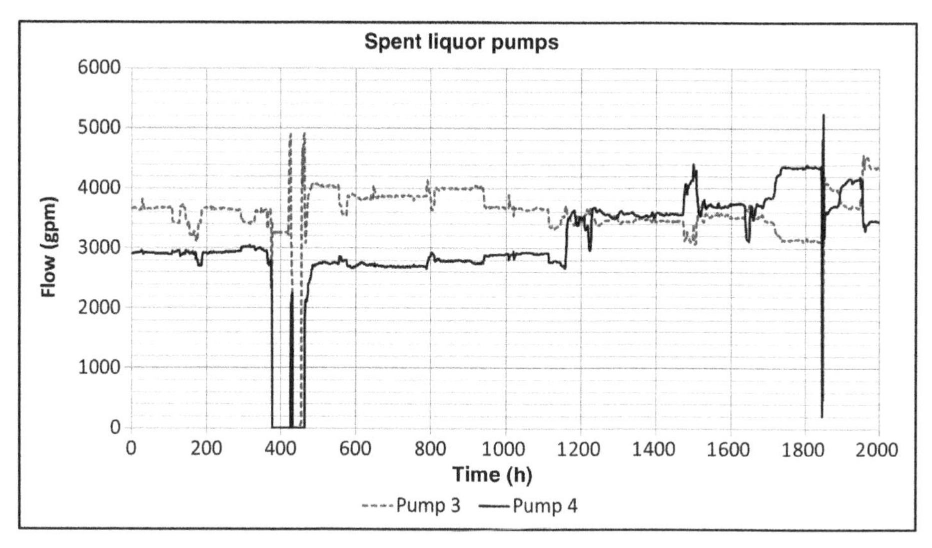

Figure 15.1. Flow rates for two 1250 hp refinery pumps with 6″ recirculation lines.

a higher head, and the pump efficiency decreases to about 50%. Thus, not all of the cost of pumping the recirculation flow can be saved by installing control valves. Actual savings considering the increase in pump head and a decrease in pumping efficiency are: power: 311 kW, consumption: 2479 MWh, and cost: $150,000. The estimated cost for adding automatic control valves for each of these two 6″ diameter recirculation lines is $30,000, for a total project cost of $60,000. The simple payback on this project is expected to be about 5 months. The alternative of adding VFDs to these pumps was dismissed due to the high cost of VFDs on medium-voltage motors. The total project cost is estimated to be $600,000 to install VFDs for both pumps, and this would yield a payback of about 2.4 years.

Centrifugal Pumps with Throttling Valve Control

The most common form of centrifugal pump flow control is with throttling valves. Pumps are often oversized for their desired service. Static balancing valves are normally used for systems with little or no variation in required flow, while automatic modulating control valves are employed with dynamic systems where flow requirements change.

A butterfly valve set at about 40% open is shown in Figure 15.2. If the handle is parallel with the pipe, the valve is 100% open, and if the handle is perpendicular to the pipe it is 100% closed. Most flow control valves have some sort of position indicator showing the valve position relative to wide open and fully closed. This valve supplied a parts washer and is a good example of a flow control system that is set for a particular downstream pressure and flow. This valve was not adjusted frequently, and its upstream pump always used more energy than was needed to provide the required flow to the

Figure 15.2. Throttled butterfly valve controlling flow on a parts washer.

washer. VFDs were installed on all the seven pump motors on this washer and yielded significant energy savings.

When pumping systems are screened for potential energy-saving projects, the easiest source of savings is a system where the entire flow from a pump goes through a single control valve that is often significantly closed. Data from a heavily throttled refinery pump are shown in Figure 15.3. Note the valve position never

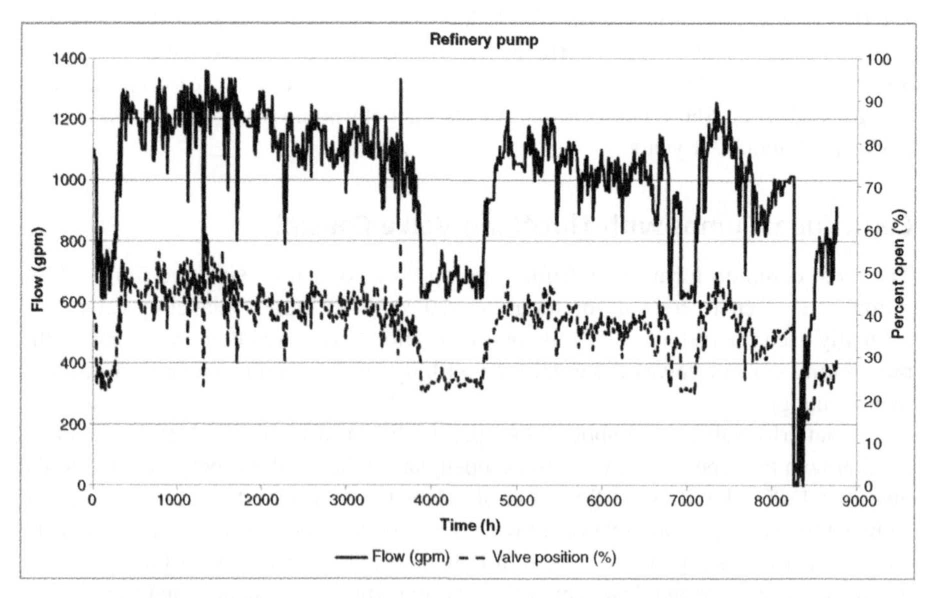

Figure 15.3. One year of hourly data on a heavily throttled refinery pump.

exceeds 60% open over a 1-year period. This is a good indication there are significant energy savings possible by replacing the throttling valve control with the installation of a variable frequency drive to utilize motor speed control to provide the desired flow.

Centrifugal Pumping System Analysis

Pumping systems operate at the intersection of the pump curve and the system curve. System curves are of the form: $\text{head}_{\text{total}} = \text{head}_{\text{static}} + K' \ Q^{1.9}$, where static head is the pressure contribution (positive or negative) due to elevation changes and gas overpressures from the fluid source to its destination [1]. For a closed loop system the static head is zero. Coefficient K' is a constant that defines the frictional losses when multiplied by the flow rate Q (GPM) raised to the power 1.9. Sometimes the exponent is taken to be 2.0. A typical throttled pump analysis is illustrated in Figure 15.4.

Pump affinity laws (Equation 15.2) are often used in pumping analysis to recast a published pump curve to different operating speed. An optical strobe instrument can be used to measure the rotating speed of a pump. Once the actual rotational speed is known, the manufacturer's published pump curve should be adjusted for the measured speed using the pump laws, shown below. Pump affinity laws can also

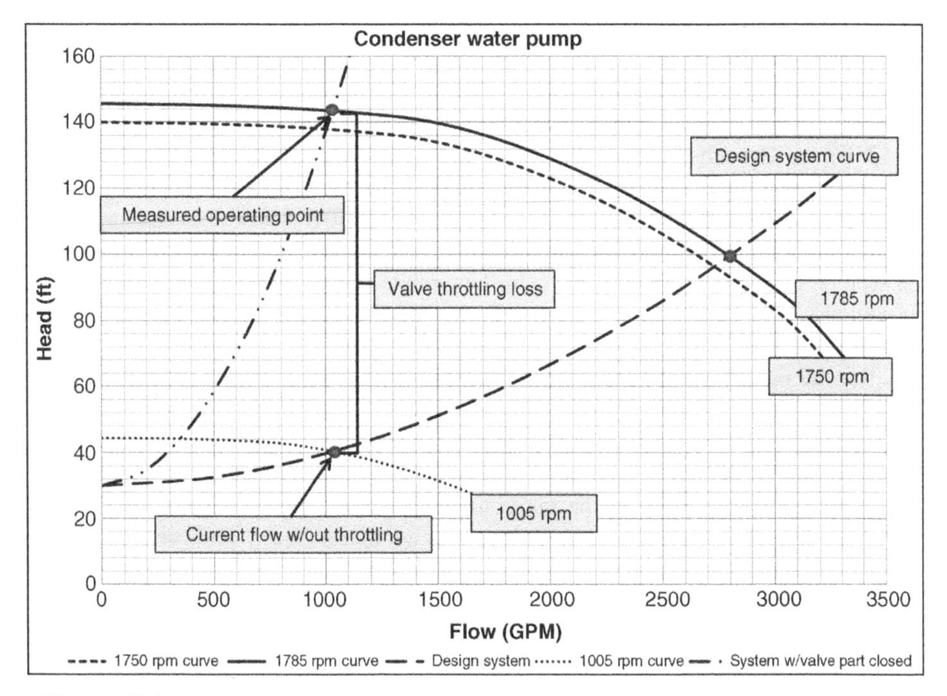

Figure 15.4. Analysis of throttled condenser water pump with reduced speed operation.

be used to redraw a pump curve for a different impeller diameter operating at the same speed.

$$\frac{Q_1}{Q_2} = \frac{N_1}{N_2} \text{ or } \frac{Q_1}{Q_2} = \frac{D_1}{D_2}$$

$$\frac{H_1}{H_2} = \left(\frac{N_1}{N_2}\right)^2 \text{ or } \frac{H_1}{H_2} = \left(\frac{D_1}{D_2}\right)^2 \tag{15.2}$$

$$\frac{BHP_1}{BHP_2} = \left(\frac{N_1}{N_2}\right)^3 \text{ or } \frac{BHP_1}{BHP_2} = \left(\frac{D_1}{D_2}\right)^3$$

where Q = flow rate, D = impeller diameter, N = speed, H = head (TDH), and BHP = brake horsepower.

Centrifugal Pumps with Throttling Valve Control: Example System

The pumping system described by Figure 15.4 is a condenser water system where one of two original chillers had been removed from service. The condenser water pump was not replaced or the impeller trimmed to reduce the pump's capacity. Instead, two separate valves were significantly throttled to move the pump operating point far enough to the left on the pump curve for the remaining chiller to operate with the desired flow. Both valves required frequent maintenance.

In the above analysis the manufacturer's pump curve was published at a pump speed of 1750 rpm. The actual measured pump speed was 1785 rpm and the pump laws were used to "expand" the published curve to the higher operating speed. The piping system included a cooling tower with the basin water level 30 ft below the discharge pipe distributing water at the top of the tower, giving rise to a static head of 30 ft. This head can be seen on the system curves at the zero flow point. The actual operating point for this system was 1030 GPM at 144 ft of head. Pump efficiency at this throttled condition was not good, at 55%.

Several energy-saving projects are possible:

VFD: If the same flow was achieved with speed control and the control valve was locked in the open position, the operating point is predicted to be 1030 GPM at 41 ft of head. Slowing the pump to roughly 1005 rpm (34 Hertz on the VFD output power) will shrink the pump curve to intersect the system curve at the new operating point. An additional benefit is that the pumping efficiency improves to about 67%. Calculations show a power savings of 41.6 kW for a VFD on this 75 hp pump. At an electric power cost of $80/MWh and operating hours of 8000/year, the savings are $26,600 annually. Installing a VFD, estimated to cost about $15,000, would pay back in about 7 months.

Trimmed impeller: Normally it would be a good idea to investigate trimming the impeller because this application does not require flow modulation. However, this pump is so oversized that the impeller could not be trimmed enough to meet the desired operating point.

New pump: Replacing the existing pump with a pump properly sized for operation with a single chiller is also a good option. A new pump selected for this operating point will have a 20 hp motor and operate at 81% pump efficiency. Unlike the VFD option, though, the new pump eliminates the possibility of returning to higher flow rates. The costs and risks of the two feasible options can be compared to find the best project.

Energy Losses in Throttling Valves

Analyzing pressure letdowns in throttled pumping systems helps to identify energy-saving opportunities. Pump discharge pressure is often measured close to the control valve, but downstream pressure is rarely measured. The control valve equation, Equation 15.3, provides a helpful tool for determining the valve pressure loss.

$$Q = C_v \sqrt{\frac{\Delta P}{(\text{specific gravity})}} \tag{15.3}$$

where C_v = valve flow coefficient, ΔP = pressure drop in $lb_f/in.^2$, and Q = flow in GPM.

Valve manufacturers publish curves of valve flow coefficient versus position for their products (Figure 15.5). With the proper valve C_v curve and valve position reading, the pressure drop can be estimated from Equation 15.3 if the flow rate is known. Ultrasonic strap-on flow meters are often used to measure the flow rate when a permanently installed

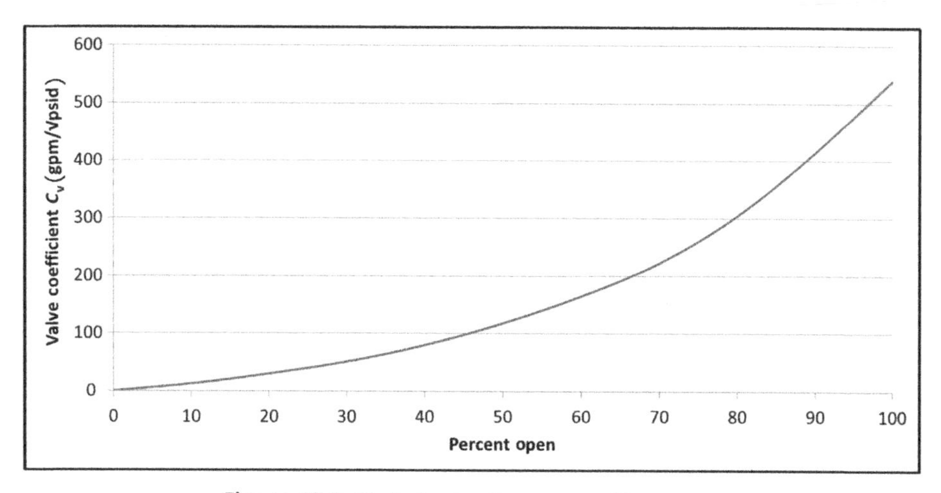

Figure 15.5. Typical valve C_v versus position curve.

meter is not available. A significant pressure drop across a throttling valve is an indication that energy savings are possible from an alternative control scheme.

PUMPING SYSTEM CONSIDERATIONS

Centrifugal Pumps Operating in Parallel

Parallel pumping is commonly employed throughout a wide variety of industries when the systems are required to provide relatively high flow rates at moderate to low head and the system curve is static head dominated. It is not unusual to have three or more identical pumps operating in parallel in a given application, so that if a single pump fails the remaining pumps can provide a large percentage of the system flow at a slightly lower head and prevent a shutdown of the process equipment.

Studying the system behavior of a parallel pumping system provides the insight that with more than three pumps operating in parallel, the last pump started does not typically add a large amount of additional flow to the system. Often the last pump on the system is there more for insurance against shutdown rather than for actual flow requirements. Operating the system differently can provide energy benefits without endangering the process.

Centrifugal Pumps Operating in Parallel: Example Systems

A parallel system operating five 100 hp pumps is illustrated in Figure 15.6. It can be noted that the operation of the fifth pump adds just over 1000 GPM of flow to the system.

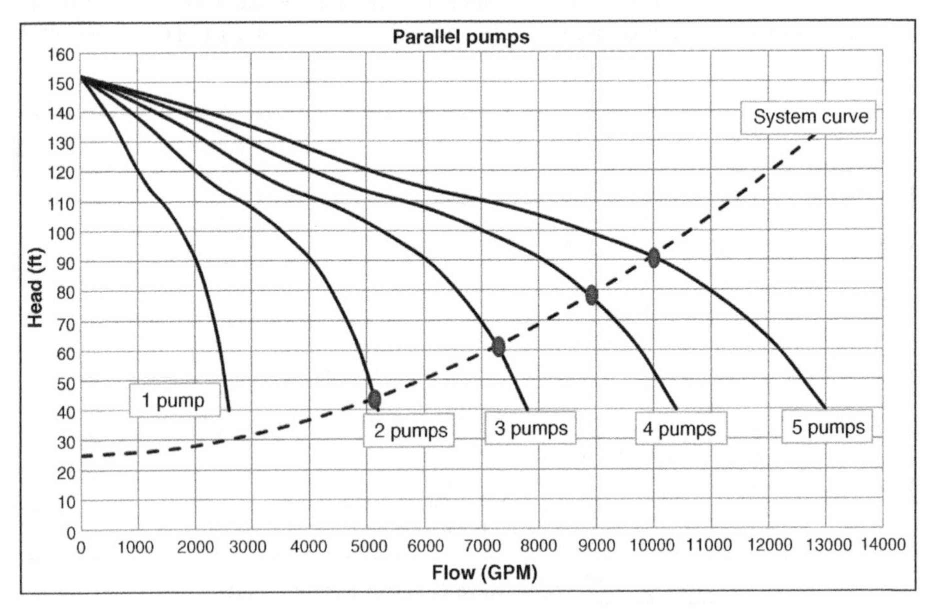

Figure 15.6. Five identical 100 hp pumps operating in parallel.

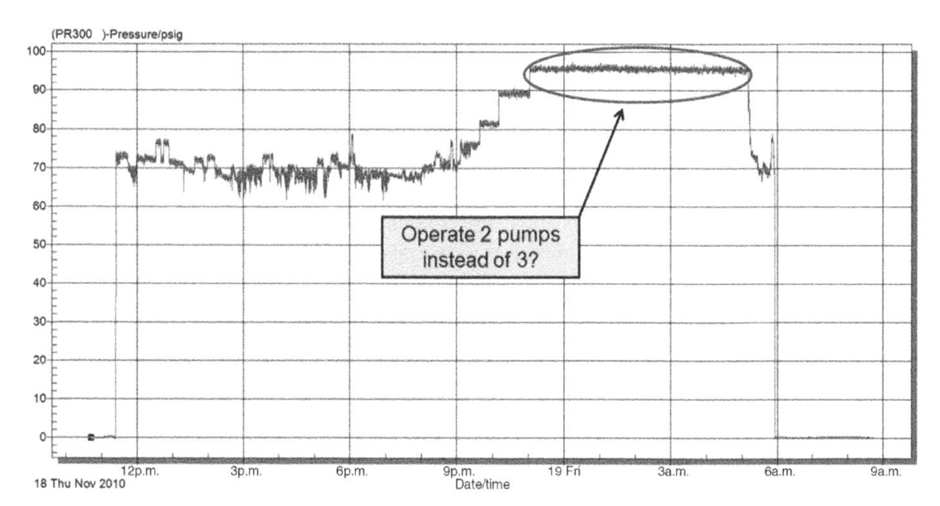

Figure 15.7. Header pressure with three 100 hp pumps operating in parallel all the time.

Often with a system like this, the 8900 GPM provided by four pumps is sufficient for proper operation. The fifth pump runs as an "insurance policy" in case one of the pumps fails. In this case, the "insurance policy" costs the company $35,000/year in electrical energy. "Insurance" could be provided instead by installing automatic controls capable of starting an additional pump if the header pressure falls, allowing the plant to operate four pumps instead of five and reduce operating cost.

A second situation can occur with parallel pumping relating to systems experiencing a reduced demand for flow during some portions of the day. If the operation is 24 h/day, but some process equipment is not operated on some shifts and the same number of pumps is operated all the time, there may be an opportunity to stop one or more pumps during these periods of the day with reduced demand.

A recording pressure logger was installed overnight on the supply header of a coolant loop operating three 100 hp pumps in parallel all the time. Data from the pressure logger is shown in Figure 15.7. As can be seen, starting at 9:00 p.m. some equipment is shut down and the header pressure begins to increase. From 11:00 p.m. until 5:00 a.m. the header pressure peaks and is consistent at 96 psig. Required pressure during the day shift with all equipment in operation is between 60 and 70 psig. This coolant loop must be circulated at night for filtration, but at least one and possibly two 100 hp pumps can be shut off for at least 6 h overnight. This can be accomplished manually at no cost by having a worker turn off pumps at night and other workers start them back in the morning, or automatically with the installation of controls.

CENTRIFUGAL FANS

Many industrial facilities employ centrifugal fans to move air and other gases through ductwork and processing equipment. Identifying energy waste in these applications usually centers on the system as a whole. The proper selection of fan wheel type (airfoil, backward inclined, radial, etc.) and a well-designed duct configuration are important. In some cases, improper design of fan inlet and exit conditions can give rise to large system effect factors that cause significant extra pressure loss and instability in fan operation. In these cases, projects can be initiated to correct the configuration. A discussion of system effect factors is beyond the scope of this publication.

Cost-effective energy projects most often center on optimizing the fan's capacity control technology. Surge control, which is a special case, is discussed in Appendix 15A. Standard centrifugal fan control techniques typically involve one of the following:

- Outlet dampers, or discharge dampers, are adjustable dampers located near the fan discharge. As the dampers close they add resistance to the duct system and move the operating point up the fan curve to the left, reducing the flow rate. Outlet dampers are very inefficient and should not be used unless they are absolutely necessary. The relative efficiencies of different fan capacity control technologies are compared in Figure 15.8. It is clear that either inlet guide vanes (IGVs) or variable speed control are much more efficient than outlet dampers.
- IGVs can be installed at the fan inlet to impart a prerotation of the incoming air in the same direction the fan wheel is turning. The more the vanes close the greater the prerotation and the more the fan's capacity is reduced. Every change in position of the IGVs alters the shape of the fan curve. Fan control with IGVs is very efficient in the higher load ranges (from 80 to 100% of full capacity), as

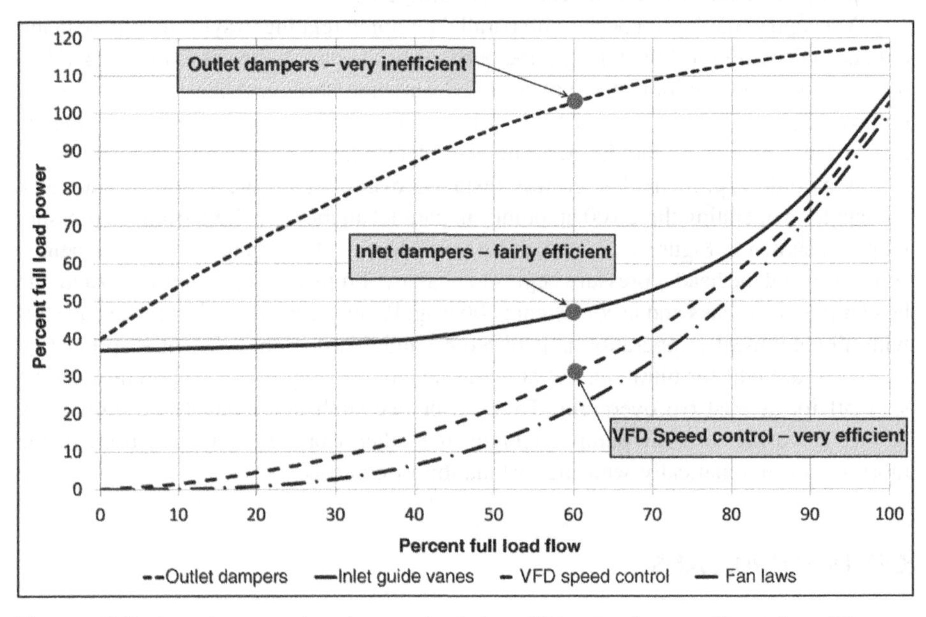

Figure 15.8. Capacity control options and relative efficiencies for centrifugal fans. ([*Source:* U.S. Department of Energy [2]].)

illustrated in Figure 15.8. In the lower load ranges speed control, usually with a VFD, becomes a better option. IGVs are not expensive to implement as compared with VFDs and can be a good choice if the fan stays loaded between 70 and 100% most of the time. Outlet dampers are so poor in terms of energy performance that retrofitting to either IGVs or VFDs usually produces a cost-effective project.

- Speed control is a more efficient capacity control approach as compared with IGVs, especially at loads below about 80%. VFDs are typically used to achieve speed control of fans. The fan affinity laws (Equation 15.4) relate how fan performance is affected by changes in fan rotational speed, fan wheel diameter, and gas density changes. They can be used to predict fan input power as speed changes. The fan law curve, also included in Figure 15.8, is an ideal limit, as an actual VFD has internal losses equal to between 2 and 5% of its full load power rating.

The relative energy efficiency of different fan control schemes operating at 60% capacity is also compared in Figure 15.8. The least efficient means of fan capacity control is the use of outlet dampers on the discharge side of the fan, showing a relative energy use of 105%. This is because the dampers themselves cause an additional pressure loss within the duct system as opposed to other control options. The next least efficient control method is inlet dampers or IGVs, with power consumption at 48% of full load power at 60% capacity. The best option is with VFD speed control where the power usage is 31% of full load power at 60% capacity.

- Fan affinity laws at constant density:

$$\frac{\mathrm{CFM}_1}{\mathrm{CFM}_2} = \frac{N_1}{N_2} \text{ or } \frac{\mathrm{CFM}_1}{\mathrm{CFM}_2} = \frac{D_1}{D_2}$$

$$\frac{P_1}{P_2} = \left(\frac{N_1}{N_2}\right)^2 \text{ or } \frac{P_1}{P_2} = \left(\frac{D_1}{D_2}\right)^2 \qquad (15.4)$$

$$\frac{\mathrm{BHP}_1}{\mathrm{BHP}_2} = \left(\frac{N_1}{N_2}\right)^3 \text{ or } \frac{\mathrm{BHP}_1}{\mathrm{BHP}_2} = \left(\frac{D_1}{D_2}\right)^3$$

where CFM = flow, D = impeller diameter, N = speed, P = pressure, and BHP = brake horsepower.

- Effect of a change in density at constant speed and wheel diameter (ρ = density):

$$\frac{P_1}{P_2} = \frac{\rho_1}{\rho_2}$$

$$\frac{\mathrm{BHP}_1}{\mathrm{BHP}_2} = \frac{\rho_1}{\rho_2}$$

The fan laws predict a cubic reduction in motor power in relation to the speed ratio as a centrifugal fan is slowed. Because of adjustable speed drive electrical losses and other effects, the speed control curve in Figure 15.8 is not as good as the fan laws predict.

However, speed control is a great improvement over outlet damper control and provides significant savings when compared to inlet vane control in the lower load ranges. The speed control curve shows power consumption at 60% fan flow capacity as 31% of full load power. Longer periods of fan operation at low loads yield increased savings using speed control as compared to inlet guide vane control.

Electrical power into a fan motor can be calculated from Equation 15.5 (Motor input power from fan pressure, flow, compressibility factor, fan efficiency, motor efficiency, and drive efficiency).

$$\text{Motor}_{\text{kW}} = \frac{(P_{\text{total}})(Q)(\text{compressibility factor})}{(8528)(\eta_{\text{fan,total}})(\eta_{\text{motor}})(\eta_{\text{drive}})} \tag{15.5}$$

where Q = flow (CFM) and P_{total} = total fan pressure developed (inches of water).

The drive efficiency in above Equation 15.5 is the efficiency of the VFD if the fan or pump is direct-coupled to the motor shaft. The efficiency of a VFD is typically 96–98%. If the fan or pump is coupled to the motor by belts, the belt losses must be estimated and $\eta_{\text{drive}} = 1 - \text{belt losses}$. The Air Movement and Control Association (AMCA) publishes data for standard V-belt losses, shown in Figure 15.9 [3]. The data in Figure 15.9 are often used to estimate belt losses when more specific data is not available.

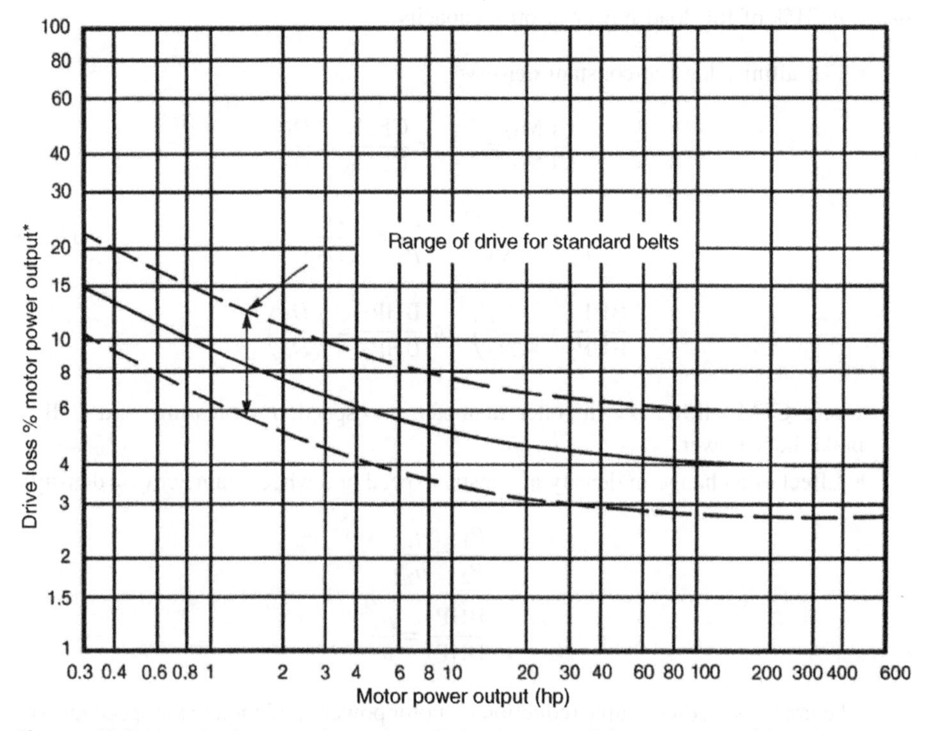

Figure 15.9. Standard energy losses by V-belts. (Reprinted from AMCA Publication 203-90 (R2011): Field Performance Measurement of Fan Systems [3], www.amca.org.)

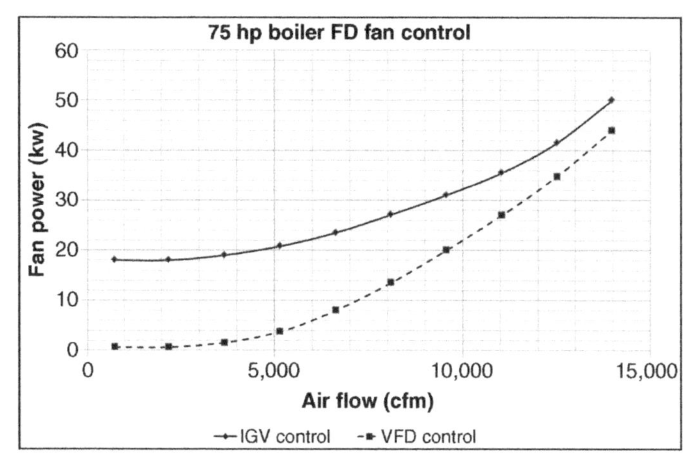

Figure 15.10. 75 hp boiler FD combustion air fan power using IGVs and a VFD.

It is clear from Equations 15.1 and 15.5 that motor efficiency is an important factor in the energy use of both pumps and fans. Motor efficiency is discussed further in Appendix 15B.

Comparison of VFD Speed Control to Inlet Guide Vane Control: Fan Example

A boiler forced-draft combustion air fan provides an example of upgrading capacity control technology from IGVs to VFD speed control. Fan power consumption with both IGVs and a VFD is shown in Figure 15.10. About 85% of the operating hours (8400 h/year) for this boiler are at loads between 30 and 60% of full load capacity. Annual energy usage with IGV control is 216.0 MWh, costing just over $14,000 per year. After the VFD was installed annual energy usage dropped to 95.4 MWh (a reduction of 55.8%), costing $6200 annually. Thus, the savings are 120.5 MWh, valued at $7800/year. The purchase and installation of the VFD cost $15,000, resulting in a simple payback of 1.9 years.

PROCESS COMPRESSOR AND TURBINE EFFICIENCY IMPROVEMENTS

Although the main focus of this chapter is capacity control schemes, it should also be noted that equipment upgrades can sometimes also yield cost-effective energy savings. Improvements in compressor and turbine manufacturing technology have resulted in significant efficiency increases over the last 15–20 years. The process industries often employ very large and high-horsepower centrifugal compressors, and many of the compressors and drivers in use date from earlier stages of manufacturing capabilities.

Large rotating machines are generally overhauled during major process turnarounds to restore the equipment to near-design conditions. In some cases, though, it can be economical to revamp compressors and turbines so that they approximate current technology by replacing their internal elements. These machines must be evaluated on a case-by-case basis, but efficiency improvements on the order of 3–5% can be possible in both the compressor and the turbine. It is worth contacting the original equipment manufacturer(s) to investigate the prospects for specific pieces of equipment. An efficiency improvement of only a few percentage points in such a significant energy user can make a step change in the specific energy use of a process unit, and in many cases it can also provide debottlenecking benefits.

CLOSING THOUGHTS

Energy efficiency projects are subject to the same financial hurdles as other projects. It is important to consider all the benefits from energy projects in their economic justifications. Process reliability and/or maintenance costs are often improved by replacing improper control schemes or piping and ductwork configurations, inappropriate pump or fan selection, or the replacement of equipment with newer and more efficient technologies.

The least costly way to significantly reduce the energy cost associated with a pumping or fan system is often by eliminating the energy waste associated with the capacity control mechanism. The application of techniques to "make the pump or fan smaller" all the time or to efficiently modulate capacity in systems with variable flow requirements often lead to cost-effective projects. Pump impellers can be trimmed and fan pulleys changed on systems with constant flow requirements. For variable capacity systems the installation of a VFD for speed control is the most common option.

The application of variable frequency drives to existing pump and fan systems is common today. Proper engineering of the VFD installation by a qualified electrical engineer is critical to the success of the project. Potential problems and pitfalls of improperly engineered VFD applications are well documented and must be considered for each new VFD installation.

APPENDIX 15A. CENTRIFUGAL COMPRESSOR SURGE CONSIDERATIONS

Compressor Surge

Surge is defined as the operating point at which centrifugal compressor peak head capability and minimum flow limits are reached. The working principle of a centrifugal compressor is to initially increase the kinetic energy of the fluid with a rotating impeller. The fluid is then slowed down after it has exited the impeller into a volume called the plenum, where the kinetic energy is converted into potential energy, causing a pressure increase.

At low flow rates when the plenum pressure behind the compressor can become higher than the outlet pressure from the impeller, the fluid tends to reverse flow direction and move back into the compressor. As a consequence of this flow reversal, the discharge plenum pressure will decrease and the compressor inlet pressure will increase, leading to a second flow reversal. This phenomenon, called surge, repeats and occurs in cycles with frequencies varying from 1 to 2 Hz. The compressor loses the ability to maintain peak head when surge occurs and the entire system becomes unstable.

During low flow situations like start-up or emergency shutdown, the compressor operating point will move toward the surge line. If conditions are such that the operating point closely approaches the surge line, flow recirculation can occur between the impeller and the exit diffuser. Flow separation will eventually cause a decrease in the discharge pressure, and flow from suction to discharge will resume. Surging can cause the compressor to overheat to the point at which its maximum allowable temperature is exceeded. Surging can also cause damage to the thrust bearing due to the rotor shifting back and forth from the active to the inactive side [4].

Surge Control in Process Compressors

Surge is an important issue in the operation of centrifugal compressors in process applications, including process refrigeration applications. The surge limit is a key constraint, along with process limits, compressor power, and in the case of variable speed machines, maximum and minimum speed limits. Compressor control systems are designed to maintain conditions in the stable zone of operation, which is bounded by these constraints, together with a control margin.

As demand on the compressor falls, surge controls will adjust the IGVs or, if the machine has variable speed capability, slow the rotation. However, when the conditions reach the control margin for the surge limit, a valve will open to recycle a portion of the compressed gas back to the inlet of the compressor in order to prevent the machine from going into surge. This "antisurge recycle" represents a significant energy debit, as it creates a continuous recompression loop.

The trend in modern advanced compressor control systems is to reduce control margins by using application-specific algorithms that minimize antisurge recycle without sacrificing reliability. The result can be significant efficiency improvements, often with the added benefits of improved productivity and reduced process disturbances.

Centrifugal Air Compressor Control Implications

Surge control operates somewhat differently in compressed air systems. As in process gas compressors, when demand on a centrifugal air compressor drops from 100% full load capacity, the first control action taken is typically to slowly close the IGVs to the minimum position. Closing the IGVs to the minimum position will reduce the capacity of the compressor to 70–75% of full load production, depending on the manufacturer and discharge pressure. Some compressors employ an inlet butterfly valve (IBV) instead of IGVs. An inlet butterfly valve is not as efficient as inlet guide vanes and sometimes a cost effective project can be implemented to replace an IBV with IGVs.

Typical centrifugal air compressor control curves are shown in Figure 15A.1. If compressor air production is still greater than the demand with IGVs in their minimum position, the system pressure will continue to increase. The next control action is for a blow-off valve to begin to open, allowing some compressed air to vent to atmosphere. This keeps the inlet vanes from closing further and ensures that the machine does not

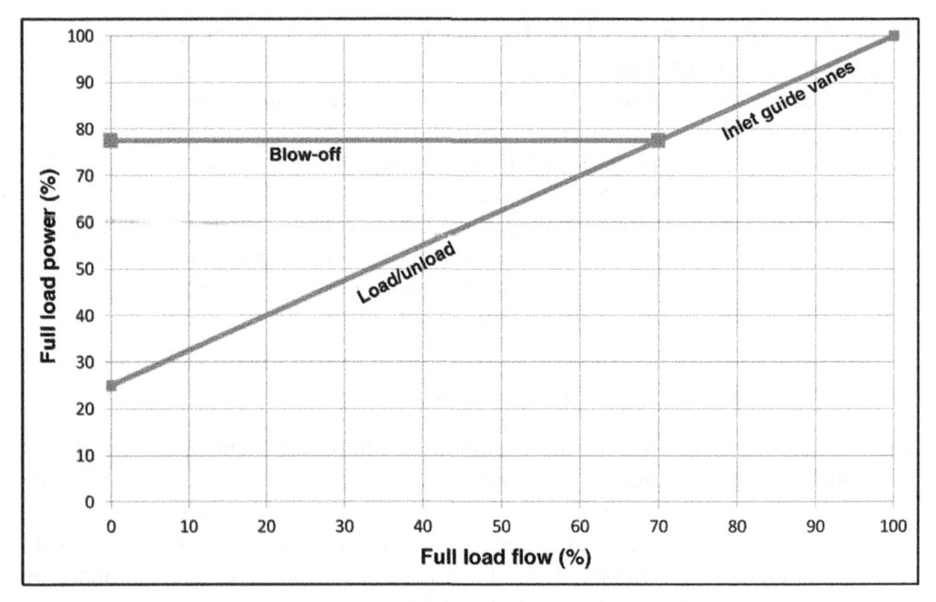

Figure 15A.1. Centrifugal compressor control curves.

operate in surge. Operation with the blow-off valve open is expensive as compressed air is produced and then immediately vented.

If the air system has enough storage capacity, it is possible for a trim centrifugal compressor to shift to load/unload control mode. Under load/unload control the compressor operates in two modes: producing maximum airflow with IGVs in their minimum position or no air with the compressor motor still turning. Load/unload operation requires storage capacity from 7 to 10 gallons/CFM of trim compressor capacity. Electric current and system pressure data for a centrifugal compressor operating with load/unload control are shown in Figure 15A.2. After the compressor operates in

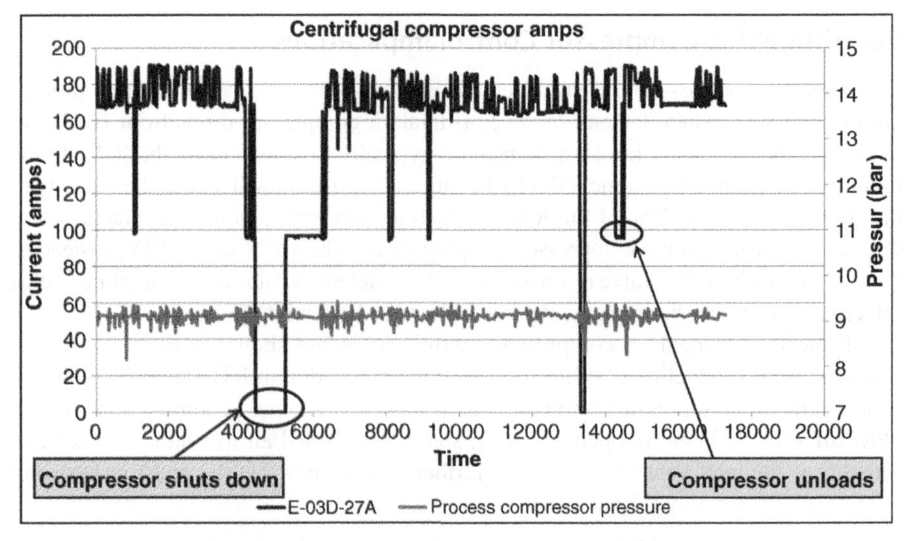

Figure 15A.2. Logged amps and system pressure.

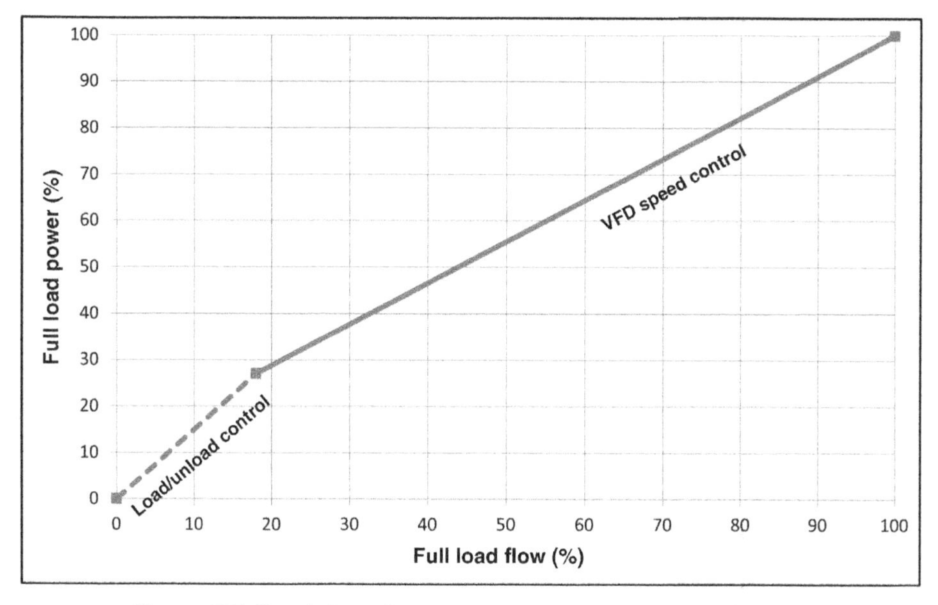

Figure 15A.3. Lubricant free rotary screw compressor control curves.

vent mode with its blow-off valve partially open for a pre-set period of time the compressor will completely unload. Sufficient compressed air storage is required to allow the compressor to remain off for a reasonable length of time before reloading. In the unloaded condition the motor continues to operate but the compressor produces no compressed air. Controls can be set to shut the compressor down completely after it operates unloaded for a preset period of time. Figure 15A.2 shows such an operation. Current usage by this compressor in the unloaded condition is about 90 A.

Heavy industrial sites often have compressed air systems comprised of a number of large centrifugal compressors. Some sites have master controllers coordinating the operation of these machines. It is common to observe three or more centrifugal compressors serving the same system with several machines operating with blow-off valves open and IGVs set in random positions. This lack of coordinated compressor control is inefficient and increases the cost of operation for the compressed air system.

One approach to improving the efficiency of a compressed air system dominated by centrifugal compressors is to install a large variable speed lubricant free rotary screw compressor to serve as the trim machine[1]. VFD compressors can turn down efficiently over their entire load range (illustrated in Figure 15A.3). The addition of the

[1] Variable speed lubricant-free rotary screw compressors typically have a wider operating range than centrifugal compressors, and are therefore better suited to handling variable loads.

VFD screw compressor with other centrifugal machines allows total system air production to be modulated with a combination of inlet guide vane control and the VFD compressor's capacity. Keeping the blow-off valves closed is the key objective of this arrangement. For example, assume a system is operating three 1000 hp centrifugal compressors and one 1000 hp VFD rotary screw compressor. If each centrifugal can reduce air production by 25% with IGVs the system can modulate air production by 75% of one compressor with IGVs. Including the VFD screw compressor increases the turndown capacity to 175% of one compressor in a four compressor system. If air demand varies by more than 175% a master controller would be needed to implement load/unload and start/stop control of one or more centrifugal compressors.

For a second example, assume a lubricant-free rotary screw air compressor operates at 25% capacity. From Figure 15A.3 the power requirement is around 30% of full load power. If a centrifugal air compressor operates at the same 25% compressed air production rate, by Figure 15A.1, the power usage is about 75% full load power (with blow-off valve control). The difference in power is about 493 kW for a 1096 kW compressor. If this condition existed for one full year and the cost of electricity is $80/MWh, the difference in operating cost between the lubricant-free rotary screw air compressor and the centrifugal compressor would be about $345,500 annually.

APPENDIX 15B. EFFICIENCY OF ELECTRIC MOTORS

Motor efficiency is an important consideration in the overall efficiency of rotating equipment. The full-load (nameplate) efficiency of induction motors above about 20 hp used in process plants is typically between 90 and 96%. Efficiency generally increases as the size of the motor increases. However, for any size there is a variety of motors available and "high-efficiency" or "premium" motors are somewhat more expensive than standard ones. The difference in full-load efficiency between high-efficiency and standard motors is typically about 2.5%, but this difference tends to increase at part load conditions, which is where many motors run. As noted in Chapter 3, electricity costs make up about 96% of the total life-cycle cost of a motor, so in most applications the incremental cost of a high-efficiency motor can easily be justified.

The operating efficiency of an electric induction motor depends on several things:

- *The design of the motor:* newer motors are typically more efficient than older motors.
- *The size of the motor:* larger motors are generally more efficient than smaller motors.
- *The operating speed of the motor:* 1800 rpm motors are slightly more efficient than 1200 rpm and 3600 rpm motors of the same size.
- *The motor's load:* load affects operating efficiency.

TABLE 15B.1. NEMA Premium Efficiency Motors [5]

NEMA Premium Efficiency Motors						
	Open			Enclosed		
Motor Size (HP)	1200 (rpm)	1800 (rpm)	3600 (rpm)	1200 (rpm)	1800 (rpm)	3600 (rpm)
1	82.5	85.5	77.0	82.5	85.5	77.0
5	89.5	89.5	86.5	89.5	89.5	88.5
10	91.7	91.7	89.5	91.0	91.7	90.2
20	92.4	93.0	91.0	91.7	93.0	91.0
50	94.1	94.5	93.0	94.1	94.5	93.0
100	95.0	95.4	93.6	95.0	95.4	94.1
250	95.4	95.8	95.0	95.8	96.2	95.8
500	96.2	96.2	95.8	95.8	96.2	95.8

Courtesy National Electrical Manufacturers Association (NEMA).

Table 15B.1 shows the current peak efficiencies for premium efficiency motors for open and enclosed motor types of different sizes [5]. Larger motors are clearly more efficient than small ones. For each motor type and size the 1800 rpm motor has the highest efficiency and the 3600 rpm motor has the lowest efficiency.

Motor efficiency is fairly constant with load from 100% rated capacity down to about 40%. Below 40% load the efficiency of the motor begins to drop off significantly. At loads below 25% motor efficiencies are well below the full load value and drop quickly with further reductions in load. Table 15B.2 and Figure 15B.1 illustrate average motor efficiencies for different sizes of motors at different loads. The data were generated from the motor performance characteristics calculator included with the US DOE Pumping System Assessment Tool (PSAT) [6]. Power factor also declines rapidly as motor load decreases. Lightly loaded motors are a prime contributor to the low power factor problems experienced at many facilities.

TABLE 15B.2. Efficiency vs. Load for NEMA Premium Efficiency Motors [6]

Average Efficiency Motors—1800 rpm				
Motor Size (hp)	25% Load	50% Load	75% Load	100% Load
1	73.3	80.7	82.2	83.2
5	78.3	85.3	86.8	86.6
10	82.0	88.1	89.3	88.8
20	85.4	90.2	91.1	90.2
50	88.2	92.1	92.8	92.6
100	89.7	93.2	94.0	93.9
250	92.0	94.2	95.0	95.0
500	93.6	94.9	95.5	95.5

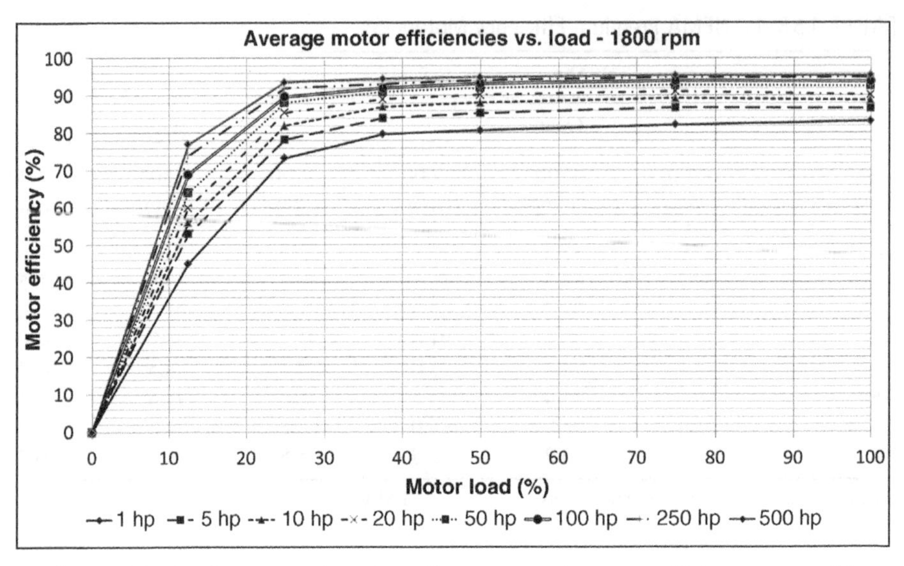

Figure 15B.1. Efficiency vs. load for different sizes of average efficiency motors [6].

REFERENCES

1. Casada, D. (2008) Overview of the Pumping System Assessment Tool (PSAT), U.S. Department of Energy Industrial Technologies Program, December 15. Available at http://www1.eere.energy.gov/manufacturing/tech_assistance/pdfs/psat_webcast.pdf (accessed June 15, 2014).
2. Improving Fan System Performance: A Sourcebook for Industry (2003), *U.S. Department of Energy Industrial Technology Program.* Available at http://www1.eere.energy.gov/manufacturing/tech_assistance/pdfs/fan_sourcebook.pdf.
3. AMCA Publication *203-90 (R2011): Field Performance Measurement of Fan Systems*, Air Movement and Control Association International, Inc. (AMCA), Arlington Heights, Illinois. Available at www.amca.org.
4. Ghanbariannaeeni, A. and Ghazanfarihashemi, G. (2012) Protecting a centrifugal compressor from surge. *Pipeline and Gas Journal*, 239 (3).
5. NEMA Standards Publication MG 1 – 2006, National Electrical Manufacturers Association (NEMA), Rosslyn, VA, 2006, Tables 12.12 and 12.13.
6. U.S. Department of Energy, *Advanced Manufacturing Office, Pumping System Assessment Tool (PSAT).* Software download available at http://www1.eere.energy.gov/manufacturing/tech_assistance/software_psat.html (accessed June 15, 2014).

16

INDUSTRIAL INSULATION

Mike Carlson

LyondellBasell Industries, Houston, TX, USA

INTRODUCTION

In this section, we will discuss various types of industrial insulation and their applications. The intent is not to give a complete list of all available insulation materials, but to discuss the more commonly used varieties at the time of this writing. Technology and research are always coming up with better products. Developing accurate costs and economic justifications for insulation projects, large or small, will also be discussed, as well as insulating specialty items like rotating equipment and bolted connections.

Usually, insulating hot services for the purposes of personnel protection and reducing energy loss is what comes to mind when we think of insulation, but insulating equipment in cryogenic services is equally important because the cost of refrigeration is higher than the cost of heating. The chapter concludes with a discussion of project justification for cold service insulation.

COMMON INDUSTRIAL INSULATION TYPES AND APPLICATIONS

Figure 16.1 shows the basic insulation properties of a number of common insulating materials. Each type of insulation has its own value for thermal conductivity, range of

Energy Management and Efficiency for the Process Industries, First Edition. Edited by Alan P. Rossiter and Beth P. Jones.

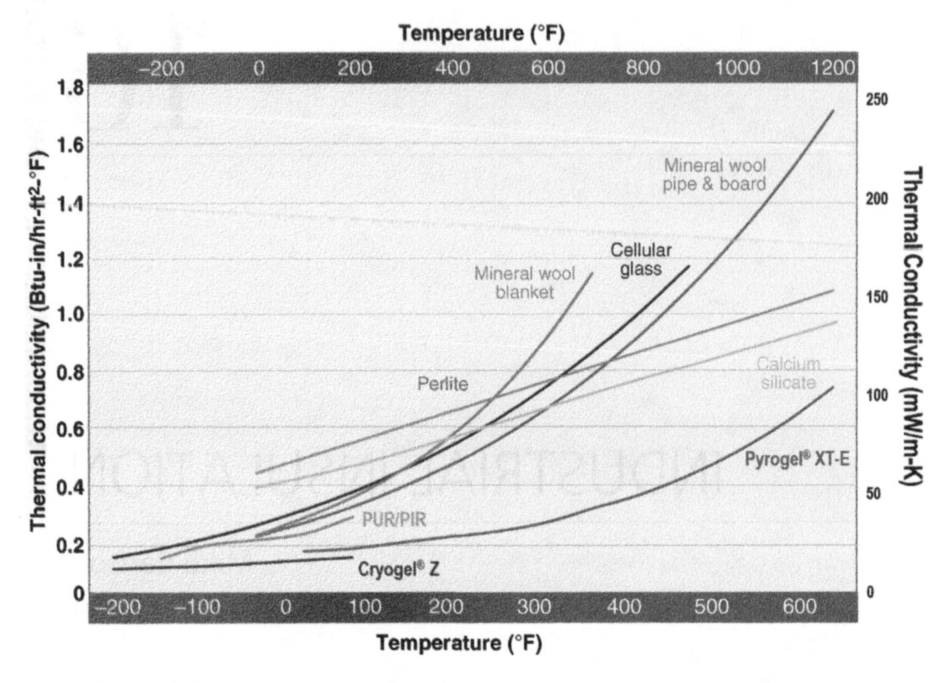

Figure 16.1. Insulation materials and properties. (Courtesy Aspen Aerogels, Inc.)

temperature operation, durability, material cost, and installation costs. Some are suited for hydrocarbon service and others are not. Several kinds of insulation can absorb hydrocarbons, which can lead to autoignition of the insulation. Others can absorb water, which leads to corrosion under the insulation. Some are suited for cryogenic services and others are not.

Various insulation types, benefits, and concerns are summarized in the following sections.

Mineral Wool

Mineral wool is the least expensive industrial insulation material. It has good insulation qualities at lower temperatures, but at temperatures above 500 °F, perlite and calcium silicate perform better. Mineral wool is not durable and needs a substantial jacket to prevent crushing. It is also porous and can absorb water through anomalies in the jacket, which can reduce its effectiveness and even support corrosion under insulation.

Perlite

Perlite is often the insulating material of choice. It is durable, does not absorb water, has good insulation qualities, and is relatively inexpensive. For higher temperatures, above 800 °F, calcium silicate has slightly better insulation qualities.

Calcium Silicate

Calcium silicate has been replaced with perlite in lower temperature applications, below 800 °F, because it hydrates easily. Calcium silicate can even absorb quite a bit of water as it is removed from its shipping packaging, before it is installed. Once installed in lower temperature applications, the outer portion of the insulation does not get hot enough to drive the water out, and therefore its effectiveness is reduced. But for operating temperatures above 800 °F, "Calsil" is the go-to material because it has excellent durability and insulation characteristics, and the higher temperatures drive water out of the insulation.

Cellular Glass

Until recently, cellular glass was the best material for insulating cryogenic services. Cellular glass is manufactured by baking it in ovens like bread, in relatively small pieces. These pieces then are cut, shaped, and glued together to form whatever shape the client desires. This manufacturing process is expensive, and its installation in cryogenic services is also expensive. The pieces of glass are held onto the cold surface with bands and then are glued together. Then the whole surface is wrapped in a vapor barrier and then cladded with jacketing. One distinct advantage of cellular glass in this application is that it does not absorb hydrocarbons, which could lead to autoignition as with other insulation materials.

Aerogel

Aerogel insulation (Cryogel® for low-temperature applications, and Pyrogel® for high-temperature applications) is made in a flexible blanket. While the cost of aerogel insulation is high, its insulation qualities are superior, so less aerogel is required for the same insulation properties than any of the other insulation materials listed earlier. Aerogel is also very easy to apply, which makes this material competitive in some applications like cryogenic services. The surface to be insulated can simply be wrapped like a blanket rather than having custom pieces fabricated and glued together as with cellular glass. Small insulation blankets made with aerogel materials are also cost-effective compared with other blanketing materials. This is because the cost of manufacturing any blanket is relatively high, due to associated labor, and less aerogel material is required due to its superior insulation properties.

ECONOMIC JUSTIFICATION FOR INSULATION PROJECTS

Everything done in business has to be economically justified, including insulation installation and repair. When it comes to insulation, the cost of installation must be compared with the value of the associated energy savings. We usually divide the two to get a simple payback for the project in years and complete the projects that meet the company's economic threshold. To do this, we just need two things: the value of energy savings from insulation and the cost to install it. But we want to be accurate in our estimates so that we complete the right projects, because all projects within the company are competing for available funds based on economic returns. So, the emphasis presented will be on accurate project development.

The industry-standard program for calculating the energy savings produced by insulation is 3E Plus [1]. The program is free, comprehensive, very easy to use, and is sponsored by the North American Insulation Manufacturers Association (NAIMA). Substantial research has gone into the development of the heat-loss calculations used in the program.

With 3E Plus, the user compares the heat loss of the bare surface with that of various insulation thicknesses for a variety of types of insulation. The heat-loss savings with the cost of energy will give the value of the energy savings.

One watch-out in using the program is the value used for average wind velocity. It is easy to find this for a particular city or area, but when insulation is installed in an operating unit there can be substantial blanketing of the wind, so using a much lower wind velocity may lead to a more accurate result. Wind velocities can be measured within the operating unit with handheld devices. This is recommended because wind velocity has a significant effect on the heat-loss calculations.

The next step in the process is getting a good cost estimate for installing the insulation. To do this, an experienced insulation estimator with in-plant experience is required. The estimator should be proficient in a variety of disciplines:

1. The estimator should be competent and experienced in the use of a thermal imaging camera so that surface temperatures can be accurately determined. Training is required to accurately calibrate the camera with respect to surface and background materials.
2. The insulation cost estimator should understand the work process and safety requirements within the plant so that they are accurately reflected in the cost of the project.
3. The estimator should have experience working with different types of insulation, so he or she understands the material cost differences and construction cost associated with each type of insulation.
4. The estimator should have a good working knowledge of the plant engineering standards and guidelines for insulation selection and use.
5. The estimator should also have construction experience, so he or she knows how to complete the project scope most efficiently. Will scaffolding be required or can a ladder or a man lift be used instead? If a man lift is used, which projects can be completed in series using the man lift so that it can be removed from the job after the selected projects are completed to keep costs down? If scaffolding is required, can a rolling scaffold be used? This may significantly reduce the cost of scaffolding up to each location. What crew size will be required, and in turn what amount of supervision will be needed? What size break facilities will be required, and how far away from the work site are they? Safety breaks, lunch breaks, ingress and egress into the plant for the work crew and equipment, permitting, work schedules, fatigue factors, quality, and inspection requirements

should all be included in the estimate to obtain accurate costs. By using a top-level estimator, the accuracy of the cost estimate and savings will be more reliable.

Once the installation costs are established for the insulation project, they simply need to be divided by the energy savings to determine the simple payback in years. Often, all the projects or components will be sorted on a spreadsheet from best to worst payback and all the projects above a particular cutoff that complies with company guidelines or the budget available for the work will be selected for installation.

RULES OF THUMB

- Normally any surface operating at 250 °F or above that can be accessed at ground level is worth insulating. Individual applications' values can be verified using 3E Plus calculations and current energy and construction costs.
- Although insulating surfaces above ground level requires scaffolding, which adds to the cost of the project, insulating elevated surfaces at or above 250 °F is still usually economical.
- Surfaces above 140 °F that are accessible to personnel are usually insulated for personnel protection.
- Surfaces between freezing and 250 °F should be investigated for the possibility of corrosion under insulation, but these applications do not generally pay back as energy reduction projects.
- Replacing crushed insulation usually does not pay. It is surprising, but it is usually not economical to replace insulation where the jacket has been damaged and the insulation crushed. From the results of 3E Plus calculations, it is obvious that a little bit of insulation does a lot of good, so the expense of removing old, crushed jacketing and insulation, and replacing it with new usually does not pay.
- Surfaces that operate below −25 °F are usually worth insulating.
- An insulation survey using a calibrated thermal imaging camera should be completed by a qualified insulation estimator every 3–5 years to identify opportunities to repair insulation that has deteriorated substantially or has been removed and not replaced during equipment maintenance.
- It is well worth the trouble to ensure that maintenance procedures include replacing any insulation that is removed during maintenance work.

COLD SERVICE INSULATION

In the past, it has been difficult to justify replacing cold service insulation, because the application is very labor intensive and the materials are expensive.

As mentioned previously, NAIMA provides a standard software program called 3E Plus used for calculating energy losses for insulation. This method determines energy

Figure 16.2. Octavio Torres, Energy Solutions Group Leader, measures condensation from iced-up surfaces in a chemical plant.

loss based on radiant and convective forces. But when insulation fails in applications below freezing temperatures, ice forms. Ice is a very poor insulator and, just like on a glass of iced tea left outside on a hot summer afternoon, water condenses from the air on iced pipe and equipment insulation and drips onto the ground (Figure 16.2). It takes energy to condense the water—energy that is supplied by the refrigeration system. By measuring the water condensing from various iced surfaces, it was determined that condensation accounts for an additional 25% energy loss during summer months beyond what is calculated using the standard NAIMA method.

It takes as much energy to condense water out of the air as it does to boil it in a boiler, but the cost of the refrigeration energy to condense the water is far greater than the heat energy to boil it. The colder the refrigeration level, the more expensive energy loss becomes. In some plants, refrigerant is produced at various temperatures from 50 °F all the way down to −150 °F, and the coldest levels require up to seven times more power to produce than the warmer levels.

So, what does all this mean? Replacing cold service insulation is very expensive. Even with the added incentives for condensation and refrigerant level multipliers, it is still not economical to replace all the failed insulation in the cold sections of plants—but there are significant opportunities at the refrigerant levels below −25 °F.

The economic evaluation for replacing insulation in cold services is more involved than for hot services [2]. In hot service, basically any bare surface over 250 °F at ground level is worth insulating. In cold service, there are several factors to consider in the analysis:

1. Ice: where insulation is failed or missing in cryogenic service, ice will form on the piping or equipment. Ice has insulating properties, although poor. The colder the

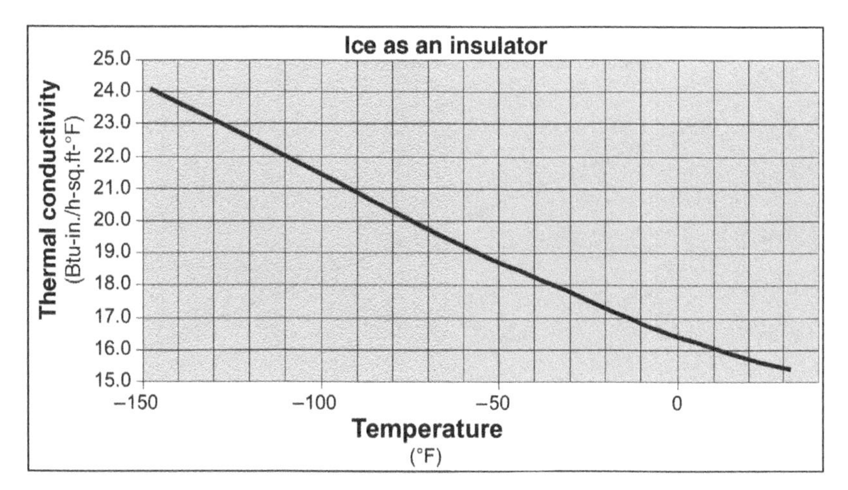

Figure 16.3. Thermal conductivity of ice.

ice, the worse insulator it becomes (Figure 16.3). Ice conducts heat at anywhere from 60 to 120 times the rate of a good insulation system, depending on its temperature. As ice forms, surface area increases, which offsets most of its insulating effects, so the heat gain compared with bare steel only decreases from 5 to 7%. Therefore, 3E Plus calculations are run twice for iced surfaces; once for the piping or equipment using ice as insulation and again for the future repaired scenario with proper insulation and thickness per company engineering standards. The difference in the results is the economic opportunity from radiant and convective heat transfer.

2. Condensation: on cryogenic surfaces with failed or missing insulation, condensation forms on the surface of the insulated equipment or ice. A little more than 1000 Btu of energy is required to condense every pound of water. This energy is provided by the refrigeration system. To quantify the amount of water condensing on iced surfaces, measurements were taken on simple geometric iced surfaces three times a day for a week in the Houston area during the month of July 2008. From meteorological data and observations, a conservative estimate was used to annualize the condensation factor. The results are shown in Figure 16.4. In general, condensation increases the heat gain through an iced surface in the Houston area from 10 to 40%, with an average of 25%, above the 3E Plus calculation results. 3E Plus only calculates radiant and convective heat transfer.
3. Refrigerant level: finally, the colder the desired process temperature, the more it costs to maintain, since more stages of compression are required to produce lower refrigerant levels. The heat gain due to condensation is added to the 3E Plus calculation results, and this sum is multiplied by the refrigerant factor to compute the fuel required to offset the heat gain due to the failed insulation. Lower heating values are used to calculate the fuel quantity.

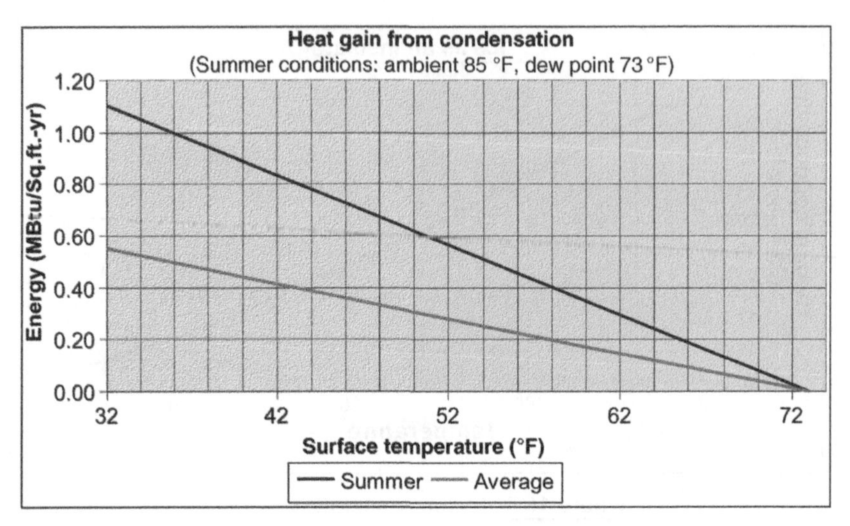

Figure 16.4. Condensation heat gain.

ROTATING EQUIPMENT INSULATION

Insulation blankets can be made for nearly any shape and size of equipment (Figure 16.5). The advantage of blankets is that they are easily removed and reinstalled for maintenance. The insulation properties of new blankets are essentially the same as those for hard insulation, but blankets deteriorate with time. Maintenance workers tend to remove the

Figure 16.5. Insulation blanket on rotating equipment.

blankets to do work and then fail to reinstall them. Insulation surveys should be carried out periodically throughout the plant to identify areas with missing blankets. Before designing blankets for rotating equipment, machinery and maintenance engineers should be consulted, so allowances can be made to fit around drains and instrumentation where the addition of blankets would interfere with operations.

FLANGE INSULATION

Insulation of flanges in hot service (Figure 16.6) is controversial. There are obvious energy savings, but there is a possibility of weakening the compression of the gasket supplied by the bolts as the bolts heat up and lengthen under the insulation. In practice, the flanges as well as the bolts heat up with insulation, and the expansion should be the same, keeping the pressure on the gasket the same as before the insulation was applied. The risk comes with abrupt cooling of the process. It is suspected that the bolts will remain hotter longer than the flange material, and therefore exert less compressive force on the gasket during abrupt cool-down periods. Stress engineers can analyze this problem for a particular process. Some companies are comfortable with the results of these analyses and insulate the entire flanged joint with piping.

To account for the possibility of the flange material cooling off faster than the nuts and studs, one way to capture most of the heat loss from flanged connections is to insulate

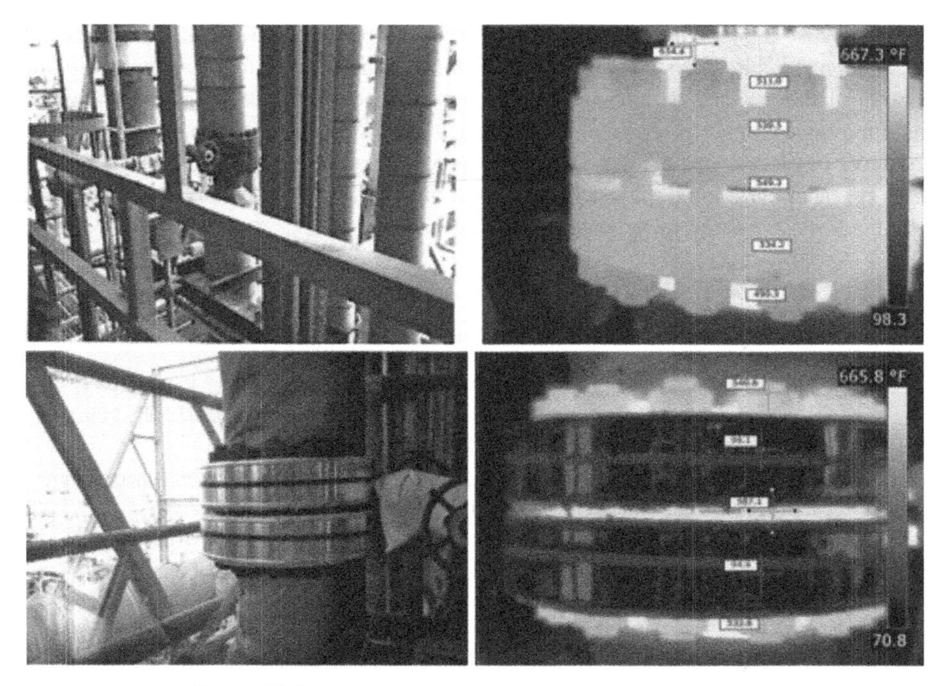

Figure 16.6. Flange temperatures and insulation methods.

the flange edges but not to insulate the nuts and bolts as they protrude from the flanges. In this way, approximately 70% of the heat loss from the flange is captured. The nuts and studs remain cooler than the process for all conditions and the joint maintains adequate compression.

Another risk in insulating flanges in hydrocarbon service is that if a small leak falls on porous insulation materials, the escaping hydrocarbons can autoignite. This can be mitigated using cellular glass insulation to insulate the entire flange joint.

MAINTENANCE AND RENEWAL

Once new insulation is installed in a plant, it does not always stay there. Just as blankets are frequently removed from rotating equipment, maintenance, new projects, or operational needs sometimes cause insulation to be removed and not replaced. Therefore, insulation surveys within a plant should be completed from every 3–5 years to locate missing insulation or new uninsulated areas. These repair projects can also be justified economically based on current energy prices, just as the initial installation was justified.

FINAL THOUGHTS

In addition to its importance in process safety, thermal insulation can also be used to reduce energy losses from both hot and cold surfaces. Good tools are available to estimate the savings and costs, and hence also the paybacks, for insulation projects. Insulation should always be considered when new plants are being designed and built, and insulation projects within existing facilities can also yield excellent returns. While it is often uneconomic to repair old or damaged insulation, there are many uninsulated surfaces in process plants today, and projects to insulate these surfaces can offer large savings and very good paybacks. Thermal insulation is an essential part of any comprehensive energy management program.

REFERENCES

1. North American Insulation Manufacturers Association, "3E Plus Computer Program (CI219)," Available at: http://www.naima.org/insulation-resources/installation-application/3e-plus-computer-program-ci219.html, (accessed February 17, 2014).
2. Carlson, M. (2009) *"Energy Maintenance."* 2009 NPRA Reliability & Maintenance Conference & Exhibition, Grapevine, Texas, May 19–22, 2009.

UTILITY SYSTEMS

17

HEAT, POWER, AND THE PRICE OF STEAM

Alan P. Rossiter[1] and Joe L. Davis[2]

[1]*Rossiter & Associates, Bellaire, TX, USA*
[2]*PSC Industrial Outsourcing, LP, Houston, TX, USA*

Process plants require both thermal energy and mechanical energy to operate. Thermal energy (heat) can of course be provided directly by burning fuel, and mechanical energy (work) is often provided by importing electric power. However, significant efficiencies can be obtained by integrating the systems for heat and power. This fact follows directly from the laws of classical thermodynamics, so we start this chapter with a brief thermodynamics review.

THERMODYNAMICS RECAP

The *first law of thermodynamics* states that energy can be neither created nor destroyed. It can, however, change in form. Within the environment of process facilities, the main forms of energy that we see include the following:

1. *Chemical energy*, within both fuels and process streams. Combustion of fuel and exothermic reactions of chemicals both release heat (thermal energy). This heat can be recovered for use in other parts of the process or facility. Some chemical

Energy Management and Efficiency for the Process Industries, First Edition. Edited by Alan P. Rossiter and Beth P. Jones.

reactions are endothermic, meaning they consume heat. Naphtha reforming is a common endothermic process in typical refineries.

2. *Thermal energy:* Virtually all processes require heating and cooling in some areas, such as distillation and other separation operations.
3. *Mechanical energy* is needed primarily for transporting materials, for example, in pumps and compressors.
4. *Electrical energy:* Some processes (e.g., chloralkali) are electrochemical, so electricity is explicitly required for the process. Electricity is also used for lighting, monitoring, metering, and control, and, of course, to provide the mechanical energy required by many of the pumps and compressors mentioned above.

The *second law of thermodynamics* focuses on the quality, or value, of energy. There are several different statements or definitions of the second law. For our present purposes, Kelvin's definition is the most useful:

> "No heat engine, reversible or irreversible,[1] operating in a cycle, can take in thermal energy from its surroundings and convert all of this thermal energy into work." ([1], p. 590)

This principle is the key to understanding the relative values of heat and mechanical work (or mechanical energy) within a process facility. Almost all mechanical energy is derived from some type of heat engine, whether this is an on-site steam turbine or a remote combined cycle electric power station. As no heat engine cycle can convert all the incoming heat into mechanical energy, mechanical energy is inherently more valuable than heat since for every 1 unit of mechanical energy, $1+x$ units of heat energy are required (see discussion below on efficiency). Moreover, as electricity and mechanical energy can be interconverted at very high efficiencies by generators and motors, electrical energy is essentially equivalent to mechanical energy, and it is therefore also inherently more valuable than heat.

We can quantify this inherent difference in value by considering the efficiency of heat engines (see Figure 17.1). Sadi Carnot examined this topic theoretically and published his results in 1824. He demonstrated ([1], p. 592) that of all heat engine cycles operating between a heat supply temperature of T_h and a heat rejection temperature of T_c, the most efficient is a reversible engine taking in all of its supply heat Q_h at fixed temperature T_h and discharging all of its reject heat Q_c at fixed temperature T_c. This is known as the Carnot cycle.

We can define the "first-law efficiency" or "thermal efficiency" (ε_{th}) of a heat engine as $\varepsilon_{th} = W/Q_h$, where W is the work output. By the first law, $W = Q_h - Q_c$ (assuming no frictional losses). So, $\varepsilon_{th} = (Q_h - Q_c)/Q_h = 1 - Q_c/Q_h$.

[1] A reversible process can be made to run backward simply by reversing the direction in which the thermodynamic variables change. In addition, (i) it must run without energy being degraded such that it loses any of its potential to perform mechanical work, and (ii) the entire process must be at thermodynamic equilibrium. If any of these conditions are not satisfied, the process is irreversible. The requirement that reversible processes must be entirely at equilibrium necessitates that they must run infinitesimally slowly, so all practical processes have at least some irreversibility.

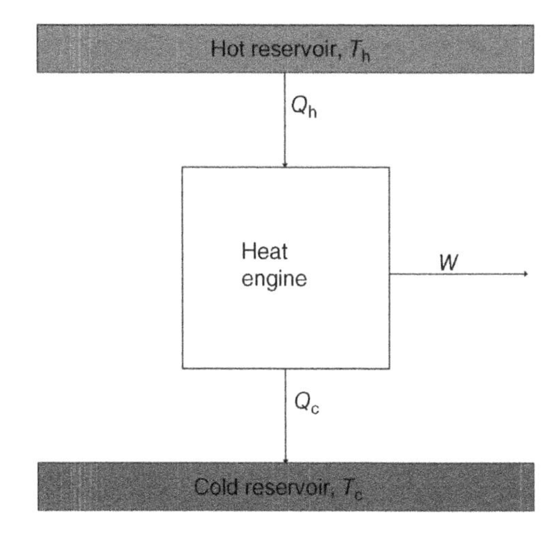

Figure 17.1. Generalized heat engine, showing heat flows Q_h and Q_c and work output W.

In the case of the Carnot cycle, by definition ([1], p. 594), $Q_c/Q_h = T_c/T_h$, so $\varepsilon_{th} = 1 - T_c/T_h$, where T_c and T_h are absolute thermodynamic temperatures. Thus, for the Carnot cycle the thermal efficiency is a function purely of the temperatures at which heat is supplied and rejected. The efficiency is greatest when the supply temperature is maximized and the rejection temperature is minimized. This is also generally true for "real-world" non-Carnot engines such as the steam turbines, gas turbines, and reciprocating engines that are commonly used commercially. The Carnot cycle provides the theoretical limit for the efficiency of a heat engine. However, for real-world applications, $\varepsilon_{th} < 1 - T_c/T_h$; that is, for any given heat supply and rejection temperatures, non-Carnot engines always have a lower thermal efficiency than Carnot engines.

We now turn to the practical question of how we define the efficiency of real-world equipment. The performance of commercial, non-Carnot heat engines, especially steam turbines, is commonly measured in terms of "second-law efficiency" or "isentropic efficiency" (ε_{is}). To explain ε_{is}, we first need to define entropy, and then consider another statement of the second law of thermodynamics:

> Entropy, S, is defined by $dS = dQ/T$, where Q is heat flow and T is absolute temperature. The change in entropy, ΔS, for a given process is therefore given by $\Delta S = \int dQ/T$, between appropriate limits.
>
> The alternative statement of the second law of thermodynamics is that ΔS is always greater than or equal to zero (mathematically, $\Delta S \geqslant 0$) for any isolated system—that is, the total entropy of an isolated system either stays constant or increases with time. ([1], p. 602)

Consider two steam turbines, one of them ideal and reversible and the other one an actual steam turbine that has some irreversibility.[2] The equality in the above expression

[2] The irreversibility of the actual steam turbine is not simply because of mechanical friction. Some of the processes that take place in any real steam turbine are inherently irreversible, for example, passing hot, high-pressure steam through a nozzle to convert thermal energy into kinetic energy. This irreversibility can be reduced by design improvements, for example, adding stages to the turbine, but it can not be totally eliminated.

for the second law of thermodynamics ($\Delta S = 0$, no change in entropy or "isentropic") applies to the ideal, reversible steam turbine, and the inequality ($\Delta S > 0$) applies to the actual steam turbine. If the pressure and temperature of the steam entering both of the turbines are the same, and their exhaust pressures are also the same, then the work output per unit of steam flow will always be greater for the ideal turbine (W_i) than that for the actual turbine (W_a). As more thermal energy is converted to mechanical work in the ideal turbine, it follows that the exhaust steam temperature for the ideal turbine must be lower than the exhaust steam temperature of the actual turbine. We define the isentropic or second-law efficiency for the actual steam turbine as $\varepsilon_{is} = W_a/W_i$.

STEAM SYSTEM TYPES

Figure 17.2a–d shows simplified diagrams of four different types of steam systems [2]. Figure 17.2a is a standard steam turbine Rankine cycle—a very common heat engine cycle that is used in many power stations to generate electricity. High-pressure (HP) steam (2000 psig in this example) is produced in a boiler, using heat provided by burning fuel. From the boiler the steam enters a turbine where it expands and transfers some of its thermal energy to the rotor in the form of mechanical energy, which in turn is converted to power in a generator. The exhaust steam from the turbine is under vacuum, to maximize the pressure difference across the steam turbine and thus maximize its power output. Typically, the vacuum is provided using eductors (steam ejectors). The exhaust steam is condensed and pumped to the boiler to repeat the cycle. In this example, the amount of fuel consumed is 307 MBtu/h, and the amount of power produced (in the same energy units) is 110 MBtu/h, so the overall first-law efficiency is 110/307 or 35.8%. (This neglects the energy consumed in the pump and any losses from the system apart from stack losses and heat rejection in the condenser.)

In addition to its role in power generation, steam is also very commonly used as the main heat transfer medium in process plants. Water has a very high latent heat of vaporization, so steam has a high energy density. Steam also condenses at a constant temperature for any given pressure, which facilitates temperature control. In addition, water and steam give rise to high heat transfer coefficients when they pass through heat exchangers, and they are safe and nonpolluting, as well as being (generally) inexpensive and readily available.

Figure 17.2b–d illustrates energy distribution using three different types of steam systems, with and without associated power generation. Key parameters for the different systems are given in Table 17.1. In all cases, pump power is neglected, as are any losses from the system apart from stack losses and, in the case of Figure 17.2a, heat rejection in the condenser.

The system in Figure 17.2b provides steam from a boiler at 150 psig to satisfy a process heating requirement. 100 MBtu/h of fuel is burned in the boiler to generate steam at 150 psig. 15 MBtu/h of heat leaves in the stack gases, and 85 MBtu/h of heat is recovered in steam generation and subsequently discharged in the process heater. The boiler therefore has a first-law efficiency of $85/100 \times 100$, or 85%. There is no power generation.

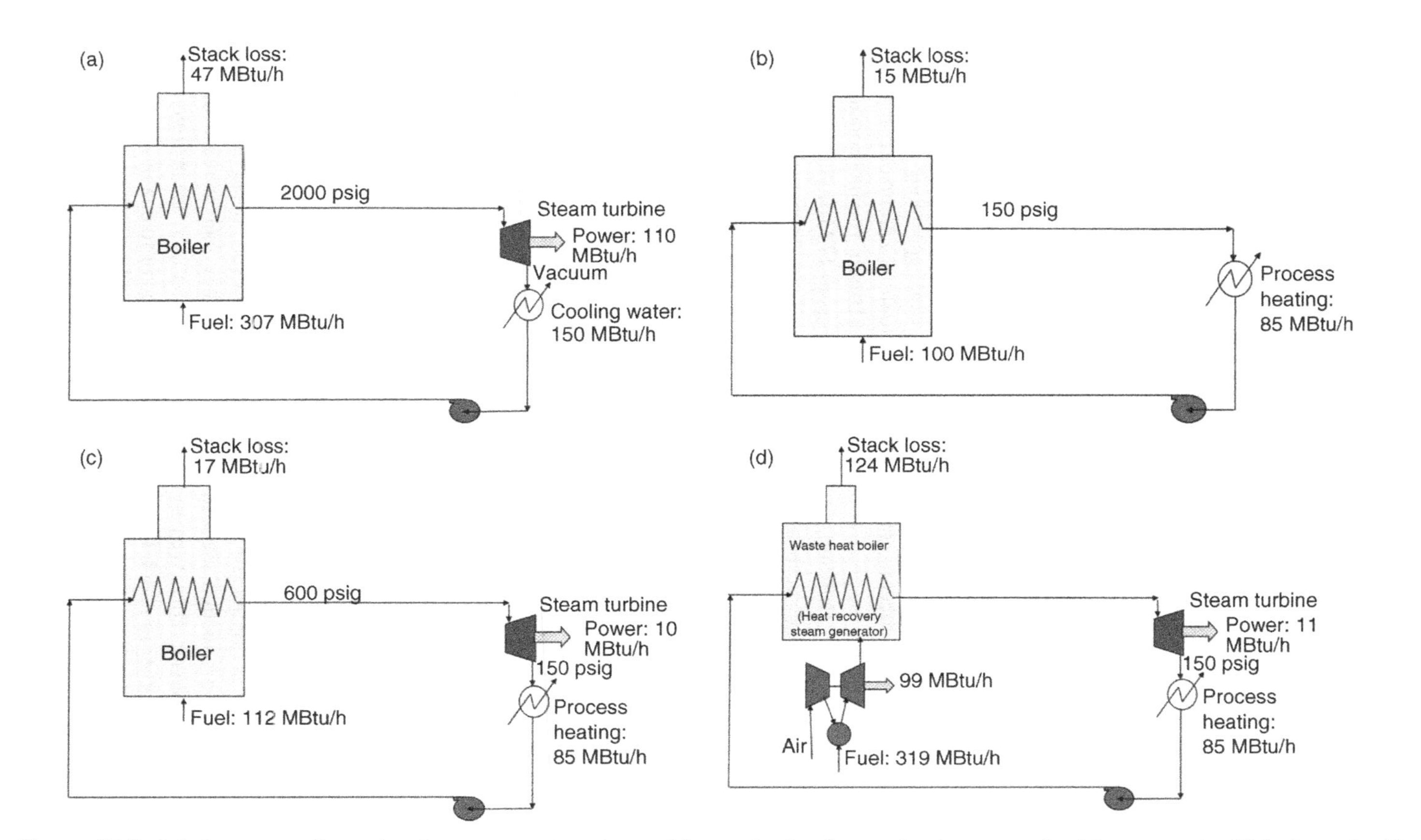

Figure 17.2. Relative energy flows showing power generation and heat rejection for condensing steam for (a) power only, (b) boiler only, (c) boiler + steam turbine, and (d) combined cycle employing gas turbine. Energy flows are shown in MBtu/h.

TABLE 17.1. Key Parameters for Typical Steam Systems in Figure 17. 2

	(a) Condensing Turbine	(b) Boiler Only	(c) Boiler and Steam Turbine	(d) Combined Cycle
Fuel	307	100	112	319
Incremental fuel	307	0	12	219
Power	110	0	10	110
Power/ incremental fuel	0.36		0.85	0.5
Power/heat to process		0	0.12	1.29

Source: Reproduced with permission of John Wiley & Sons, Inc. [2].

In Figure 17.2c, the boiler pressure is increased to 600 psig and a backpressure steam turbine is added. This drops the steam pressure to 150 psig while generating 10 MBtu/h of power, which can be used either to generate electricity or to drive a pump or compressor. The exhaust steam provides the process heating duty of 85 MBtu/h. The fuel consumption in the boiler increases to 112 MBtu/h, and the stack loss rises to 17 MBtu/h, so the boiler efficiency remains at 85% (subject to a small rounding error). If we compare the energy use in Figure 17.2c against that in Figure 17.2b, we note that the fuel consumption increases from 100 to 112 MBtu/h—a change of 12 MBtu/h. Power generation, on the other hand, increases from 0 to 10 MBtu/h. Thus, for an incremental 12 MBtu/h of fuel we generate 10 MBtu/h of power, giving an incremental first-law efficiency of $10/12 \times 100$, or 83.3%, for power generation. This corresponds approximately to the efficiency of the boiler, and it is 2.3 times greater than the efficiency of the stand-alone power generating system shown in Figure 17.2a. This comparison illustrates the value of cogeneration, or combined heat and power systems. By integrating power generation and heat delivery in the same steam system, we can generate power incrementally at efficiencies that are much higher than those of conventional power stations.

However, a backpressure steam turbine cogeneration system is limited in its useful power generation capability by the size of the steam demand on the exhaust side of the turbine. If too much steam passes through the turbine, the excess will have to be vented or condensed with rejection of heat to ambient, and this will result in very inefficient power generation. We can increase the amount of efficient, cogenerated power that can be made by raising the steam inlet pressure or lowering the exhaust pressure, subject to process constraints. We can also make more power with a given steam flow if we use a steam turbine with a higher isentropic efficiency. However, these increases in cogenerated power production are usually modest, and they can increase the cost of the installation considerably. We can achieve a much larger increase in cogenerated power production, either to match the process power demand or to enable us to export electricity, by adding a gas turbine to create a "combined cycle cogeneration system," as shown in Figure 17.2d.

The combined cycle first fires fuel into a gas turbine. Hot exhaust gas from the turbine then goes to a waste heat boiler or "heat recovery steam generator" (HRSG), where a large portion of the exhaust heat is recovered in the form of steam generation. Steam from the HRSG then passes through a backpressure steam turbine to the process heater, as in Figure 17.2c.

The combined power production from the gas turbine and steam turbine in Figure 17.2d is much greater than that from the steam turbine alone in Figure 17.2c. Systems of this type have allowed many petrochemical plants to become exporters of cogenerated power to utilities. However, as the numbers in Table 17.1 show, the simple backpressure steam turbine (Figure 17.2c) gives by-product power at the lowest incremental energy use.

Most gas turbine applications in the process industries are linked to steam cycles, but gas turbines can be integrated anywhere in the process where there is a large requirement for heating, especially at high temperature levels. For example, gas turbine exhaust has been used to preheat air for ethylene cracking furnaces [3] and crude oil in an oil refinery [4]. On the power side, gas turbines can be used to generate electricity, or they can be coupled to compressors or pumps.

The economics for developing new combined cycle projects can be compelling. However, it is important to evaluate all aspects of the project. Several refineries that installed new combined cycle units with the intention of gaining a cheaper, more reliable supply of steam and electricity discovered after start-up that they had major problems with their fuel gas balances. When they shut down their old, often much less efficient steam boilers, they suddenly had no home for a significant portion of refinery-produced fuel gas, resulting in flaring. Chapter 20 discusses refinery fuel gas containment in more detail.

The combined cycle is also applicable to dedicated power production. When the steam from the waste heat boiler is fed to a condensing turbine, it is possible to achieve overall conversion efficiencies of fuel to electricity of 50% or more.

PRICE EQUIVALENT EFFICIENCY

The concept of price equivalent efficiency (PEE) provides another, very simple way to quantify the energy conversion benefit of cogeneration. This treatment of PEE was presented at the 27th Industrial Energy Technology Conference, New Orleans, LA, May 10–13, 2005 [5], and is adapted with permission.

In the discussion of steam systems in the previous section, both heat flows and power are expressed in the same units, MBtu/h, in order to more easily illustrate how energy is distributed. These units are commonly used in the United States for fuel and heat flows, but power is usually expressed in watts or multiples of watts (kW, MW), where $1\,W = 3.414\,Btu/h$.

Consider a situation where a refinery or chemical plant uses natural gas as its marginal fuel, at a price of $3.25/MBtu. The plant imports marginal electric power at a price of $85/MWh. The PEE is the ratio of the cost of marginal fuel to the cost of marginal power, in consistent units. In this case, the PEE is 3.25/(85/3.414), or 13%. This means that the cost of the imported power can be considered to be equivalent to power generated from a heat engine with a first-law efficiency of 13%. Any power production system with a marginal efficiency above 13%, such as systems shown in Figure 17.2c and d, reduces the net cost of providing heat and power to the plant. Of course, some investment is needed to install the turbines and other equipment in these systems, and this must be compared against the energy savings to determine whether power generation is economically justified.

SIMPLE MARGINAL STEAM PRICING

When considering energy-saving opportunities that impact steam loads, it is important to know the true value of steam. When considering improvements in existing facilities, the steam system itself is already in place. In this case, the cost of installing the steam system is already "sunk," and we do not need to consider it when pricing steam. Moreover, most opportunities for improving energy efficiency, such as reducing stripping steam flows or small heat integration projects, involve only modest changes in steam flows. For these reasons, most commonly we are interested in *marginal* steam costs, that is, the incremental cost of providing or eliminating small amounts of steam production in real time. While a detailed discussion on this topic is outside the scope of this chapter, a brief discussion is warranted. This treatment of simple marginal steam pricing was presented at the 27th Industrial Energy Technology Conference, New Orleans, LA, May 10–13, 2005 [5], and is adapted with permission.

Establishing the marginal cost of high-pressure steam in terms of $/klb is straightforward; it is the enthalpy change across the boiler, divided by boiler cycle efficiency, multiplied by the cost of fuel. Establishing the marginal cost of medium- and low-pressure (LP) steam, however, is more difficult. Many simply assign an enthalpy-based cost to these steam headers. This method will not represent the true cost of the steam, and in fact will often overstate the value. Unless the steam is provided via letdown valves from the high-pressure header, the actual fuel equivalent cost of medium- and low-pressure steam is a function of the marginal fuel to power price ratio and the efficiency of the backpressure turbines that provided the exhaust steam.

Figure 17.3 illustrates the proper method for determining the marginal cost of steam at each pressure level. This example illustrates a simple, two-header steam system where an 85% first law efficient boiler generates high-pressure steam at 150 psig. The HP steam is then used to drive a 50% isentropic efficient backpressure turbine that generates 19 kWh for every 1000 lb of steam. The turbine exhausts steam at 15 psig into the LP header. The marginal fuel price is $3.25/MBtu, and the plant imports marginal electric power at a price of $85/MWh.

The marginal cost of HP steam is simply the enthalpy change across the boiler, divided by the boiler efficiency: (1201 − 250)/0.85 = 1120 Btu/lb (or 1.12 MBtu/klb), and then multiplied by the fuel cost ($3.25/MBtu) to get $3.64/klb.

The marginal cost of LP steam must take into account the power generated by the turbine. Consider the following example: Low-pressure steam is used on a stripping tower. The current stripping steam rate is high compared with industry standards and a target is set to save energy by reducing the amount of stripping steam. In this simple example, saving LP steam would result in less power being generated by the turbine. This power would have to be replaced by power import. Therefore, the value of the LP

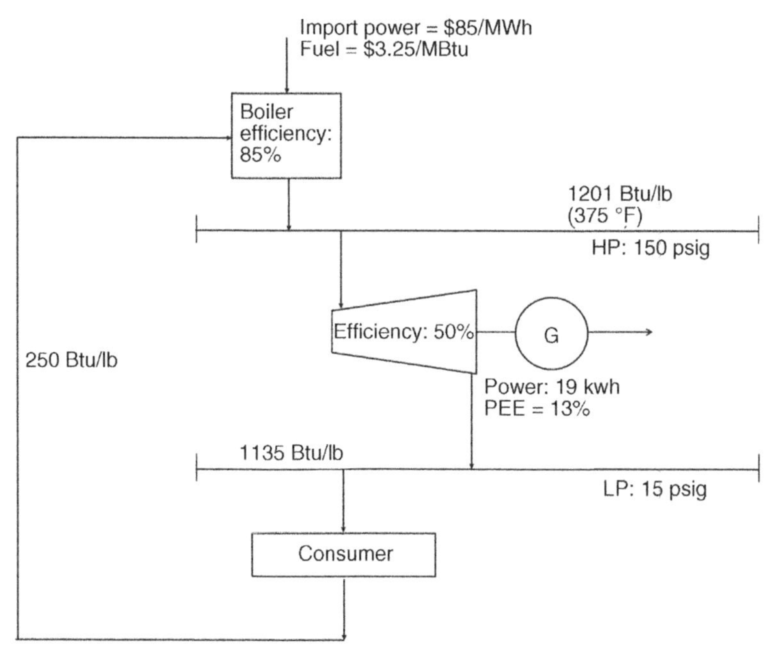

Figure 17.3. Marginal steam costing example. (*Source*: Ref. 5.)

steam must take into account the "power credit" obtained by running the backpressure turbine.

Thus, the marginal cost of the LP steam is the marginal cost of HP steam minus the power credit. The power credit is simply the amount of power generated multiplied by the power price. In the example above, the power credit per 1000 lb (1.0 klb) of steam is $0.019 \times 85 = \$1.62$/klb. The marginal cost of the LP steam then becomes $(3.64 - 1.62) = \$2.02$/klb. Alternatively, the power credit can be calculated by dividing the amount of power generated by the PEE. In this example, the power credit is $0.019 \times 3.414/0.13 = 0.5$ MBtu/klb. The cost of the LP steam then becomes $(1.12 - 0.5) \times 3.25 = \$2.02$/klb, which agrees with the earlier calculation.

Note that if the import power price increases relative to fuel, and/or if the efficiency of the turbine increases, the marginal cost of LP steam is reduced. In extreme circumstances, the marginal cost of the LP steam can become negative, which means that venting LP steam can make money since the alternative is backing down the backpressure turbine and buying more expensive imported power.

ENHANCED MARGINAL STEAM PRICING AND EFFICIENCY IMPROVEMENTS

The above example includes only one steam turbine operating between two pressure levels. In many plants, there are three or more pressure levels with many steam turbines operating between them, and the calculation of marginal steam costs can be difficult to carry out manually. Furthermore, the approach described above for estimating marginal costs focuses only the two main factors: boiler efficiency and power generation. If it is necessary to calculate more rigorous marginal costs, additional factors need to be incorporated. These include the following:

1. *Water and chemicals:* A source of water is needed to produce steam, and there are costs associated with providing this. Depending on how the water is obtained, typical costs might include municipal charges or fees for extracting from a river or other natural water source. In addition, treatment is required to ensure that the water meets boiler quality standards, and this requires water treatment facilities that consume chemicals. The amount and cost of fresh water makeup can be significantly reduced if steam condensate is recovered and recycled.
2. *Pumping:* Fresh water, treated boiler feed water, and condensate all require pumping, which adds to the power demand and costs of operating a steam system.
3. *Deaerator steam usage:* Most steam systems employ thermal deaerators to remove oxygen and other gases from boiler feed water. The deaerators use a significant amount of steam—often more than 10% of the total boiler steam production—and this increases the energy requirement and cost to deliver steam to the process. Opportunities to reduce deaerator steam use are discussed further in the next chapter.
4. *Boiler blowdown:* Boiler feed water contains dissolved solids, and they concentrate in the liquid phase as water evaporates in the boiler. If they are allowed to concentrate too far, solids can precipitate, forming scale deposits that thermally insulate heat transfer surfaces and can lead to boiler failures. In order to prevent the concentration of dissolved solids to rise excessively, a portion of the water—typically between 2 and 10% of the feed rate—is drawn off or "blown down" from within the boiler.

   Although it is necessary for the safe operation of the boiler, the blowdown constitutes a loss of both water and energy, and it increases the cost of producing steam. There are several steps that can be taken to mitigate the resulting inefficiencies, of which the most common are the following:

   - Route the blowdown through a flash vessel. The flash steam can then go either to a low-pressure header or directly to the deaerator, while the liquid phase goes to a drain.
   - Add a heat exchanger to recover heat from the blowdown into boiler feed water or some other heat sink.
   - Reduce the blowdown. Almost all boilers use treated boiler feed water from which some of the naturally occurring dissolved solids have been removed. However, it may be possible to upgrade the water treatment and remove additional dissolved solids, and thus reduce the amount of blowdown that is required.
5. Distribution losses, including steam leaks, steam trap performance (see Chapter 13), and heat losses from piping and equipment (see Chapter 16).

In order to address all of these complexities, an energy management best practice is to have a detailed steam and power system model in place to calculate accurate marginal costs and to evaluate energy efficiency opportunities. Without such a model, accurate steam costing is difficult to achieve, and this can lead to misplaced priorities. Steam system models are discussed further in Chapter 18.

REFERENCES

1. Weidner, R.T. and Sells, R.L. (1965) *Elementary Classical Physics*, Vol. 1, Allyn & Bacon, Inc., Boston, MA.
2. Rossiter, A. (2004) Energy management, in *Kirk-Othmer Encyclopedia of Chemical Technology*, Vol. 10, 5th edition, John Wiley & Sons, Inc., pp. 133–168.
3. Kenney, W.F. (1983) Combustion air preheat on steam cracker furnaces. *Proceedings of the 1983 Industrial Energy Conservation Technology Conference*, Texas Industrial Commission, p. 595.
4. *New ExxonMobil cogeneration plant presented at COGEN Europe's Annual Conference*. March 27, 2009. Available at http://www.cogeneurope.eu/medialibrary/2011/05/27/82444a1a/270309%20COGEN%20Europe%20Press%20Release%20-%20FINAL.pdf. Accessed February 13, 2015.
5. Davis, J.L., Jr. and Knight, N. (2005) Integrating process unit energy metrics into plant energy management systems. *27th Industrial Energy Technology Conference*, New Orleans, LA, May 10–13, 2005.

18

BALANCING STEAM HEADERS AND MANAGING STEAM/ POWER SYSTEM OPERATIONS

Alan P. Rossiter[1] and Ven V. Venkatesan[2]

[1]*Rossiter & Associates, Bellaire, TX, USA*
[2]*VGA Engineering Consultants Inc., Orlando, FL, USA*

Steam systems provide a core energy conduit in most process plants. They are used to generate and distribute not only steam but also power in many cases. Typical equipment includes boilers, steam turbines, and deaerators. Many larger systems also incorporate gas turbines to increase power production, and heat recovery steam generators (HRSGs) to recover heat in turbine exhaust gases. The performance of the individual equipment items is important, and is covered elsewhere. This chapter focuses on overall considerations for steam/power system efficiency, specifically steam header balances and operational optimization.

STEAM BALANCES

There are three very common inefficiencies in steam balances: steam *vents*, *letdown* of steam through pressure reduction (letdown) valves rather than steam turbines, and excessive use of steam in *deaerators*. These are illustrated in Figures 18.1–18.3 in the context of a generic steam system "ladder diagram." The discussion that follows on ladder diagram is adapted with the permission of World Scientific Publishing Company [1].

Energy Management and Efficiency for the Process Industries, First Edition. Edited by Alan P. Rossiter and Beth P. Jones.

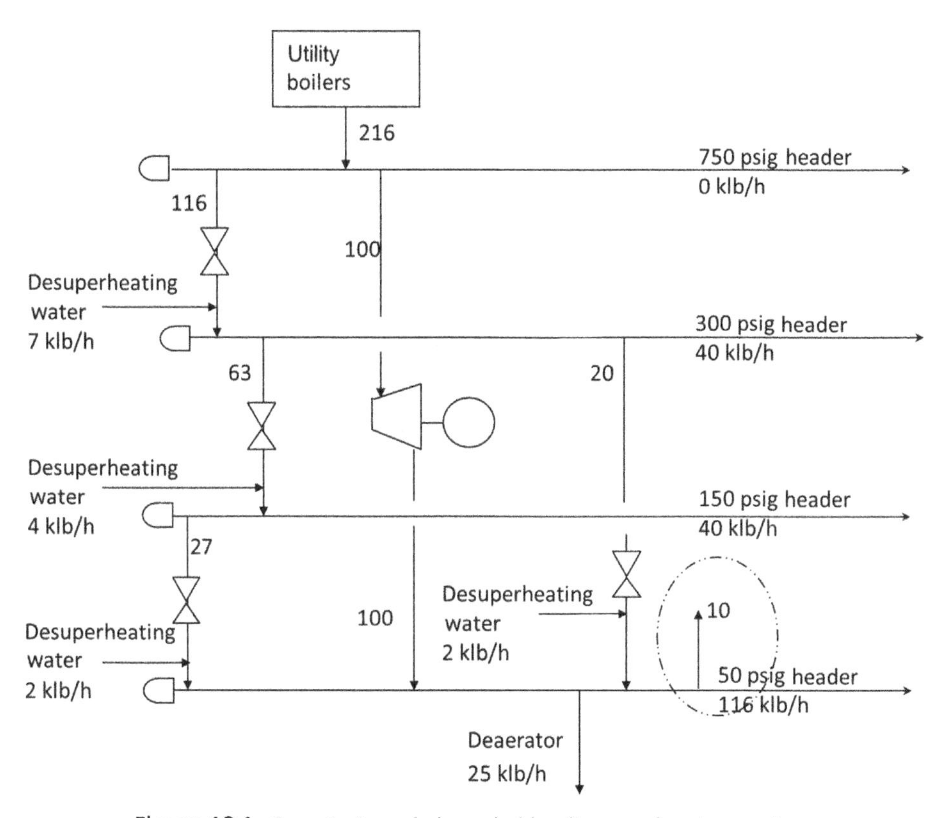

Figure 18.1. Generic steam balance ladder diagram showing venting.

The ladder diagram is a simplified representation that provides a convenient way to show how steam cascades from higher pressure levels to lower pressure levels through steam turbines and letdown valves.

- Steam enters from boilers. In this simplified, generic example (see Figure 18.1), there is only one boiler, linked to the highest-pressure header. Flows are shown in klb/h, and header pressures in psig.
- Steam demands for process use are shown on the right side of the diagram, and steam for deaeration is taken from the lowest-pressure header at the bottom of the diagram.
- Steam flows between headers through steam turbines and letdown valves.
- The steam vent (if present) is shown leaving the lowest-pressure header.

Steam Venting

Figure 18.1 illustrates a steam vent. There is excess steam in the lowest-pressure header, and the excess must be removed from the system. Working back up through the steam balance, too much steam passes through turbines and letdown valves and there is an excessive boiler load, which implies an unnecessarily high-energy input (fuel firing), and also additional boiler feedwater (BFW). Venting is thus a very visible loss of both energy and water.

Venting can arise due to a number of reasons:

a. Often there is too much flow through backpressure steam turbines, which creates more low-pressure (LP) steam than the system can accommodate. This can often be corrected by running a switchable motor (see next section).
b. In other cases, large amounts of LP steam are produced in waste heat boilers (WHBs) or by recovering the flash from condensate collection tanks (not shown in Figure 18.1). If there are not enough LP steam consumers, excess LP steam will vent.
c. Boilers typically have a limited turndown capability. If the steam demand is less than the "minimum turndown" on the boiler(s), it often leads to venting. This can sometimes be corrected by modifying the boiler(s) to enable operation at lower loads.
d. In yet other cases, the steam vent and the letdowns are in different areas, under the control of different personnel. The simplified diagram in Figure 18.1 shows an overall balance, but individual operators are typically aware of only their own area. Thus, it is very easy to open a letdown valve to satisfy a process demand in one process unit without realizing that excess steam at the same pressure level is being vented in another part of the facility. Situations like this can often be corrected by improving monitoring and control systems.
e. Undersized piping between steam sources and consumers at different locations of the plant can create a hydraulic constraint. As mentioned in point (d), the result can be venting in one area and letdown in another. This situation can generally be corrected by adding a parallel header or replacing the existing header with larger diameter piping.

Figure 18.2 shows the same generic steam system with the vent corrected by reducing flows through letdown valves.

Steam Letdown Versus Steam Turbines

As previously noted, steam can flow from higher pressure headers to lower pressure headers either through letdown valves or through steam turbines. The turbines consume some of the heat content in the steam to produce power; if the steam passes through letdown valves, there is no consumption of heat, but also no power production. Water may be injected to cool (desuperheat) the steam as it flows to a lower pressure header to control the header temperature. As power (in the form of electrical import) is typically much more expensive than heat (from imported fuels), it is generally desirable to maximize the flow of steam through turbines and minimize the flow through letdown valves. This is illustrated in Figure 18.3, where a steam turbine (shown shaded) is added between the two highest-pressure headers.

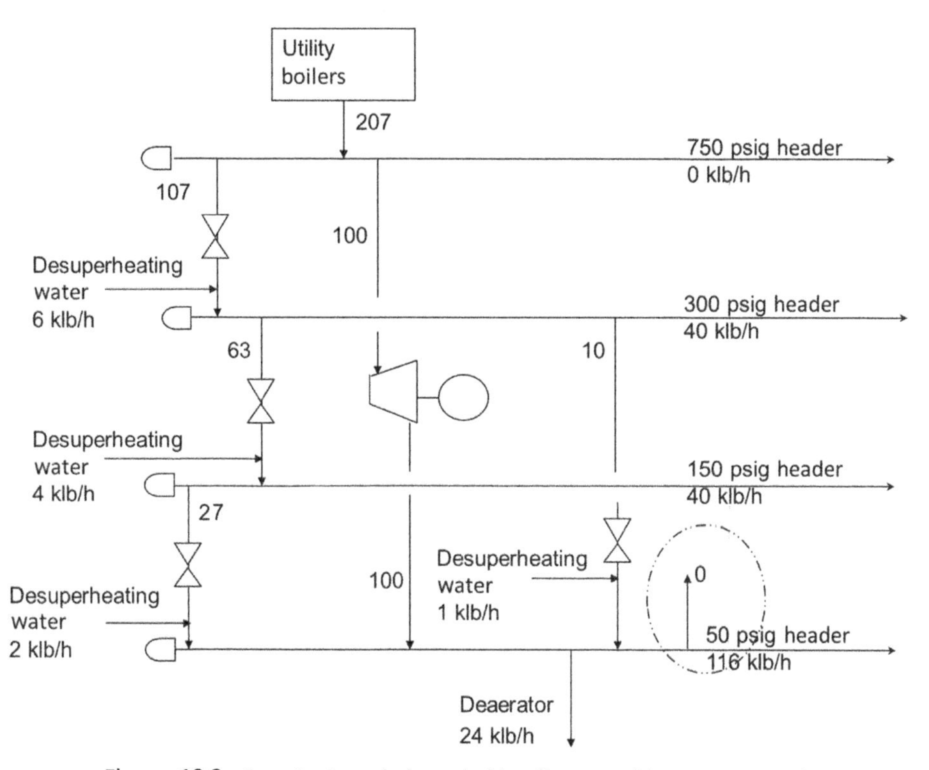

Figure 18.2. Generic steam balance ladder diagram with vent corrected.

There are two common ways for using the power that is produced:

a. If the turbine is coupled to a *generator* (shown as a circle in Figure 18.3), it produces electricity, which can either be used by electricity consumers on the site or exported to the local electric grid. In this configuration, the flow of steam through the turbine can be adjusted subject to its design limits to balance the steam headers and minimize letdown flows, while avoiding excessively high steam flows that would lead to venting.
b. Alternatively, the steam turbine may be "directly coupled" to a *pump* or a *compressor* (shown as a square attached to the new steam turbine in Figure 18.3). In this case, the power output (and hence the steam demand) of the turbine is dictated by the requirements of the pump or compressor, and it cannot be adjusted without impacting plant performance.

Steam balance changes continuously for many reasons—for example, changes in ambient conditions or bringing equipment online or off-line—and as the steam balance changes, a direct-coupled steam turbine can cause venting. This can be corrected with "switchable pairs" or "switchable sets" of equipment. In a switchable pair, there are two

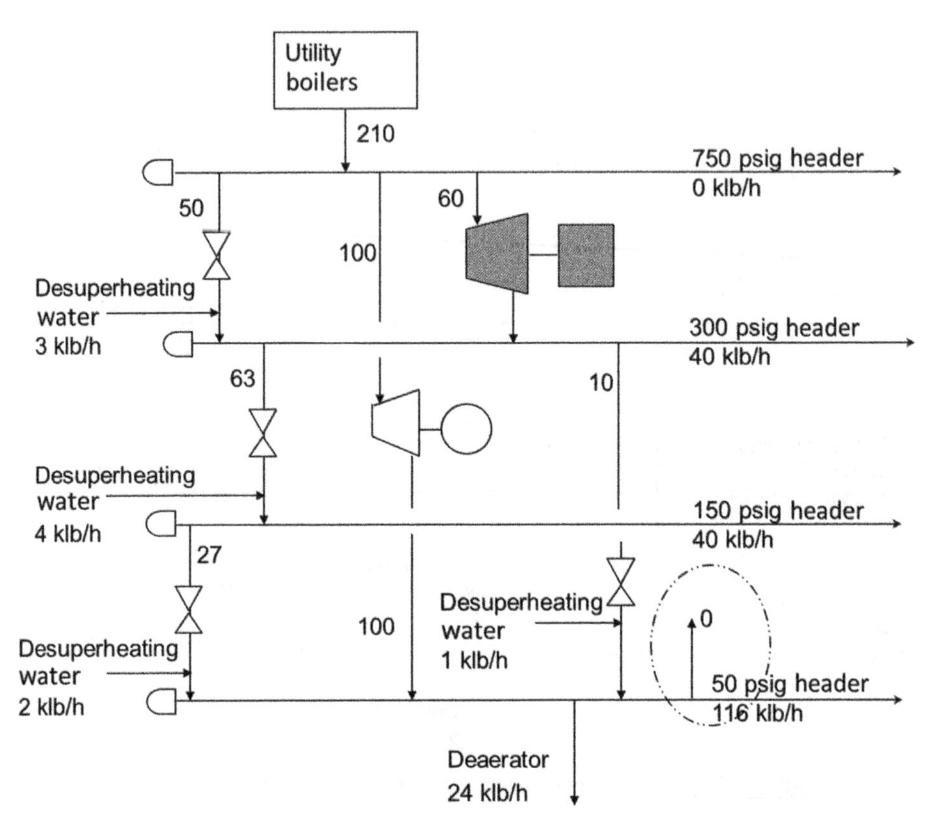

Figure 18.3. Generic steam balance ladder diagram with steam turbine added.

pumps, one driven by an electric motor and the other by a steam turbine, and the operators can choose which one to operate based on current steam balance requirements. A switchable set may consist of, say, three pumps, one with a steam turbine and two with electric drives. Assuming only two of the pumps are needed at any given time, the operators can decide, based on steam balance requirements, whether to run two electric pumps or one electric pump and one steam-driven pump.

In addition to letdown flows, steam temperature is also an important factor in determining how much power can be produced within a steam system. This example is from a commodity chemical plant in Europe [2]. In the plant, two steam turbines drive a large compressor and an electric generator in a single-shaft arrangement. The first (high-pressure (HP)) steam turbine is a backpressure machine, supplied with steam at 930 psig and exhausting at 290 psig. The second (low-pressure) steam turbine is supplied with steam at 290 psig, and it has two exhausts—steam extraction at 73 psig and a vacuum exhaust that goes to a condenser (see Figure 18.4).

In the original design, only the exhaust steam from the HP steam turbine was used in the LP steam turbine. The design specifications of the LP machine limit its maximum

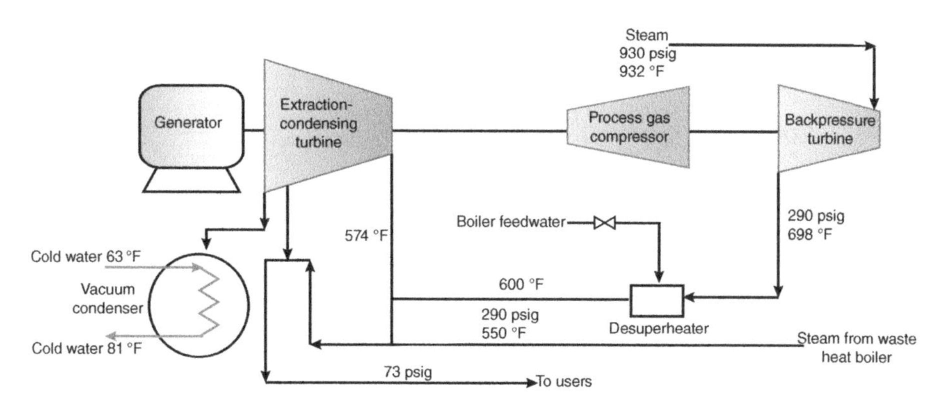

Figure 18.4. Turbine/compressor/generator configuration before modification. (*Source:* Reprinted with permission from Ref. [2]. Copyright 2012, American Institute of Chemical Engineers.)

inlet steam temperature to 660 °F, and as the exhaust from the HP machine was hotter than this, the design included a desuperheater. Here boiler feedwater is injected into the steam to reduce its temperature. The desuperheater outlet temperature was controlled at 600 °F to provide a margin of safety. The specific power generation in a steam turbine falls as the inlet temperature goes down, so the desuperheater causes some loss of power generation in the LP machine.

Later the steam from a waste heat boiler was added to the exhaust steam to increase the steam flow to the LP turbine and thus increase its power output. The steam from the waste heat boiler is at 550 °F—significantly cooler than the exhaust steam from the HP steam turbine. However, the operation of the desuperheater was not reevaluated when the steam from the WHB was added. Typical operating conditions are shown in Figure 18.4.

A subsequent study of this system by one of the authors found that the maximum temperature that can be reached with the combined steam flow to the LP steam turbine is below the turbine inlet temperature limit, and desuperheating of the HP turbine exhaust steam can be safely eliminated. Based on this finding, a bypass line was installed around the desuperheater and the water supply to the desuperheater was shut off. The flow diagram after the modification is shown in Figure 18.5. This change increases electricity generation by 500 kW, resulting in annual energy cost savings of $400,000.

This example shows the need to challenge existing operating practices and also to reevaluate conditions when process changes are made. Although it is always essential to operate within design limits, excessively large margins of safety can result in unnecessary losses of energy efficiency.

Deaerator Steam

Most steam systems use thermal deaerators (Figure 18.6) to drive off oxygen and other dissolved gases from boiler feedwater. In principle, only a small amount of steam is needed to do this. However, as the incoming water is often far below the saturation

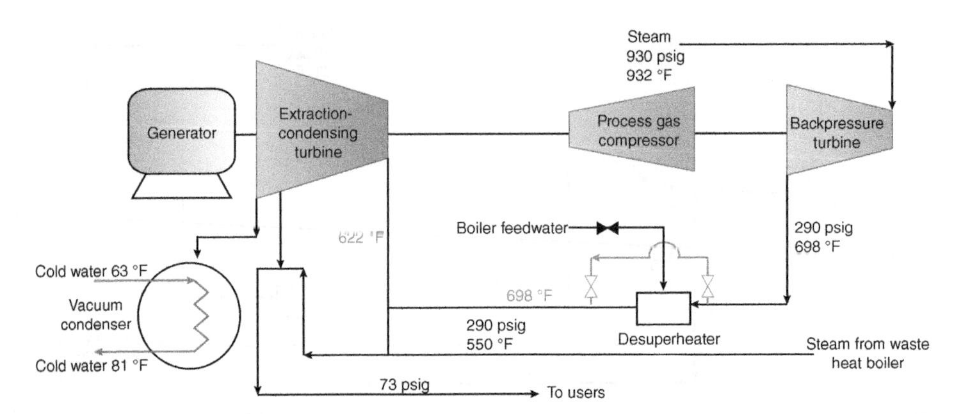

Figure 18.5. Turbine/compressor/generator configuration after modification. (*Source:* Reprinted with permission from Ref. [2]. Copyright 2012, American Institute of Chemical Engineers.)

temperature in the deaerator, a substantial amount of additional steam is consumed in preheating the water. It is not uncommon to consume 10% or even 15% of the total steam made in heating deaerators, as shown in Figures 18.1–18.3.

There are two common options for managing deaerator steam demand to improve energy efficiency:

a. It is sometimes possible to improve energy efficiency simply by adjusting the operating pressure of a deaerator. For example, if excess low-pressure steam is creating a steam vent, raising the deaerator pressure can save energy. Increasing the deaerator pressure also raises its temperature and therefore increases its steam demand, which provides a consumer for additional low-pressure steam, reducing the vent. As the boiler feedwater from the deaerator is now hotter, less fuel is

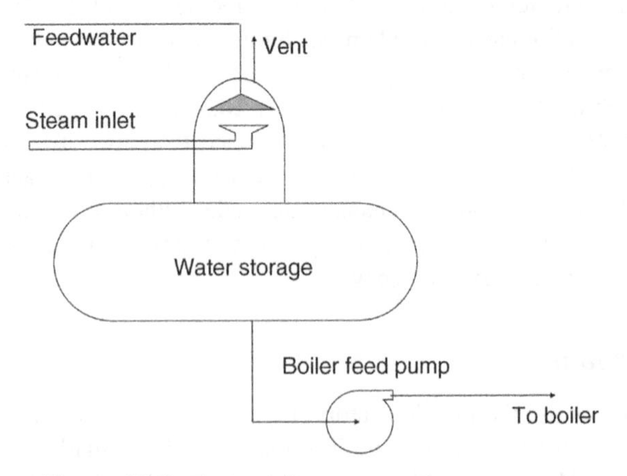

Figure 18.6. Thermal deaerator: main components.

needed in the boiler to generate steam, and this constitutes the energy saving.[1] However, it is important in all cases like this to consider the entire steam system, as there can be other changes that offset the savings. Steam balance models, which are discussed later in this chapter, are useful for this purpose.

b. Many facilities install deaerator feedwater preheat projects that recover waste heat to reduce the steam load. This is illustrated in the following example [2].

The deaerator in a chemical plant processed a combination of warm returned condensate and cold softened water makeup. There was no preheater for the softened water.

Within the boiler house, there were also several water-cooled air compressors (Figure 18.7), with one of them experiencing chronic maintenance problems in its cooling tower. A project to replace the cooling tower was under consideration. However, it was observed that the average quantity of makeup water flow (115–150 gpm) was almost identical to the cooling water flow requirement of the air compressor (120 gpm). Based on this observation, a new project was proposed to route the makeup water through the air compressor and isolate the cooling tower with blinds (Figure 18.8). The new proposal was evaluated and accepted, and the piping modifications were completed within 2 months.

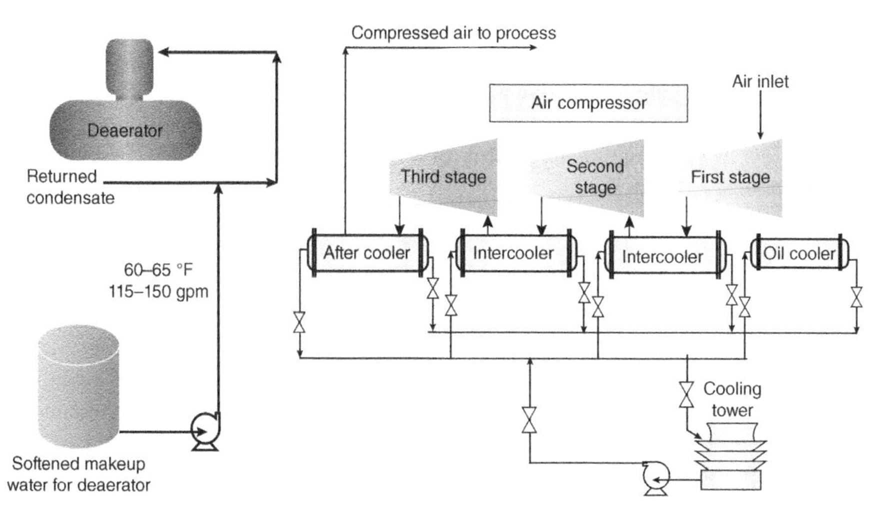

Figure 18.7. Air compressor cooling before modification. (*Source:* Reprinted with permission from Ref. [2]. Copyright 2012, American Institute of Chemical Engineers.)

[1] Deaerator pressure is also sometimes increased to combat cold-end corrosion in boiler economizers. The boiler feedwater from the deaerator raises the temperature at the cold end of the economizer, and the higher temperature reduces or eliminates the condensation of moisture and acid gases on the exhaust gas side of the economizer.

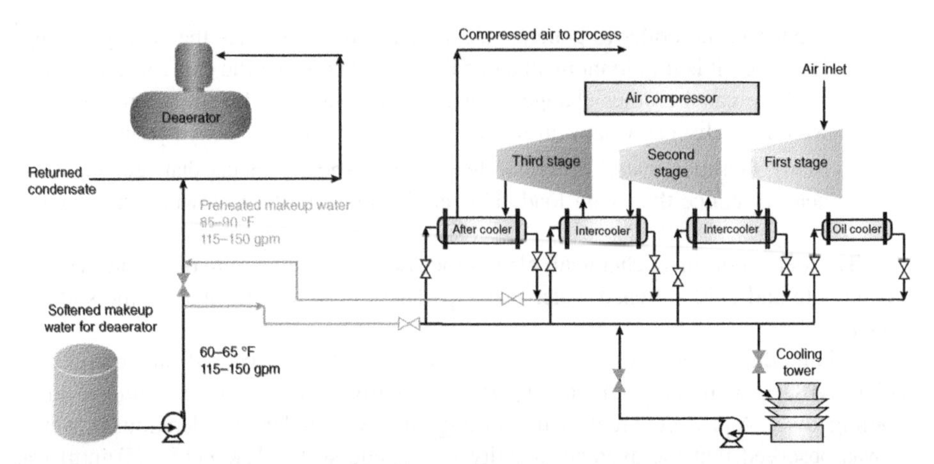

Figure 18.8. Air compressor cooling after modification with heat recovery. (*Source:* Reprinted with permission from Ref. [2]. Copyright 2012, American Institute of Chemical Engineers.)

As a result of this project, heat from the air compressor is now recovered in the softened water, saving $80,000/year in deaerator steam usage. Implementation was very inexpensive, as it required only local piping changes; and the project removed the need to maintain or replace the cooling tower, thus eliminating a significant cost.

This example illustrates the benefits of preheating deaerator feedwater. Perhaps even more important, it shows how a single project can achieve multiple objectives—in this case, saving energy and eliminating a chronic maintenance problem. It also illustrates the importance of looking for creative ways to redeploy existing equipment *in situ* to develop low-cost projects to save energy. As always, it is essential to check equipment limitations and follow appropriate management of change procedures when making process modifications.

COMPUTING AND OPTIMIZING STEAM BALANCES

There are several ways to establish steam balances and populate ladder diagrams. For example, data from pressure and temperature gauges and flow meters in a steam system can be linked directly to a display in a boiler plant console to provide operators and engineers with a current snapshot of the steam system. This can be used to identify operating problems and initiate corrective action. Similarly, data from a plant historian can also be imported to provide profiles of steam system performance at previous times.

However, most steam systems have only limited instrumentation, especially on the lower pressure headers, and it is rare to be able to construct a complete steam balance from measured data. Moreover, a steam balance derived purely from measured data has no predictive power. In contrast, computer models can be used to develop complete, internally consistent steam balances, which in turn can check current operating data for consistency and predict performance for different scenarios such as higher and lower production rates, start-up and shutdown, and summer or winter weather conditions.

Building the Model. To the extent possible, computer-based steam balances should be based on an item-by-item buildup of steam flows into and out of each steam header. Flow meters should be used where available. Equipment design sheets can supplement the balance where meters are missing. Inevitably, there will be some inconsistencies in the input data, and the computer model can be used to highlight and resolve these and thus improve the accuracy of the steam balance.

Simple models can be built using spreadsheets with no special features, simply by constructing heat and material balances. Alternatively, commercially available software can be used to facilitate the development of more rigorous and comprehensive steam balances. These programs incorporate functions for physical properties (enthalpy, entropy, phase equilibria, etc.) and calculations for the various components of the steam system (boilers, deaerators, steam turbines, letdown valves, gas turbines, heat recovery steam generators, etc.).

Outputs from the Model. Commercial software packages generally model both the steam balance and the power balance, and thus provide a realistic simulation of the site heat and power balance. These models carry out a mass and energy balance for each header. In addition, they calculate the specific energy for each pressure level of steam, the shaft work generated between each of the levels, and the marginal fuel and power prices. This is particularly important for low-pressure steam, as its value is often overstated. At many facilities, the low-pressure steam value is calculated based on enthalpy alone and it does not take into account the power that was produced from the higher pressure steam that feeds the low-pressure header via backpressure turbines.

A good steam model will provide a diagrammatic representation of the steam system, and will include the following [3]:

- *Steam and condensate distribution system:* This includes balances, flows, and conditions throughout the facility, as well as logic to match supply with demand. This can be used to check meter readings for consistency.
- *Steam generation equipment:* Package boilers can be modeled according to manufacturer's data or operating data as appropriate. Process waste heat boiler output should be related to production rate.
- *Gas turbines:* These are modeled to relate fuel consumption to power output and exhaust gas characteristics.
- *Waste heat boilers and supplementary firing:* Typically, these are modeled based on vendor data.
- *Steam turbines (backpressure or condensing):* These are modeled to relate steam flow to operating load and include switches where a turbine's use is optional.
- *Fuel balance:* This is integrated with the steam system model so that changes in boiler loading and gas turbine operation can be related to changes in the fuel balance.
- *Deaerators, desuperheaters, letdowns, flash drums, and so on:* These are included to ensure that a complete heat and material balance of the facility is calculated.

- *Total operating cost of the utility system:* This is based on the costs of providing fuel and power to the site, as well as the raw water costs, together with the heat and power balance derived from the model. Total operating cost information can be used to calculate the marginal cost of steam in each header, and it is also extremely valuable in performing "what–if" cases. For example, it can address the following questions: If I have to take this boiler out for maintenance, what will be the cost of operating the utility system? What mitigating steps can I take to minimize the cost impact while keeping the system in balance?

The following are the other features that are provided in some commercial modeling systems:

- A display of key operating parameters calculated by the model, for example, fuel consumption, power generation, and efficiency of each item.
- Emission tables (CO_2 and SO_2) calculated from each source of combustion.
- Tables of supplementary data used in calculations, lists of equipment performance parameters, mechanical constraints, and so on.

Some Modeling Systems. Some free tools for modeling simple steam systems are available on the Internet – for example, the U.S. Department of Energy's Steam System Modeler. Several other modeling tools with greater flexibility are available commercially. These include ProSteam^{TM} from KBC, Aspen Utilities Operations^{TM} from Aspen Technology, and Visual MESA^{TM} from Soteica Visual MESA.

Modeling and Optimization. As discussed earlier, the function of most utility systems is to provide both steam and electric power to the operating site. Typically, there are many different ways in which the equipment can be run to meet the steam and power demands, but the cost of meeting these demands may vary considerably depending on which items of equipment are used and how they are loaded (e.g., selection of steam turbines or motors in switchable pairs, steam production rates from each boiler, and flows through each steam turbine). Process conditions are constantly varying due to factors such as production rates, feed quality, and weather conditions. Furthermore, the prices of imported fuel and electric power also change, sometimes on time frames as short as a few minutes. It follows that the combination of equipment selection and loading that minimizes operating cost is also constantly changing.

The major commercial modeling software packages include optimization capabilities, which allow the models to determine the minimum-cost operating mode for any given set of steam and power requirements. Models can be set up for real-time optimization (RTO) by interfacing them with data from a site's instrumentation or distributed control system. This enables the model to verify a consistent heat and material balance and determine the optimum operating conditions for the steam/power equipment to satisfy the site's current need for steam and electric power.

Most RTO systems are reported to save 1–3% of the operating cost of a steam/power plant. While this may not seem like a large saving, the cost of supplying steam and power to a large refinery or a chemical plant is typically in the hundreds of millions of dollars per year, so the absolute savings are often on the order of several million dollars per year.

Real-time steam/power system optimization is discussed in more detail in Chapter 19.

REFERENCES

1. Rossiter, A. (2012) Energy management for the process industries, in *Recent Advances in Sustainable Process Development* (eds D.C.Y. Foo, M.M. El-Halwagi, and R.R. Tan), World Scientific, Singapore, pp. 609–628.
2. Rossiter, A.P. and Venkatesan, V. (2012) Easy ways to improve energy efficiency. *Chemical Engineering Progress*, 108 (12), 16–20.
3. Davis, J.L. Jr. (2004) Overcoming fuel gas containment limitations to energy improvement. *Proceedings of the Twenty-Sixth Industrial Energy Technology Conference*, Houston, TX, April 20–23, 2004.

19

REAL-TIME OPTIMIZATION OF STEAM AND POWER SYSTEMS

R. Tyler Reitmeier

Soteica Visual MESA LLC, Houston, TX, USA

Industrial operations have historically focused on improving performance on the process side, but improving the performance of utility systems (steam, power, fuel, and water) that support the industrial process is often a much lower priority, and in some cases, unfortunately, it is forgotten entirely. But in many cases, significant improvement and ongoing savings can be delivered through a structured approach to utility system energy management. For example, a refining company in the United States has captured upward of $25,000,000 in savings with a return of $2 per year for every $1 originally invested in such systems.

Utility systems must provide the heat and power required by the industrial process at all times. The safety and reliability of this supply is paramount. Optimizing the cost of these systems is accomplished through adjusting manually controlled elements without affecting regulatory controls, such as steam header pressure controllers or horsepower control for steam turbines in process service.

Since utility systems are not the focus of industrial process production, these systems are often lacking in metering, and this can lead to a significant lack of real-time measurement information available to the operators who are charged with manipulating manual set points to maximize efficiency and minimize cost. As a result of limited

Energy Management and Efficiency for the Process Industries, First Edition. Edited by Alan P. Rossiter and Beth P. Jones.

real-time advice, operators take a conservative approach to operation of utility systems, prioritizing "excess reliability" over efficiency and cost concerns.

"Excess reliability" is system capacity exceeding that which is needed to meet the demand and to maintain system operation in case of an unexpected shutdown of the largest contributor to a utility balance. Therefore, excess reliability is not only unneeded but can also prove very costly. For example, an extra boiler might be run to provide additional reliability (often called $N+1$). The actual load on the system may be low enough, however, that two of a facility's boilers must be run at minimum capacity, instead of shutting down one boiler and running the remaining boiler at higher capacities. The difference in efficiency between the lower end of a boiler's operating envelope and the middle of that envelope can be as much as 20–25%. Excess reliability, intended to prevent infrequent, high-cost issues, can lead to long-term operation at reduced efficiency, which can increase annual steam production costs by $1,000,000 or more at a large facility.

In order to remain competitive, industrial facilities with heavy energy use must now consider not only the cost of excess reliability but also of every potential operational inefficiency. Operators must understand the true incremental cost of every real-time decision. This information can be provided effectively by a rigorous thermodynamic model of the site-wide utility system, which accurately accounts for the incremental price of fuel and electricity purchase and/or sale in order to calculate the cost of every potential operating mode that satisfies the heat and power needs of the process in real-time.

However, the complexity of most modern utility systems yields hundreds, if not thousands, of potential operating modes among which the operator must choose in order to achieve the goal of minimum cost operation. A site-wide thermodynamic model can be leveraged by applying an optimization engine to calculate the lowest-cost operational mode currently available.

A key component in successful application of such an optimization method is proper constraining of the limits the optimizer must recognize. A mathematical optimizer will by definition seek for lower cost until it encounters a constraint that will not allow it to find a feasible lower cost operating mode. The optimization, of course, must not violate any reliability, environmental, safety, or contractual constraint.

The limits to optimization must be realistic in consideration of actual operation and decided with alignment from management to engineering to operations staff. Management must define the expectations of the energy management framework under which operators are to act in order to measure success.

The ISO 50001 Energy Management Standard (Chapter 6) addresses this need and defines specific principles that allow an organization to design, implement, and continuously improve energy management. The standard utilizes the "Plan–Do–Check–Act" approach to continuous improvement. Figure 19.1 illustrates the cycle from management planning through local execution back to audit and adjustment to planning required for energy management process improvement.

Real-time optimization and management systems for utilities can provide a rigorous basis for the Plan–Do–Check–Act cycle by aligning recommended operator behavior with the expectations of management regarding energy use at an industrial facility. In the process, they can drive significant operational savings that can be sustained over time.

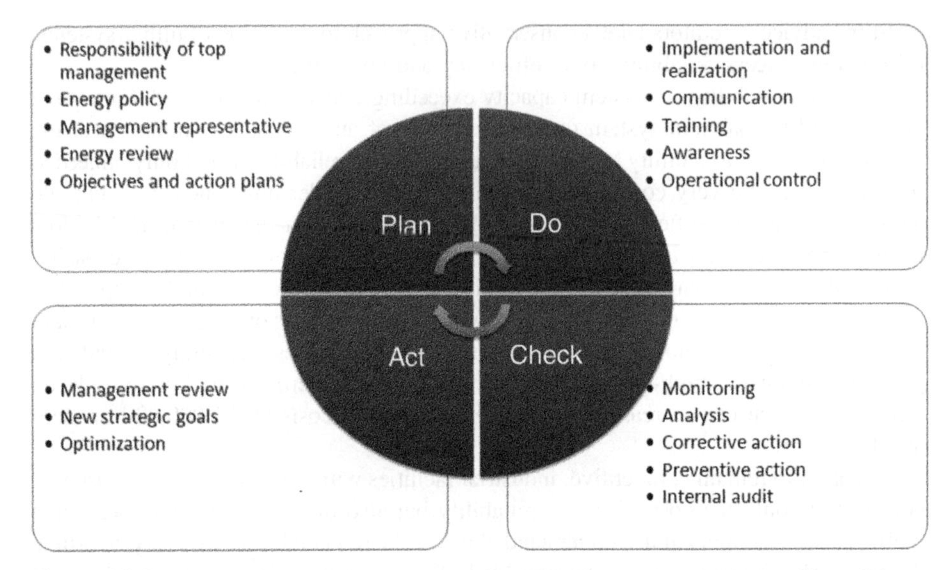

Figure 19.1. The Plan–Do–Check–Act approach to continuous improvement. (Copyright © 2012, FW8100 [CC-BY-SA-3.0 (http://creativecommons.org/licenses/by-sa/3.0)], via Wikimedia Commons, unaltered.)

HISTORICAL CONTEXT AND BACKGROUND FOR REAL-TIME ENERGY OPTIMIZATION TECHNOLOGY DEVELOPMENT

As energy prices increased during the 1980s and 1990s, along with deregulation of electricity markets and a new focus on limiting emissions, industrial facilities recognized the need to reduce waste in their utility systems. Many began to seek methods to more effectively manage steam balances, often employing spreadsheets to perform the task manually. Utility simulation systems were designed to perform steam and other utility balances in order to understand real-time operation more effectively than the spreadsheet methods allowed and these systems began to employ real-time data connections to plant information systems.

Once real-time operation and energy balances were understood, the next logical step for operations was to focus on understanding real-time efficiency in order to utilize available data and simulations to drive increased efficiency in utility system operation.

However, maximum efficiency is not always correlated to lowest cost. Optimization necessarily involves consideration of the "marginal cost" of utilities, as was discussed for steam, in particular, in Chapter 17. Optimization model results sometimes challenge long-held assumptions about utility system operation. Industrial facilities with multiple fuel sources of different cost may find that firing lower cost fuel to produce steam, even in a lower efficiency boiler, is more cost-effective than burning higher cost natural gas in more efficient boilers, even though the overall efficiency of steam production is lower. Since the operators' job is ultimately to provide utilities to the industrial process at the lowest cost within all reliability, contractual, and other constraints defined by management, industrial utility optimization systems were developed to address the ongoing need for real-time minimum cost operation.

To achieve maximum benefit, these systems should be integrated into standard operating procedures and included in the facility's process for management of change, ensuring that the system is maintained over time as the utility system is improved or the process changes. Some optimization systems issue recommendations for control system set points to be implemented by operators. Even more benefit can be gained by "closing the loop" so that the system communicates set points directly to the control system. Closed-loop systems achieve lowest cost operation more frequently than is possible in an "open-loop" configuration requiring operator intervention.

KEY ELEMENTS OF REAL-TIME ENERGY OPTIMIZATION SYSTEMS

This section will describe important functions that should be considered in designing a real-time utility optimization system. In all cases, real-time information indicating status of operation based upon local measurements throughout the utility system is a requirement for optimization.

Most industrial facilities require the purchase of fuel and/or power from third parties, for example, the purchase of electricity from the local grid. In addition, the proliferation of cogeneration integrated with industrial plant operations, where steam and power are produced internally, has given some facilities the opportunity to export power to the grid.

Real-time price for third party power and fuel purchases and sales is an essential driver in finding lowest cost operation. Incremental natural gas and other fuel pricing, if not known, can be estimated from web-based sources, such as www.theice.com. Transportation and distribution charges should also be estimated and included in the price used in the optimization for natural gas. Similarly, electricity prices can be determined by modeling applicable tariffs for facilities operating in regulated power markets, or from real-time pricing available on the grid operator (or independent system operator) website for those operating in deregulated markets.

For steam systems, accurate measurement of flow for letdowns and vents, which were discussed in Chapter 18, is critical for optimization, as reduction of vents and letdown of steam are key handles that the optimizer uses to save cost on an ongoing basis. Where flow meters are absent, steam flows can often be inferred from other sources, such as estimating the flow through a valve using the real-time measurement of valve position.

Data reconciliation is an important part of any solution calculating mass and/or heat balance, as a real-time energy optimization solution necessarily does. Utility systems are historically lacking in effective metering, and measurement errors are widespread. As a result, the optimization system must find ways to account for imbalances, which will always be present, while still finding the lowest cost operation effectively. One way to do this is through header "balloons," as shown in Figure 19.2, which comes from the commercial steam system optimizer Visual MESA™.

In the above-mentioned example, the header balance is held constant during the optimization, so all incremental changes to operation are fully accounted for in the

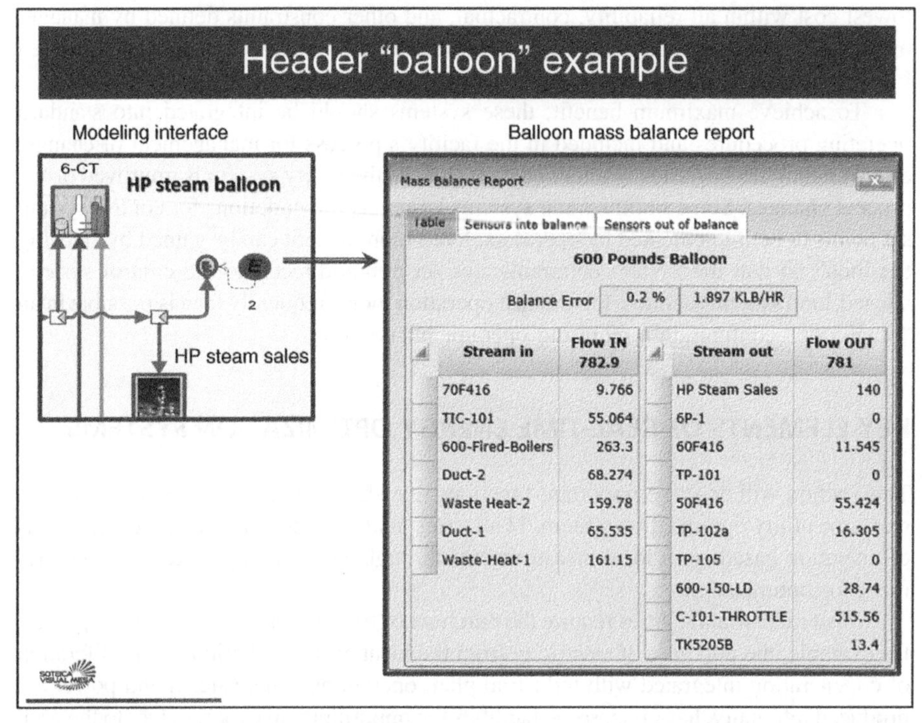

Figure 19.2. Header balloons provide a mechanism to track header imbalances. (*Source:* Courtesy of Soteica Visual MESA LLC.)

calculation of cost. This offers a very effective solution method, with details where needed for the purpose of optimization and simplicity in the elements not essential to this goal. This makes the solution less resource intensive to build and then to maintain over time.

One essential aspect of the solution is that it should be "site-wide" in nature, meaning the entire utility system should be included in the model. The reasoning is that savings found in a "local" optimization may be at the expense of increased cost in portions of the utility system that are not included in the "local" optimization scope. This can lead to increased operational cost overall, defeating the ultimate goal of optimization.

Optimization systems should account for all applicable constraints to operation, including regulatory, environmental, contractual, and reliability constraints. An effective solution will never recommend changes to operators that would be unsafe, unreliable, or illegal. Care should be taken in design and testing to ensure that all of the constraints are well understood and properly considered by the model. The Plan–Do–Check–Act method of continuous improvement suggested by ISO 50001 is applicable and can be helpful if applied effectively during model testing.

STEAM AND POWER SYSTEM EQUIPMENT OPTIMIZATION

Most utility systems have common elements, including similar "handles" or variables for optimization. In particular, many steam system elements are common from site to site.

Steam venting and steam letdown from a higher to a lower pressure header are typical and are often useful optimization handles that can provide significant savings. Since letdowns and vents are not typically under the direct control of operators, changes to other equipment under the operators' direct control are required in order to affect them.

A common way to change the letdowns and vents is through switching between pairs or groups of steam-driven and electric motor-driven pumps in shared services. By selecting the most appropriate of these machines to run at any given time, the operator can do much to minimize vents and letdowns. For instance, starting a steam turbine-driven pump to replace a motor-driven pump can reduce an open steam letdown, if the inlet and exhaust headers are the same for both the turbine and letdown. The benefit is seen in reduced electric power consumption. This concept is discussed in more detail in Chapter 18.

Figure 19.3 illustrates the reduction in letdown and vent provided by site-wide optimization. The numbers next to the bold gray lines indicate the difference in flow between the starting point and the optimized solution recommendations. The turbine-driven

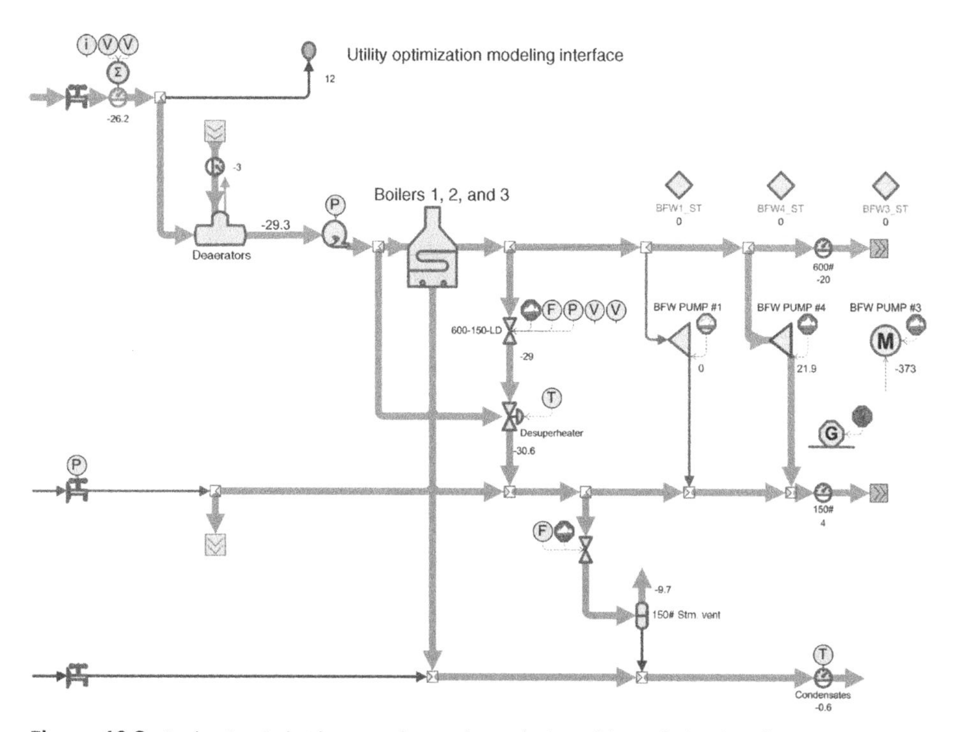

Figure 19.3. Reduction in letdown and vent through site-wide optimization. (*Source:* Courtesy of Soteica Visual MESA LLC.)

BFW Pump 4 replaces motor-driven Pump 3. Net 600 psig steam export goes down by 20 klb/h, and 600-150 psig steam letdown is reduced by 29 klb/h.

Other common optimization "handles" include steam producers, fuel sources, steam turbines, and a number of cogeneration plant variables.

On-demand steam producers, such as auxiliary boilers and duct firing on heat recovery steam generators, can provide excellent handles for reducing cost site-wide through boiler-to-boiler rebalancing for maximum system efficiency. When multiple fuel options such as refinery fuel gas and natural gas are available, the optimization can negotiate the complexities of differential pricing and constraints to find real-time operational savings. Even complex constraints are respected, such as the requirement to consume all refinery fuel gas (discussed in Chapter 20) to avoid flaring.

Steam turbines can provide excellent handles for optimization if their operation can be adjusted. For example, a steam turbine generator with a manual set point provides the optimizer with an opportunity to react to changes in the incremental fuel versus electricity pricing. So, if the price of electricity increases in comparison to the price of fuel, the steam turbine may be a cost-effective option for either reducing site-wide power purchases or increasing power sales.

Even extraction/condensing turbines with a constant horsepower requirement to the process can provide savings by rebalancing the horsepower between the high- and low-pressure sections of the turbine with the adjustment of other steam equipment in manual control.

Cogeneration systems with gas turbines and heat recovery steam generators can react to changes in fuel versus electricity purchase and sales pricing if the power generation is in manual control. Excess steam or water injection for power augmentation, inlet air chilling, and combustion air recirculation all provide additional flexibility in cogeneration operation and handles the optimization system can use to increase the savings it provides through real-time recommendations.

Common Barriers to Real-Time Optimization Effectiveness

Before a site initiates development of an optimization solution, goals for the optimization system should be understood and aligned from operations to engineering to management.

For operators, the priority is reliable and safe operations in their daily decisions. Since the information available to them for real-time utility systems is commonly limited, there is seldom a standard against which the effectiveness of past operations can be measured. The introduction of an optimization system can require a significant change in the operators' approach, which often requires training and ongoing oversight to ensure effectiveness.

For engineers, a "project" focus is often prioritized over day-to-day operations improvement. In addition, engineering staff may lack understanding and alignment with how operators make daily decisions, due to their focus on other priorities. The implementation of an optimization solution may be viewed as another "project" rather than an ongoing "process" requiring consistent, ongoing engineering oversight and adjustment to remain effective. A lack of committed engineering support can also undermine the solution's effectiveness over time.

Management may sometimes approve an optimization solution without full understanding of the basis of the optimization system's value delivery. As a result, the message of the optimization solution's priority to the organization is not communicated from the top in some cases. Time constraints may not allow a continued focus on the optimization initiative among all other priorities competing for the manager's time and attention. In the end, lack of management sponsorship or "buy-in" results in underprioritizing, and underfunding, the staff and resources necessary to maintain a cost-effective optimization solution.

Transition between internal "champions" for the optimization system can lead to optimization value deterioration due to lack of ownership and daily oversight of the model and implementation of recommendations. In addition, if "management of change" procedures do not address the completion of needed changes to the optimization system as process and utility system equipment and operation are modified over time, the model of the system will not be kept up-to-date, and will give inaccurate results.

Deficient or ineffective measurement can also prevent the optimization opportunity from being realized. What is not measured cannot be optimized, so the value of the solution is negatively affected by both the lack of important measurements and the measurement inaccuracies.

CHARACTERISTICS OF SUCCESSFUL REAL-TIME OPTIMIZATION SOLUTIONS

Alignment of management to operator is crucial to avoid the pitfalls discussed previously. First, the commitment of the site "champion" and support of upper management are requirements for success and ongoing value creation. The operators "buying in" to the value of the solution is of key importance, as they must implement the recommended changes in order to produce the savings.

The system must give accurate and direct advice in a format that fits existing operator systems and preferences. Operators are motivated to implement the optimizer's recommendations when they see a clear indication of the financial benefit alongside the recommended changes to manual set points.

Another important consideration in the success of a real-time optimization system is the technical approach. The optimization model should be based upon rigorous thermodynamics and an effective optimization algorithm. A mass balance alone will provide suboptimal solutions. A thermodynamic balance is essential to completely describe the system and achieve a true optimum.

The system must also be capable of considering equipment starts and stops in addition to "continuous" variables in order to maximize savings opportunities. And, even with the additional complexity of such "on/off" decisions, the calculation engine should solve within a few minutes in order to provide timely advice to the operators.

In addition to the importance of the site-wide approach discussed earlier in this chapter, the system should run at a regularly scheduled frequency in order to provide the operators with automatically updated recommendations, and the regular use of the system should be included in operational procedures for each shift.

As with any system utilizing a customized model, the ongoing maintenance and upkeep of the solution must be prioritized by the organization. Whether a shared ownership model with the solution provider or an internal "center of excellence" model is chosen, the organization must commit adequate resources to ensure ongoing success. This includes providing support and training during the transition of internal ownership to new staff and integration of model update and upkeep into "management of change" procedures in order that the model remains accurate and useful in driving savings over time.

Regular review of past performance of the system is also critical. The gap between recommended and actual performance should be analyzed. Constraints to the optimization should be adjusted, as appropriate, to accurately account for the actual allowable operating limits, in line with the continuous improvement recommended by the Plan–Do–Check–Act method of ISO 50001. If certain constraints are always active, the organization should question whether the constraint is valid (can the system run closer to a limit?) or the constraint can be removed through equipment or process changes. The optimizer can provide the economic justification for the changes.

Finally, the organization should regularly review, document, publish, and consider rewards for saving goals achieved. When a "team" approach is combined with a rigorous technical basis and sustained support for the solution over time, the site reaps the ongoing benefits of cost savings, improved reliability through standardization of operations, and assurance of environmental compliance based upon the rigorous utility model.

CLOSING THOUGHTS

In today's competitive marketplace, industrial companies face a growing need for energy and environmental stewardship to stay competitive. An effective real-time energy optimization solution can address this need. Ongoing savings of 3–6% of a facility's total energy bill are not uncommon, and at many industrial facilities this is worth upward of $1,000,000 per year.

ACKNOWLEDGMENTS

The author wishes to thank his colleagues at Soteica Visual Mesa LLC, and especially Jorge Mamprin, Visual MESA Technical Lead, Houston, TX, and Carlos Ruiz, Energy Management Director, Rosario, Argentina, for providing input and reviewing material for this chapter.

20

FUEL GAS MANAGEMENT AND ENERGY EFFICIENCY IN OIL REFINERIES*

Joe L. Davis

PSC Industrial Outsourcing, LP, Houston, TX, USA

Oil refineries convert crude oil into high-value products such as gasoline, diesel, liquefied petroleum gas (LPG), and petrochemical feedstocks. After squeezing as much salable product from the crude oil as possible, there remains a light gas stream known as refinery fuel gas (RFG), which is too light for any of the high-value products. The RFG (predominantly made up of a mixture of hydrogen, methane, and ethane) is returned to the plant and used as fuel to fire the process heaters. At one time, a typical plant did not make enough RFG to supply all of its heaters and was therefore required to purchase a makeup fuel from a third-party supplier. In the United States, the purchased fuel is usually natural gas, which is typically injected directly into the RFG manifold or mix drum and represents what refiners call their "fuel gas cushion."

Changes in refined product demand, environmental requirements and specifications, and increasing emphasis on energy optimization over recent years have all changed the refinery fuel gas balance. Hydrogen supply has become more important as both refined product sulfur limits and reformer severities have been reduced, and lower emission burners are limited in their tolerance for hydrogen in the fuel system. Environmental

* This chapter was presented at the Twenty-Sixth Industrial Energy Technology Conference, Houston, TX, April 20–23, 2004 [1]. Adapted and updated with permission.

Energy Management and Efficiency for the Process Industries, First Edition. Edited by Alan P. Rossiter and Beth P. Jones.

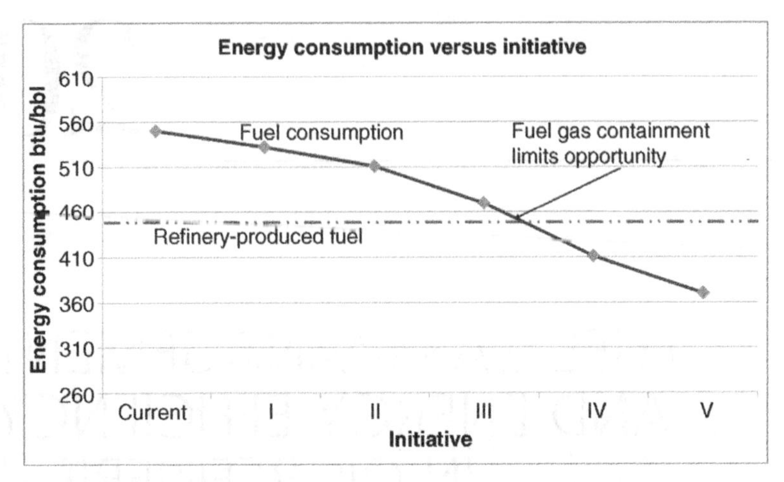

Figure 20.1. Fuel gas cushion versus energy project implementation.

restrictions and the cost of fuel prompted many flare gas recovery projects. The refinery fuel gas system has become a fertile ground for optimization to ensure that each component goes to its highest-value disposition. Combining optimization objectives—reducing overall energy use, recovering hydrogen, and minimizing emissions—can lead to interesting projects with robust justifications.

As energy prices soared and refiners investigated energy reduction projects, many found that their ability to reduce energy consumption was limited by the amount of refinery fuel gas their plants produced relative to their gas import. As energy-saving projects were implemented, the amount of marginal import fuel diminished to the point where the next energy project would result in the flaring of refinery gas (see Figure 20.1). The economics associated with wasting RFG in a flare, not to mention the environmental restrictions on flaring, made the incremental energy project wholly unattractive to the refiner. Therefore, many refiners looked for cost-effective ways to reduce the amount of fuel gas being produced in the refinery.

CAUSES OF HIGH FUEL GAS PRODUCTION

Excessive refinery fuel gas production can be attributed to a variety of factors. One that stands out is related to Fluid Catalytic Cracking Unit (FCCU) and reformer operation changes over the last 35 years. When the shift was made from leaded to unleaded gasoline, higher octane naphtha was required. The higher octane demand was met in a couple of ways that resulted in higher fuel gas production. First, FCCU severity increased through higher reactor temperatures. The side effect of this was an increase in LPG and fuel gas production from the FCCU. Second, reformer severity was increased to make higher octane reformate. As was the case on the FCCU, this more severe operation on the reformer resulted in additional fuel gas production due to cracking brought on by higher reactor temperatures. More recently, changes in both gasoline versus distillate demand and gasoline specifications have reduced severity in many reformers and FCCUs, which shifts the fuel gas balance back toward shortness. However, lower severity reformer operation also reduces the availability of hydrogen for feedstock and product desulfurization, whereas product sulfur specifications have been tightened and the quality of some feedstocks has declined. Many refiners have found it necessary to improve hydrogen recovery from refinery fuel gas and even to supplement hydrogen supply through purchase or on-purpose production.

In addition to the higher severity operations of the FCCU, simple increases in FCCU throughput can also impact the refinery's fuel balance. Hot gas from the FCCU catalyst regenerator is routed to a CO boiler, where its heat is used to generate steam. At higher throughput, even if conversion remains the same, increased amounts of steam are generated from waste heat in the CO boiler. Since FCCUs often generate steam at the highest level in the refinery, this causes the refinery's on-purpose boilers to reduce production, thus reducing the demand for fuel gas.

In addition to conversion unit fundamentals, inefficient operations can send light end molecules into the fuel system. For example, poor use of hydrogen can lead to hydrogen letdown into the fuel system. Furthermore, refineries often have a greater fuel gas containment problem in the hot summer months when higher volumes of C_{3+} material make their way to the fuel system. This can be due to poor cooling tower operations where summertime cooling water supply temperatures are as high as 90 °F.

SITE-WIDE ENERGY BALANCE

A site-wide energy balance that includes fuel, steam, and power forms the foundation of any energy study. Only after a base case energy balance is generated can one effectively evaluate the impact of energy savings and fuel optimization projects on the system as a whole, from both an economic standpoint and a pure utility balance perspective.

For example, if the steam system were not fully understood, how would one know that an increase in waste heat steam production in one area of the refinery would not simply result in the venting of steam somewhere else? Or, without a detailed fuel balance, how could one determine the impact of steam use reduction projects on the fuel gas balance when boilers are turned down, or shut down completely?

As an added benefit, the fuel, steam, and power balance discussed in this chapter becomes an extremely valuable tool for designing new facilities for the site. For example, a well-understood plant energy balance would help in the selection of major rotating equipment drivers—especially when deciding between steam turbine and electric motor drivers, which have very different impacts on all three components of the balance (fuel, steam, and power).

Fuel Balance

As previously mentioned, a reasonably accurate fuel balance is important to understanding the impact that future energy reduction and other fuel-related projects will have on the fuel system. As far as possible, the fuel balance should also take into consideration the impact of future projects, such as new process units or expansions, which are planned for the site. In developing the fuel balance, it is assumed that all consumed fuel flows are known from meters at the furnaces, produced fuel is metered at the source, and fuel gas compositions are available through either laboratory or online analysis.

The challenge is typically in the accuracy of the production meters and the availability of laboratory analysis of these streams. Depending on the fuel gas system configuration, it is often acceptable to measure fuel gas production at the outlet of the collection drum(s), where meters are normally better maintained and laboratory samples are taken more frequently. However, for predicting the impact of future operational changes, as well as for routing the right streams to the right end users, it is important to understand the contributions from individual producers. A good refinery fuel balance closes within 5–10% on a Btu basis (total internal fuel production plus imports versus total fuel consumption).

Table 20.1 shows an example of a refinery fuel balance. Note that the balance is based on an overall Btu balance rather than on a detailed component basis. A Btu balance is normally adequate for predicting the impact of future energy projects, although a rough component balance is necessary for more complex optimization. The heating value of each stream can be determined through laboratory analysis. If the stream compositions change frequently with different crude slates or other modes of operation such as summer versus winter, then the refinery should develop separate fuel balances for each mode.

If a refiner is having trouble obtaining 5% or better closure on the fuel gas balance, some of the following steps may help close the balance:

- Confirm that all gas production streams are being accounted for. Some refineries produce off-gas from units, such as the crude distillation unit, that bypass the refinery fuel collection drum and are routed directly to a process heater, thus bypassing the major meters around the fuel drum.
- Check and recalibrate all flow meters, particularly those that have not been inspected in some time.
- Verify that the process data historian is making proper meter flow corrections for temperature, pressure, and gravity.
- Verify gas sampling procedures. Confirm, for example, that the laboratory is correctly running the gas chromatographs (GCs). Use a third-party laboratory for validation.

Power Balance

By comparing the total power supply (purchased plus in-plant produced) against the total connected load, a good power balance can usually be developed. The total connected load is generated from a site-provided motor list containing rated horsepower and "on/off" status for each motor. The sum of rated horsepower for all running motors is compared with the total power supplied. Generally, this approach results in an estimate of connected load that is 10–20% higher than the measured total power supply. This is to be

TABLE 20.1. Sample Refinery Fuel Balance

Production	kscf/day	LHV Btu/scf	MBtu/h
Nat gas to mix drum	30,835	976	1254
FCCU	914	1023	39
SGRU	9940	811	336
CDU 2	7770	434	141
GU	2053	786	67
PACC	27,603	971	1117
PSA purge gas	1880	781	61
PSA bypass	2831	434	51
Benzene unit	573	781	19
EU	13,205	622	342
CU	15	781	0
Total fuel gas production	97,620	843	3427
Consumption			
GFU	915	781	30
CDU 1	13,972	781	455
DU2	12,995	781	423
HFAU	3,530	781	115
GU 1	1,109	781	36
GU2	396	781	13
Boiler House 1	12,093	969	488
Boiler House 2	18,569	747	578
Boiler House 3	14,326	781	466
Hydrocracker	2,775	971	112
Delayed Coker	9,938	971	402
H2 Plant and PSA Unit	6,439	781	210
Benzene Unit, Tank Farm	8	781	0.28
Flares	16	781	1
Total fuel gas consumption	97,082	823	3,328
Difference	538		99
%Difference	0.60%		2.90%

expected since many motors will not be operating at their full rated horsepower, but somewhere down the curve. One can simply allocate the percent difference over the entire population of motors in order to close the balance.

A good power balance will also account for "switchable" drivers—that is, pumps or compressors that can be switched between a motor and a steam turbine. When the motor is used, there is a demand for electric power. When the steam turbine is used, the electric power demand disappears, but there is an impact on the steam balance (see Chapter 18).

Steam Balance

The steam balance is normally the most challenging component of the site energy balance. This topic is discussed in Chapter 18. The steam balance should be integrated with the power balance and fuel balance so that changes in boiler loading and turbine operation can be related to changes in the power and fuel demands.

EVALUATE OPTIONS FOR OPTIMIZING FUEL GAS DISPOSITION

If the energy balance described above confirms that the refinery cannot achieve its energy reduction goals without reducing the fuel gas supply, the ideas briefly described in this section should be considered:

- *Composition of fuel gas (H_2, C_{3+}, and H_2S):* Is the fuel system swollen with hydrogen, C_{3+} material, or both? A good fuel gas system will have less than 25% hydrogen, and hydrogen likely has more value in a desulfurization unit. However, in many refineries, the fuel system contains upward of 50% hydrogen. In these cases, LPG often has to be vaporized into the fuel system to prop up the heating value of the fuel gas—further aggravating any fuel surplus.
- *Plant hydrogen demands—current and future:* Will a new steam methane reformer (SMR) be required, or might there be a project to recover hydrogen out of the fuel gas?
- *Alternate fuels (LSFO, LSR, and LPG):* What other current fuel sources could be reduced or eliminated, and do they command a higher value elsewhere?
- *Summer versus winter demands:* Is the fuel gas imbalance seasonal?
- *Steam balances—current and future:* Is a cogeneration plant planned that will shut down old package boilers, resulting in a decrease of fuel gas demand? Could any components of the fuel gas supply be used in a gas turbine?
- *Hydrogen plant feed flexibility:* If the current feed to the hydrogen plant is purchased natural gas, can this be replaced with refinery fuel gas or any of its components?
- *Long-term power generation strategies:* Will the plant continue to buy all power from a third party, or will it install on-site power generation facilities?
- *Fuel drum configuration, outlet metering, and sampling:* Can the entire pool of refinery-produced fuel gas be metered and sampled from the overhead of a common mix drum?
- *Fuel gas supply control strategy:* Most refineries control fuel on supply line pressure, but some refiners have found it worthwhile to provide advanced control to manage fuel supply and composition.
- *Third-party fuel gas purchase or sales:* Are the available options used in the most cost-effective manner? Are new options necessary?

- *Flare gas recovery:* Are recoverable fuel components routinely in the flares? Flare gas recovery projects can pay unexpected dividends—not just in achieving environmental compliance but also in avoiding significant expenditure for new flare capacity, particularly with new and tighter flare tip regulations.

Depending on the situation, several fuel optimization strategies should be considered along with their energy improvement and other benefits. Short-to-medium-term strategies include increased liquid recovery, improved hydrogen recovery, and optimized fuel system control parameters.

- *Improve liquid recovery*
  - Optimize absorber lean oil selection and rate.
  - Improve de-ethanizer operation to minimize C_3 to fuel.
  - Bypass poor recovery feeds around gas or vapor recovery unit (GRU/VRU) to reduce GRU/VRU loading and improve capacity for better feeds.
  - Improve cooling tower efficiency by upgrading the packing.
- *Improve hydrogen recovery*
  - Eliminate unmetered spills to fuel.
  - Improve cascading from high- to low-pressure hydrodesulfurization (HDS) units.
  - Optimize membrane and PSA feed selection and operation.
- *Improve fuel gas system control*
  - Check and if necessary modify minimum stop on natural gas import.
  - Review the control strategy for capability and consistency, especially if there are multiple fuel drums.
  - Review the LPG vaporization strategy. If LPG is added to the fuel system only during off-test incidents, look for ways to mitigate the off-test events.

Longer term solutions may involve the following new projects:

- *New power generation*
  - New boiler/steam turbine generator (STG) combo
  - New gas turbine/heat recovery steam generator (HRSG)/(STG) combo
- *Expanded recovery capacity*
  - New gas recovery plant
  - New pressure swing absorption (PSA) unit
  - New hydrogen compression equipment
  - New Ammonia Absorption Refrigeration (AARU) plant
- *Hydrogen plant feed conversion*
  - Swap or supplement natural gas feed with refinery fuel gas or fuel gas component feed

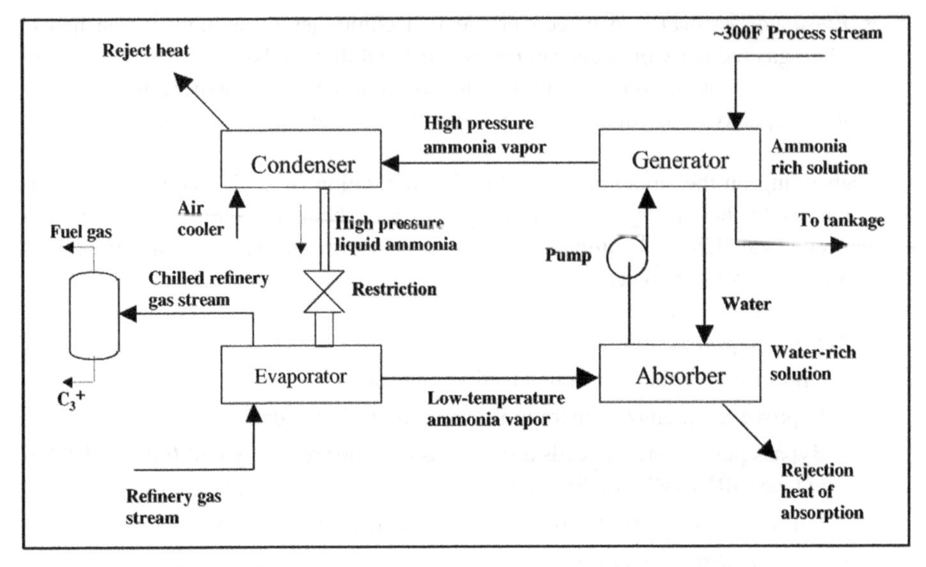

Figure 20.2. Ammonia Absorption Refrigeration Unit (AARU).

Case Study

One U.S. refinery has successfully commissioned an AARU that actually solved two problems [2]. By taking a 300 °F process stream that was currently exchanging its remaining heat with the atmosphere via a fin–fan cooler and using it instead as the heat source for the AARU process, the refinery was able to use the resulting refrigerant to cool a fuel gas stream to a point where C_{3+} material could be recovered prior to injection into the refinery fuel gas system (see Figure 20.2). The refinery now has little problem with the flaring of fuel gas since the average LPG recovery is at least 50% higher than pre-AARU. This resulted in an increase in natural gas makeup to the fuel system, which in turn made room for additional energy conservation steps.

The ammonia absorption cycle is based on the principle that absorbing ammonia in water causes the vapor pressure to decrease. The absorption cycle has a distinct advantage over other refrigeration cycles such as propane or ethylene because instead of requiring a large mechanical compressor, absorption requires only a pump and a very modest amount of electricity. In addition, the working fluid, ammonia and water, contains no ozone-depleting hydrocarbons [3].

Conclusions

All refiners continue to be under tremendous cost pressure. As energy accounts for between 30 and 50% of a typical refinery's operating costs (excluding crude oil), this is certainly an area that deserves attention. The refinery fuel gas system can be a limiting factor in how far a refiner can go on energy optimization. The key to successfully meeting energy reduction goals is found in developing a thorough understanding of current and future fuel balances and identifying and executing projects to maintain the overall fuel balance while reducing energy demand. Many energy reduction projects can be coupled with other light component and environmental constraints to improve returns and satisfy multiple objectives.

REFERENCES

1. Davis, J.L. Jr. (2004) Overcoming fuel gas containment limitations to energy improvement. *Proceedings of the Twenty-Sixth Industrial Energy Technology Conference*, Houston, TX, April 20–23, 2004.
2. Energy Concepts, Inc. (2014) *Ammonia absorption refrigeration unit provides environmentally friendly profits for an oil refinery*. Available at http://www.energy-concepts.com/_pages/app_refinery_chilling.htm (accessed March 13).
3. Energy Concepts, Inc. (2014) *The absorption process*. Available at http://www.energy-concepts.com/index.html (accessed March 13).

21

REFRIGERATION, CHILLERS, AND COOLING WATER

William (Bill) Turpish

W. J. Turpish and Associates, PC, Consulting Engineers, Shelby, NC, USA

Cooling is an important requirement in the process industries. Cooling systems can be extremely energy intensive, and poor cooling can also cause energy use in other systems to increase. For example, high cooling water temperatures can cause a distillation column's overhead temperature and pressure to rise, which in turn increases the temperature and heat load of the column reboiler. In contrast, proper cooling enables high production rates at the lowest energy input, and can also enhance product quality.

Despite these issues, cooling systems are often overlooked in energy efficiency programs, as there is a tendency to focus on heat input rather than heat removal. In this chapter, we look briefly at the main components of cooling water, chilling, and refrigeration systems, and then focus on factors that can improve their energy efficiency.

APPLICATIONS OF COOLING WATER, CHILLERS, AND REFRIGERATION SYSTEMS

Almost all process plants must reject significant amounts of heat, which is discharged in one way or another to the ambient environment. Most of this heat is generated above

Energy Management and Efficiency for the Process Industries, First Edition. Edited by Alan P. Rossiter and Beth P. Jones.

ambient temperature and can be removed, comparatively simply, in air, using air fin or fin fan coolers, or in cooling water.

However, some of the heat sources are below ambient temperature. In these cases, the temperature level of the heat must be boosted before it can be rejected to ambient, and refrigeration is required. Often, this is accomplished via a circulating fluid (most commonly water) that removes heat from the process and then rejects the heat to the refrigeration circuit. The term "chiller" is generally applied to this type of system, and chillers are widely used both in process facilities and in heating, ventilating, and air conditioning (HVAC) applications in buildings (see Chapter 24).

In many process applications, heat is removed directly from below-ambient heat sources into the working fluid of the refrigeration cycle, without the use of an intermediate circulating fluid. The term "refrigeration system" is generally applied to the equipment used in these cases. In the process industries, refrigeration systems are often used for loads at temperatures significantly below the freezing point of water—for example, in the light-ends separation section of an olefins plant, where temperatures of −250 °F or lower may be required.

Chillers and refrigeration systems depend on the same technologies and equipment types, and the terminology is often used interchangeably.

Cooling Water and Cooling Towers

There are many different types of cooling water systems, including once-through river water and seawater systems. However, most process facilities in the United States use recirculating cooling water, and the key component is the cooling tower. The tower provides a mechanism for rejecting heat to the ambient air and returning the cold water to the process, with only modest makeup water requirements.

Cooling towers are most commonly classified based on their airflow generation method, of which there are two main types:

- Natural draft, which uses buoyancy effects in a large chimney structure. These are often used in large power stations, but only rarely in refineries, chemical plants, or other process facilities.
- Mechanical draft, which uses power-driven fans to generate airflow. These are commonly seen in the process industries. Two major subclassifications are forced draft, where a fan at the inlet blows air into the cooling tower, and induced draft, where a fan at the discharge draws air through the cooling tower (Figure 21.1).

Hybrid designs (such as "fan-assisted natural draft") use both mechanisms for generating airflow.

Cooling, within cooling towers, is accomplished primarily by the evaporation of a portion of the circulating water. The wet bulb temperature[1] is the theoretical temperature limit to which the water can be cooled. In practice, most modern cooling towers

[1] The wet bulb temperature is the temperature that a mass of air would reach if cooled to saturation by evaporation of pure water into the air, the latent heat of evaporation being supplied by the air.

Figure 21.1. Induced draft cross-flow cooling tower. (Courtesy Baltimore Aircoil Company.)

are designed for an approach of 5–7 °F to the design wet bulb temperature for the local area.

Refrigeration Cycles and Water Chiller Types

There are several types of water chillers and refrigeration systems. They differ from each other based on the refrigeration cycle and the type of compressor they use.

Systems using the *vapor-compression refrigeration cycle* can use several different types of compressors, which are as follows:

- Reciprocating and scroll compressors are typically used in small chillers.
- Helical-rotary (screw) compressors are typically used in medium-sized chillers.
- Centrifugal compressors are typically used in large chillers and refrigeration systems. Note that today the difference between screw and centrifugal chillers is fading and both can be had in similar sizes.

Absorption water chillers make use of the absorption refrigeration cycle and do not use a mechanical compressor. Most commonly the working fluid is lithium bromide and water. These are uncommon in process applications, and are not described further.

Figure 21.2 shows a two-stage centrifugal refrigeration cycle with economizer. Liquid refrigerant from the condenser passes through an expansion device en route to the economizer, where it flashes at reduced pressure. This produces a two-phase vapor–liquid mix at the saturation temperature corresponding to the economizer pressure. The vapor is separated from the mixture and routed directly to the inlet of the second-stage impeller. The remaining saturated liquid refrigerant enters the second expansion device.

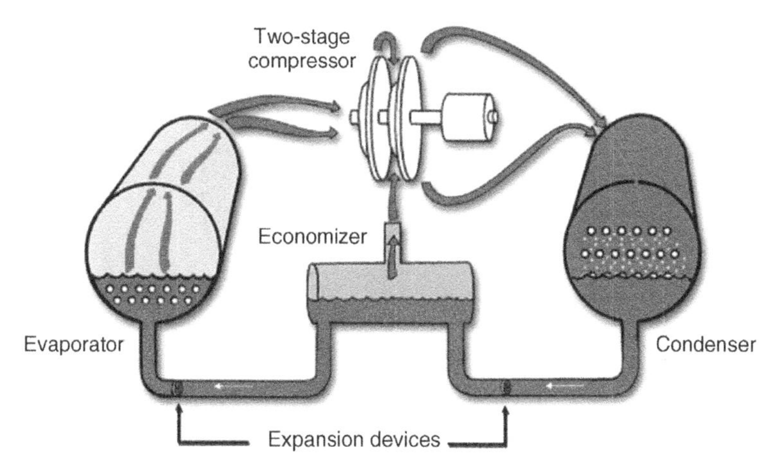

Figure 21.2. Refrigeration cycle for a two-stage centrifugal compressor with economizer. (Used by permission from Trane.)

The pressure drop created by the second expansion device causes a portion of the liquid refrigerant to vaporize, reducing the refrigerant temperature further. The resulting cooled mixture of liquid and vapor enters the evaporator.

In the evaporator, the liquid refrigerant boils at a pressure less than atmospheric as it absorbs heat from water or from a process stream, and the resulting vapor is drawn back to the compressor to repeat the cycle.

Refrigerant vapor leaves the evaporator and flows to the compressor, where it is compressed to a higher pressure and temperature. The hot, high-pressure refrigerant vapor then travels to the condenser where it rejects heat to cooling water, condenses, and then returns as a saturated liquid to the evaporator, to repeat the cycle.

Figure 21.3 shows a helical-rotary compressor refrigeration circuit. This circuit creates refrigeration in the same way as the centrifugal refrigeration cycle, but in this case it is a single stage system, and therefore does not have an economizer. The other key difference is that a substantial amount of oil is injected into the compressor with the refrigerant to provide sealing and lubrication. Oil is also used to cool and lubricate the bearings. The oil is recovered and recycled in the oil separator.

Energy Terminology for Chillers, Refrigeration Systems, and Cooling Towers

The refrigeration load of a chiller is generally expressed in tons, where a ton is the amount of heat that would melt 1 ton (2000 pounds) of ice in 24 h, or 12,000 Btu/h.

When we consider cooling towers coupled to a refrigeration or chiller system, we need to use a different definition for the ton. If we use a chiller to cool process water, the heat rejected in the cooling tower is not just the heat removed from the water, but also the heat equivalent of the energy needed to drive the chiller's compressor, typically assumed to be 3000 Btu/h (though efficient modern chillers often use

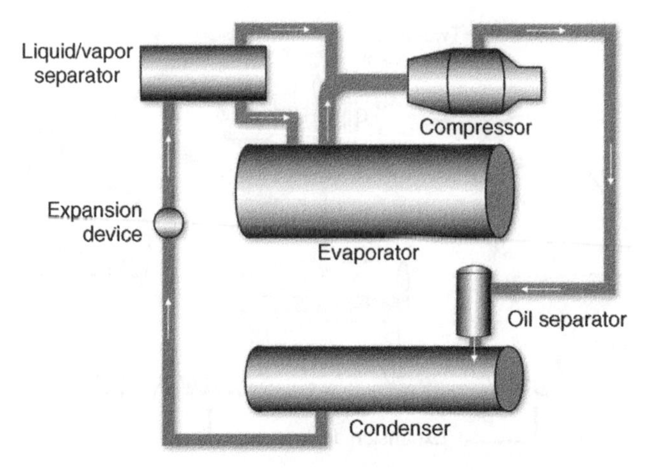

Figure 21.3. Helical-rotary water chiller refrigeration circuit. (Used by permission from Trane.)

less energy). Thus a cooling tower ton = process load + motor load = 12,000 + 3000 Btu/h = 15,000 Btu/h.

The energy efficiency of a chiller or refrigerator is often expressed in kW/ton. Another measure that is commonly used is the coefficient of performance (COP), which is the useful work divided by the energy required. This is illustrated by the following example:

1 kW = 3413 Btu/h, so 1 ton of refrigeration = (12,000 Btu/h)/3413 Btu/kWh = 3.51 kW. It follows that, if an electrically driven chiller or refrigerator consumes 1.0 kW of electricity per ton of refrigeration, its "electric COP" = 3.51/1.0 = 3.51.

EFFICIENCY IMPROVEMENTS IN COOLING WATER SYSTEMS

A cooling water system can be considered as two main sections: the cooling tower(s) and the cooling water distribution system. Opportunities for energy efficiency improvements in these two areas are discussed below.

Cooling Towers

A cooling tower provides an environment where air and cooling water are in contact, allowing evaporative cooling to occur. The key to efficiency in a cooling tower is to maximize the benefit obtained from both the air and the water moving through the tower.

The design of a cooling tower can have a significant impact on its energy efficiency. The many variables in the design include the size of the tower, air and water flow patterns, type of fill, fill spacing and height, air and water loading per square foot of fill, and spacing, type, and sizing of nozzles. Key design and maintenance issues, together with possible improvements, are discussed below.

Figure 21.4. An old cooling tower (a) is upgraded by replacing its fill with new high-efficiency fill (b). Courtesy Baltimore Aircoil Company. All rights reserved.

Cooling Tower Fill. Modern cooling towers generally use high-efficiency fills, and they are typically designed for an approach of 7 °F to wet bulb temperature at 78 °F entering wet bulb. However, designs with an approach of 4 or 5 °F may be justified in some cases where the system can benefit in terms of greater output and/or lower energy usage from the colder water temperatures.

The performance of older towers can generally be improved cost-effectively by upgrading to a high-efficiency fill (Figure 21.4). Not only can better fill reduce both overall system and tower energy costs, it can often provide other valuable benefits such as increasing plant throughput or product yield.

Water Flow Rates. Water flow rate is also an important factor in tower design. As with most energy decisions, life cycle cost should be considered in specifying flow rate. Lower flow designs (~2 gpm/ton) typically offer a lower first cost than the more common higher flow (3 gpm/ton) designs. However, the higher flow designs can offer closer approaches to the wet bulb temperature and often require lower horsepower fan motors, with associated energy savings. Higher cooling tower flow rates also increase the options to reduce noise and to use waterside free cooling.

Flow Patterns. Air- and water-flow patterns have a big impact on the efficiency of a cooling tower. The two basic configurations are cross-flow and counter-flow (see Figure 21.5). The counter-flow arrangement matches the coldest water against the coldest incoming air, which typically reduces the size of the cooling tower for a given load. However, other factors may still favor cross-flow designs in certain situations, such as

- first cost;
- dirt loading;
- certain icing conditions;
- ease of maintenance and access;

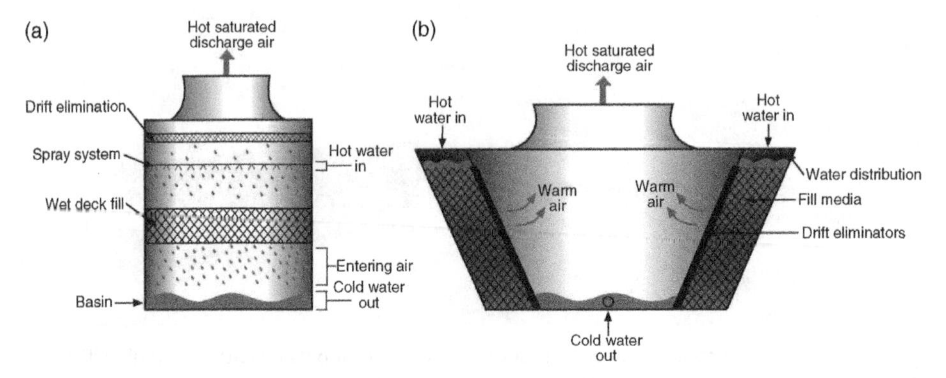

Figure 21.5. Air and water flow patterns in cooling towers: counter-flow (a) and cross-flow (b). (Courtesy Tower Components, Inc.)

- high temperature;
- location and space; and
- water distribution/loading (flow turndown).

Fans. Fans should also be chosen based on life cycle costs. Tip velocities should be kept below 12,000 fpm to increase fan life and prevent premature failure, while air discharge velocity should be above 1200 fpm to minimize recirculation of saturated exhaust air back to the air inlet.

Fan power represents a significant part of the operating cost of a cooling water system. Selecting a larger tower with a smaller fan motor for a given cooling load can result in large energy savings, often with paybacks of 2–3 years or less.

Some fans are equipped with dual speed motors or variable speed drives, which allow the airflow to be adjusted based on demand. If the demand on a cooling tower with a fixed-speed fan drops permanently (e.g., due to energy efficiency improvements in the process area, or elimination of some of the equipment served by the cooling tower), it may be cost-effective to derate the fan. If there is more than one cooling tower in the system, it may be desirable to decommission one tower and isolate it from the rest of the system, or even to demolish it entirely.

Belt drives are commonly used with fans, but these can vary significantly in efficiency. Typically synchronous belt drives are about 4% more efficient than V-belt drives over the life of the belt and offer greatly increased belt life. When properly installed, they do not require retensioning.

A further option to reduce fan power requirements (or increase capacity) is the addition of a velocity recovery (VR) stack (see Figure 21.6). The VR stack is a conical fan cowl extension, and it can reduce the discharge pressure the fan has to work against. This allows the fan to move more air for the same energy input. VR stacks can be incorporated in the initial design of a cooling tower or retrofitted. However, they are not always cost justified.

Figure 21.6. VR stacks installed on cooling towers increase airflow for the same energy input. (Courtesy Baltimore Aircoil Company. All rights reserved.)

Water Quality. Water quality is another important consideration with cooling towers. Dirt and scale in the water can cause plugging of nozzles and blockages in the fill. These problems are generally worst at low water flows (<3 gpm/ton), so adequate water flow must be maintained. Furthermore, if the water chemistry is not properly controlled, it can lead to fouling and/or corrosion in the heat exchangers of the cooling circuit, and also to growth of biological contaminants within the cooling circuit.

Water quality is generally managed by a combination of

- filtration;
- water softening;
- blow down of a portion of the recycle water flow to remove impurities; and
- chemical additions.

Water treatment professionals, either in-house or external, should be consulted to manage water treatment.

Maintenance. Failure to maintain the physical condition of a cooling tower inevitably causes degradation in performance. This may manifest itself in many different ways, including reduced cooling capacity, increased approach to wet bulb temperature, increased loss of water by drift, and rising power demand. Key components to be inspected and maintained include

- inlet louvers;
- drift eliminators;
- casing;
- fan deck;

- hot deck covers;
- structure;
- cell partitions;
- motor, gear drives, and mounts; and
- vibration switches.

Cooling Water Distribution System

The distribution system takes the cooled water from the cooling tower(s) and circulates it through heat exchangers where cooling is required and then returns it, now warmer, to the cooling tower(s). Distribution is a major energy cost in the system due to the power required for pumping. Common options for improving energy efficiency are examined in Reference 1, and the discussion is adapted and expanded below with the permission of John Wiley & Sons, Inc.:

Ensure That Water Flow Does Not Greatly Exceed the Need. Many cooling water pumps are oversized. Furthermore, the cooling water system is generally expanded as new equipment is added to a facility, and this expansion often includes additional pumps. If equipment is removed from the facility, surplus pumping capacity is left. The result is that, for much of the time, there may be a significant mismatch between the output from the pumps and the requirements of the consumers. There are many ways that this situation can be addressed—for example, derating pumps, installing variable speed drives, energizing specific pumps to match system demands, and so on. This topic is discussed in more detail in Chapter 15.

One word of caution: reducing the water flow rate can save energy, but if the flow is reduced below the cooling tower manufacturer's recommendations, this can result in scaling in the cooling towers and heat exchangers, which leads to poor performance and negates the pump energy savings.

Avoid Unnecessary Pressure Drop. In order to minimize hydraulic losses and the associated increased pumping costs, systems should be designed with relatively large pipe diameters (subject to life cycle cost analysis), and there should be no unnecessary valves and fittings. Unnecessary valves and fittings should also be eliminated from existing distribution systems.

Design to Ensure That the System Is Balanced. In most process facilities, the main pressure drop in the cooling water system is due to heat exchangers, and it is important to ensure that the system is designed to ensure that the cooling water system is pressure balanced to provide the expected cooling in each exchanger. This requires careful consideration of the design basis (design flow and design pressure drop) for each heat exchanger. However in many cases, heat exchangers are added or repurposed over time. These changes can result in overflow and underflow in portions of the cooling water system. It is not uncommon to have some heat exchangers that take twice their design cooling water flow, while others get half or less of their design flow. This imbalance can lead to poor control and capacity constraints, as well as negative impacts on energy efficiency and reliability.

Improved instrumentation (especially flow and temperature measurements) can be used to monitor and manage flow imbalances. Ideally, there should be sufficient instrumentation to calculate heat balances around all heat exchangers. Fluid flow models can also be used as a tool to evaluate and manage flow imbalances. In many cases, especially where there are complex distribution systems, models can uncover unexpected effects and find nonintuitive solutions to flow balance problems.

EFFICIENCY IMPROVEMENTS IN REFRIGERATION AND CHILLER SYSTEMS

Many energy efficiency improvements can be achieved in refrigeration and chiller systems with little or no capital investment. The 18 opportunities for savings described below are adapted courtesy of *Engineered Systems*, www.esmagazine.com [2]. They are divided into three areas:

- Component-related opportunities, which involve the proper operation and maintenance of chiller or refrigeration components, including setting optimum water temperatures and flow rates.
- System-related opportunities for multiple unit cooling plants, which involve operating the most efficient mix of equipment under varying load conditions.
- Retrofit opportunities, which include upgrading existing systems with the latest energy-saving technology and equipment.

Component-Related Opportunities

1. *Reset chilled water exit temperature.* Chilled water systems typically operate at partial load most of the year, when mild temperatures and lower outdoor humidity levels decrease the need. At reduced loads, the cooling coils can produce the required cooling at higher chilled water temperatures. Raising the chilled water temperature lowers the compressor head and decreases energy consumption.

   Studies on centrifugal chillers show that, for constant-speed chillers, this strategy saves only nominal amounts of energy at operating loads in the 40–80% range. Within this range, savings are around 0.5–0.75% per degree increase in chilled water exit temperature. Surprisingly, the efficiency of a constant-speed chiller below 40% load can be degraded to the extent that kilowatt consumption increases with reducing load (Figure 21.7).

   Centrifugal chillers equipped with variable-frequency drives (VFDs) (see item 15 discussed later) respond better to chilled water reset. At loads between 10 and 80%, a variable-speed chiller will consume 2–3% less energy, per degree of increase.

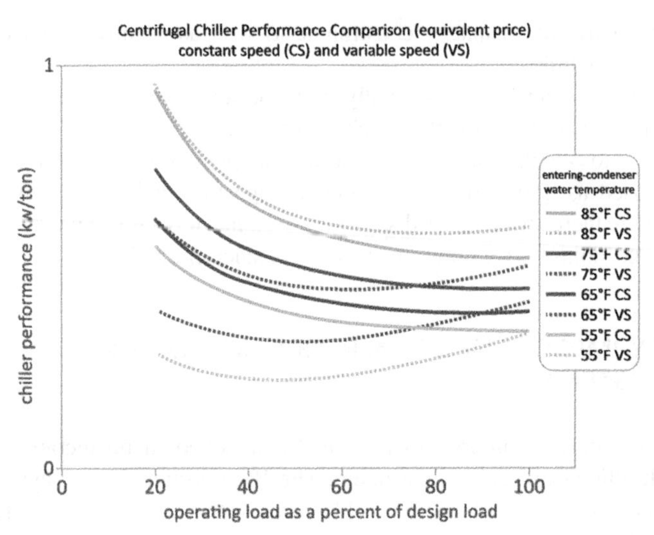

Figure 21.7. Chiller-specific power requirement variation with load. (Used by permission from Trane.)

2. *Maintain proper refrigerant charge.* Too little or too much refrigerant charge limits the heat-transfer capacity of a chiller or refrigerator and increases head pressure and energy consumption. Improper charge levels also can decrease the evaporator temperature. In a typical chiller, every 1 °F increase in evaporator temperature saves 1.5% of the full-load energy requirement. The percentage savings in refrigeration systems vary depending on its design. For centrifugal chillers, a sight glass in the evaporator shell is used to monitor refrigerant levels. For reciprocating chillers, an undercharge is indicated by bubbles in the liquid line sight glass, while an overcharge is indicated by high discharge pressure or low refrigerant temperature leaving the condenser. The system should be charged according to the manufacturer's instructions.
3. *Reduce entering condenser water temperature.* Most manufacturers specify a minimum entering condenser water temperature. The drive to conserve energy, however, has led many manufacturers to reevaluate the minimum temperature. Chiller and refrigeration energy consumption is partly a function of the condenser pressure and temperature. Lowering the condenser water temperature also lowers the refrigerant condensing temperature and condensing pressure. The result is lower head and reduced energy.

   For a typical chiller, energy savings at full load amount to 1.5% for every degree the entering condenser water temperature is reduced. Centrifugal chillers equipped with VFDs (see item 15 discussed later) respond better to condenser water reset. At loads between 10 and 80%, a variable-speed chiller will consume 2–3% less energy, per degree of reduction. The percentage savings in refrigeration systems will vary depending on design.

4. *Eliminate refrigerant and air leaks.* Leaks in the refrigerant system should be eliminated. In high-pressure chillers and refrigeration systems refrigerant will leak out, reducing refrigerant inventory and increasing energy demand (see 2 discussed previously). Air will leak into low-pressure systems and collect in the condenser, displacing refrigerant vapor and causing higher condenser pressure. In a typical chiller, energy increases about 1.5% for every 1 °F of excess refrigerant temperature leaving the condenser. The percentage savings as a result of eliminating air leaks in refrigeration systems vary, depending on design. The savings tend to be greater for a chillers equipped with a VFD.

   Low-pressure chillers and refrigeration systems use a purge unit to remove air. Problems arise when the purge unit is not functioning properly, or when the volume of leaked air surpasses the purge unit's removal capacity.

   To check for excess air, subtract the leaving condenser refrigerant temperature from the saturated condensing temperature. (Use a standard refrigerant table to convert condensing pressure to temperature.) This yields a measure of air in terms of temperature, since it is the air that accounts for the temperature difference. If the difference is greater than the difference given in the design specifications, take corrective action. Check both for leaks and for a correctly functioning purge unit.

5. *Reduce condenser tube fouling.* Fouling of the condenser tubes—including crystal or scale formation, sedimentation, slime, and algae growth—results from poor water treatment and poor maintenance of the system's waterside. Such fouling causes inefficient heat exchange, increasing the condensing temperature and creating greater head. As the head increases, the compressor motor uses more energy. To maintain the same cooling, the difference between the leaving condenser water temperature and the refrigerant condensing temperature must increase. This temperature difference is called the "small difference." For a typical chiller, every 1 °F reduction in the small difference typically reduces full-load energy consumption by 1%. The percentage savings in refrigeration systems will vary depending on design.

   Condenser tubes should be cleaned periodically to remove fouling and keep the small difference within design specifications—typically 0.5–3.0 °F. Brush cleaning often suffices, although chemical (acid) cleaning or high-pressure water blasting is sometimes necessary. Persistent fouling indicates the need for better water treatment or an automatic tube cleaning system (ATCS), sometimes called an automatic tube brushing system (ATBS).

   The application of sidestream media filtration will greatly assist in the prevention of fouling from suspended particles and dirt.

6. *Maintain the proper flow rate of condenser water.* A reduced water flow rate in the condenser increases head and energy consumption. A 20% reduction in flow rate typically increases full-load energy consumption by 3%.

   Common causes of reduced flow are partially closed valves, clogged nozzles in the cooling tower, dirty strainers and filters, mud in the condenser tubes, and air in the water piping. The flow rate usually can be adjusted to remain within design limits with the discharge valve on the condenser pump. If adjustment does not work, determine and correct the cause of the reduced flow. A rough measure of flow can be obtained from comparing the pressure drop in the heat exchanger with the design pressure drop.

7. *Control demand charges.* Most electric utilities base their demand charges on the maximum amount of energy used during any demand interval. Peak demand occurs during machine startup. The most severe demand typically occurs on a hot summer morning when several machines are started and the chilled water loop or the equipment requiring refrigeration is warm.

   Limiting demand can save significantly in demand costs. Most centrifugal chillers have either manual or automatic demand limiters. Use the limiters to limit demand to about 60% of maximum at start-up. To reduce demand charges even further, stagger the start-up of multiple machines over multiple demand intervals so that demand occurs from one machine per interval. VFDs also reduce current during machine start-up (see item 15 discussed later).

   While demand-limiting strategies can reduce energy costs, actual savings depend on your utility's rate structure.

8. *Maintain motor efficiency.* The compressor motor is the largest energy consumer in a chiller or refrigeration system. The most common degrader of motor efficiency is lack of cooling. If the operating log reveals an increase in current draw without an increase in voltage, the motor probably is not being cooled properly.

   For hermetic motors, check for restricted refrigerant flow or clogged refrigerant filters. For open motors, check for inadequate ventilation/air circulation, obstructed air intake and exhaust openings, and clogged intake filters. In either case, check for dirty oil and oil filters and for loose or corroded electrical connections.

System-Related Opportunities

9. *Operate equipment in proper sequence.* When several chillers or refrigeration compressors serve a common load, careful evaluation of each machine allows you to pick the best operating sequence for a given situation. Evaluate not only the part-load and full-load efficiencies of each machine, but also part-load efficiency of multiple machines versus a single machine. Many opportunities exist when chillers with VFDs are available.

10. *Operate condenser and evaporator pumps in proper sequence.* Sequencing pumps to stop when the chiller stops and isolating idle chillers from the chilled water circuit can save energy.

    Water may be pumped through an idle chiller, which consumes energy unnecessarily and can raise the temperature in the chilled water system by as much as 2.5 °F. Automatic shutoff valves can be added to prevent water flow through an idle chiller. Shutting off the water saves energy and maintains the circulating water temperature and flow at design conditions in the rest of the water circuit. Similarly, with a multicompressor refrigeration system, refrigerant flow should be shut off from any idled compressor.

Another method of saving pumping energy is to use two-speed or variable-speed pumps where a single pump serves several machines.

One note of caution is necessary. Modification to the pumping system can be complex and can affect the system in unexpected ways. Fluid-flow models can be used to evaluate the pumping system and to anticipate problems or facilitate troubleshooting.

Retrofit Opportunities

11. *Interconnect systems.* Many facilities have several separate refrigeration or chiller units, often as a result of facility expansions over the years. At times, some or all of the systems operate at inefficient low partial loads. Interconnecting the systems and centralizing loads may allow system efficiency to be optimized in any situation by shutting down some of the units and loading others more fully.
12. *Use power factor correction capacitors.* Power factor is the phase relation between current and voltage in an electrical system. The ideal factor is 1.0; practical factors typically range from 0.8 to 0.9. A lower factor means heavier currents must flow in the distribution circuit to deliver a given number of kilowatts to an electrical load. Improving the power factor reduces electrical demand on the utility; so many utilities offer lower rates for facilities with higher power factors. Typically, the factor can be improved to a range of 0.88–0.95 by adding correction capacitors. Capacitor installation is fairly simple and provides a potentially significant return on a small investment, depending on the utility rate structure.
13. *Install a telemetry system.* A telemetry system provides 24 h/day electronic monitoring to continuously evaluate operation of the refrigeration or chiller plant and communicate the information to a remote service office. Telemetry can detect and anticipate problems for rapid correction. Telemetry systems have three main purposes: reducing energy consumption by signaling when out-of-design conditions occur, anticipating and signaling problems (such as a dirty oil filter) that can be corrected before a safety shutdown occurs (e.g., from low oil pressure), and limiting the risk of severe problems.

    Telemetry systems do not diminish the need for accurate log keeping and periodic equipment inspections by qualified service personnel. Rather, they enhance an already good maintenance program.
14. *Add heat recovery.* Heat recovery can be applied to many situations where there is a simultaneous need for heating and cooling. In buildings with chiller plants, the applications might include such things as domestic hot water and mechanical cooling. In buildings where such needs are great, heat recovery investment costs should be recovered in three years or less.

There are also many opportunities for heat recovery and heat integration in the chilled and refrigerated portions of industrial processes—for example, heat from refrigeration condensers can be used to drive low-temperature distillation column reboilers. Pinch analysis, which is described in Chapter 26, has been widely used to identify and define opportunities of this type.

15. *Install variable-speed drives.* Guide vanes at the compressor inlet have traditionally been used to limit capacity on centrifugal chillers and refrigeration machines with constant-speed motors (see Chapter 15). The vanes act to limit refrigerant flow and reduce capacity. While this approach reduces overall electricity consumption, savings are not proportional to the capacity reduction because the power used per ton of refrigeration increases at low loads due to drive losses.

    Recent developments inVFD applications allow capacity to be controlled by motor speed, reducing energy consumption, and making better use of the inlet guide vanes. While a constant-speed machine responds to lower condenser water temperature by closing the vanes, the variable-speed machine slows the motor before restricting the vanes, thereby increasing efficiency.

    VFDs provide extremely high part-load efficiencies when the chillers are operated at low-lift conditions. In chillers or refrigeration systems, VFDs adjust the compressor's speed and inlet guide vanes to automatically match the load and operating conditions for maximum efficiency. In general, the slower the compressor's speed, the greater the energy savings from installing a VFD; however, this is only possible at low-lift conditions. As an added advantage, a VFD also controls the inrush current at start-up, reducing stress on the compressor motor.

    Certain system characteristics favor the application of a VFD, including the following:

    - A substantial proportion of operating hours at partial load.
    - Variability of process refrigeration load.
    - A significant proportion of operating hours at reduced condenser water temperature.
    - Chilled-water reset control.
    - High electrical prices.

    Energy savings for replacing a conventional motor with a VFD can approach 30%, although the exact savings depend on hours of operation, loading, and the availability of "colder-than-design" entering condenser water temperatures.

16. *Retrofit with a smaller compressor–motor driveline.* As energy conservation measures are implemented and cooling loads go down, an existing chiller or refrigeration system can become oversized for the remaining cooling load. In cases like this, retrofitting with a smaller driveline to match present loads can reduce operating costs. Downsizing the driveline brings the chiller or refrigeration system in line with load requirements, resulting in more efficient operation.

However, caution should be exercised as smaller compressors have lower flow rates at design temperature difference (ΔT), which may cause velocities in the heat exchangers to be too low for turbulent flow. In these cases, smaller ΔTs at the original flow rates may be a better solution.

Another reason to consider a driveline retrofit is to reduce maintenance costs on an aging system. Since the compressor drive system contains the majority of maintenance-intensive parts, a new driveline alone can significantly upgrade a chiller or refrigeration plant at much lower cost than purchasing an entirely new system. A new driveline also can reduce operating costs by introducing the latest advances in compressor and control design. A driveline retrofit can also be used to change refrigerants. Moreover, retaining the existing heat exchanger shells with a smaller compressor actually boosts efficiency further because the now-oversized shells result in lower compressor head.

17. *Replace chiller or refrigeration unit.* When the desired efficiency simply cannot be obtained from an antiquated system, consider replacing the entire unit. Recent improvements in technology include better tube performance, increased surface area, more efficient compressors, new refrigerants, variable speed, and higher efficiency motors.

    A careful analysis of the present operating cost per ton of cooling should be performed and compared with the anticipated costs of a new system. Then the initial capital expense must be compared with the reduced costs of operation. With the improvements in efficiency found in new centrifugal and reciprocating chillers and refrigeration systems, this changeover can be quite attractive.

18. *Install a chiller or refrigeration plant automation package.* Improved control and automation can greatly enhance the overall performance and energy efficiency of a chiller or refrigeration system. A chiller plant automation package is an energy management system designed specifically to operate the chiller plant—chillers, pumps, and towers—at peak efficiency. Similar automation packages can be applied to process refrigeration systems. A chiller package, depending on its complexity, can perform one or more of the following functions:

    - *Demand limiting*—monitoring overall load and, according to a preset strategy, automatically limiting chiller demand.
    - *Chilled water reset*—resetting water temperature automatically to reduce energy use (see items 1 and 3 discussed previously).
    - *Time of day starting/stopping*—controlling machine start-up and shut down by using outdoor temperature and other factors to anticipate the cooling requirement and operating the machine in the most efficient manner.
    - *Optimized sequencing*—operating the proper mix of chillers, pumps, and towers based on load conditions.
    - *Maintenance needs identification*—spotting and signaling maintenance needs based on performance records so that the chiller plant can be

maintained for peak efficiency. A chiller plant automation control panel permits pushbutton programming of set points, electrical demand peaks, and loading rates. The panel also displays such parameters as system temperatures, pressures, motor current, and oil pressure.

FINAL THOUGHTS

Both heating and cooling are essential for most operations in the process industries. In this chapter, we have addressed the often neglected cooling side of the equation. Applying the ideas presented here can lead to great improvements in the efficiency of your cooling systems, as well as enhancing plant capacity, product quality, and reliability.

REFERENCES

1. Rossiter, A. (2004) Energy management, in *Kirk-Othmer Encyclopedia of Chemical Technology*, 5th edition, Vol. 10, John Wiley & Sons, pp. 133–167.
2. Barr, R. (1986) 18 ways to improve chiller efficiency. *Engineered Systems*, September/October, pp. 29–35.

22

COMPRESSED AIR SYSTEM EFFICIENCY

Joe Ghislain

Ford Motor Company, Dearborn, MI, USA; Ghislain Operational Efficiency, LLC, Milford, MI, USA

Compressed air is often referred to as the "fourth utility" for good reason. In some industrial facilities, air is the largest consumer of electrical energy, and air costs can be as much as 70% of total energy costs. Even in large continuous processing sites, the cost of air is not insignificant. Air can cost from 20 to 40 cents/1000 ft^3, depending on the system and cost of electricity. Regardless of the industry, compressed air still provides meaningful opportunities for savings.

My background is in the automotive industry, which is a major user of compressed air, and I have drawn the examples in this chapter from that industry. However, the principles that are presented here are applicable across all industry sectors that use compressed air, including the process industries.

SYSTEMS APPROACH

As one of the original contributing members and an advanced instructor for the Compressed Air Challenge, www.compressedairchallenge.org/, a voluntary collaboration of stakeholders committed to the improved performance of compressed air systems,

Energy Management and Efficiency for the Process Industries, First Edition. Edited by Alan P. Rossiter and Beth P. Jones.

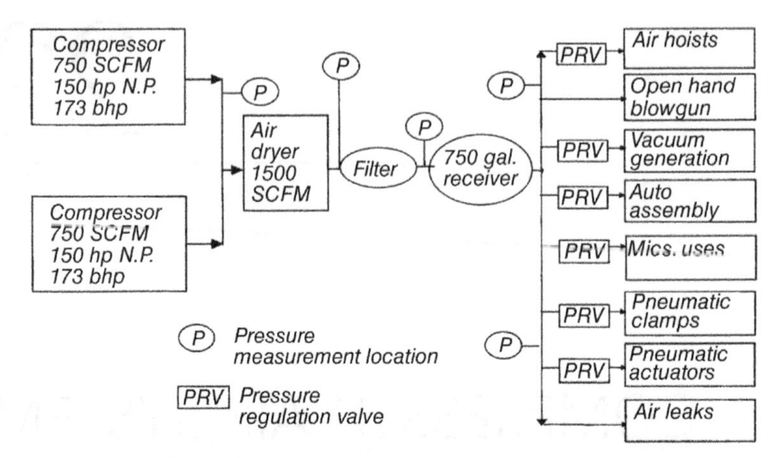

Figure 22.1. Block flow diagram of a typical compressed air system. *Note*: both compressors are lubricant-injected rotary screw. ([*Source:* [1]. Courtesy The Compressed Air Challenge®.])

I am a firm believer that the most effective way to reduce energy consumption is to take a systems approach. For many years, compressed air system efficiency was evaluated on the basis of the compressor and supply-side equipment, with little regard to what was happening on the other side of the equipment room. If the compressors were running efficiently, the compressed air system was considered to be efficient. But what if that efficient air compressor is feeding a system with 50% air leaks? How efficient would you consider that system to be? As with all utilities, the demand side drives the supply side, and compressed air systems are very dynamic. If one only looks at the supply side it limits the opportunities for improvement, so the demand side must also be a focus of attention. So, let us walk through some of the opportunities that can be identified when taking a systems approach.

The first step is to establish a basic system understanding. This can be accomplished by developing a simple block diagram of the compressed air system (Figure 22.1). The diagram starts by laying out the supply side components and then adding the demand side at a high level. This gives a good basic view and allows for better understanding and diagnosis of the system.

ESTABLISHING A BASELINE

The dynamics of the compressed air system are driven by the changing needs of the production or manufacturing process. Understanding the true demand requirements and how best to serve them is the key to an efficient and cost-effective compressed air system operation. Developing a pressure profile, establishing the usage baseline, and calculating operational costs are crucial for improving compressed air system performance. Remember, "If you cannot measure it, you cannot manage it." To develop a pressure profile, take pressure readings after the main supply components, at the beginning and at the end of the main piping distribution system, and at several critical or large points of use. Readings should be repeated over a period of time to establish the high, low, and average system demand. The magnitude of the pressure variation during these times indicates the dynamic behavior of the system demand. The greater the variation, the more dynamic the system. How the compressors and ancillary equipment react to these demands is critical.

Pressure is just one of the system parameters necessary in establishing a baseline. Other elements include electrical usage and airflow, temperature, and dew point. While temperature and dew point may influence operational efficiency, they are more useful in determining system health (maintenance requirements) than energy efficiency, and so will not be addressed here. So, pressure flow in CFM (actual cubic feet per minute), and electrical usage in kW and kWh are the measurements needed to monitor system operation, establish a base line, determine the operating cost, and evaluate and compare improvements.

The efficiency of the compressed air system is based on the relationship between the rate of flow (standard cfm) and power consumption (kW) on a real time basis. Flow meters are used to establish the compressed air usage. The type of meter used and location of the meter should be determined by the size of the system, location of the components, and estimated maximum and minimum flow. The best method for obtaining electrical usage is by using kWh meters. Data loggers are used to capture pressure, kW, kWh, and flow over time, to help draw a more complete picture of the system. For smaller systems, kW can be calculated by obtaining the volt and amperage readings. Once the kW/cfm is established, it can be converted to cost by applying the electrical cost. Because dollars are the universal language in business, converting compressed air usage into dollars is an important step, and it puts the system operation and improvements in terms that everyone can understand. Translating this into dollars per unit of production puts further emphasis on the cost of compressed air.

ENERGY-SAVING OPPORTUNITIES

Now that the baseline is established and the demand profile is known, the operational efficiency improvements can begin. The most common types of opportunities are described below.

Improved Control

The first type of opportunity to consider is improved controls—not only for the air compressors and the supply side components, but also for the end users with the greatest impact on the system.

Different types of compressors have different operational characteristics. Centrifugal and modulating control rotary screw compressors are best suited for use as base-loaded machines because they lose efficiency at partial load. In contrast, other compressors, including multistage reciprocating VSD (variable-speed drive) and variable-displacement compressors, are good "swing" machines because they have better partial-load efficiency and can be used to meet the changing demand of the system. Understanding the type, control, size, and number of compressors, along with the treatment equipment, is important, especially when trying to align supply with demand.

The key to efficient compressor operation is to run only the necessary number of compressors, base-loading (operating at full capacity) as many compressors as necessary and using only one compressor to "trim" (varying load to meet demand). Remember that the most efficient compressor is the one that is shut off. So the goal of any compressor control scheme is to shut off compressors and keep them off for as long as possible and still meet production's needs.

For small compressor systems with stable loads, a simple cascade may work fine, but most systems benefit from some type of control system. For multiple compressors of the same type, the onboard controls can be linked together or a simple sequencing control system can be used. The goal is to run all but one compressor at full capacity. These controllers will not only control the "trim" compressor's turndown but will also turn compressors off and bring them back on line based on the system demand. For compressed air systems with different types of compressors (e.g. rotary screw, reciprocating, and centrifugal), it may be beneficial to separate the control of the different compressor types, but ultimately using more sophisticated sequencing controllers and/or a "global" system (like the energy-management systems described in Chapter 21) is the most practical way to control more than one type of compressor. The sophistication required depends on the number, size, and type of compressors, the dynamics of the system, and most importantly the cost justification for installation. The bigger and more dynamic the system, the greater the need for a more advanced control system and the greater the payback will be to justify the system.

Use of Air Receivers

Controls alone will not solve all system problems, and other methods are needed to help align the supply with demand. Air receivers are used throughout the system to store compressed air to meet peak demand events. Two types, primary and secondary storage, are used to help align the supply with demand by minimizing the adverse effects of high-demand end users on the system. Primary storage is located close to air compressors and reacts to any event in the system. A primary receiver located before the dryer is known as a "Wet" or "Control" receiver, and a secondary receiver located after the dryer is a "Dry" receiver. While both have their own advantages and disadvantages the best practice, as shown in Figure 22.2, is to provide both.

Sizing of the storage vessels should be based on the volume required to meet a demand event, so that the large demand events do not require the startup of additional compressors. Traditionally, the system design rule of thumb for primary storage was 1–3 gal/cfm of compressor capacity, but that was often insufficient to meet system needs. For lubricated rotary screw compressors, an adequate amount of primary storage is particularly important because it allows the time needed to unload, reduces short cycling, and can help delay the startup of additional compressors, all of which lead to a more efficient operation and energy savings. For example, a load/unload-lubricated rotary screw compressor running at 40% capacity will use approximately 27% less power in a

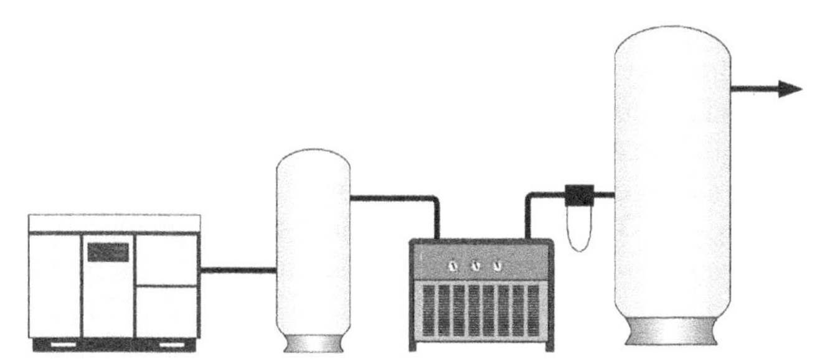

Figure 22.2. A primary or "wet" receiver is located before the dryer, and a secondary or "dry" receiver is located after the dryer. ([*Source:* [1]. Courtesy The Compressed Air Challenge®.])

system with 10 gal/cfm storage than it would in a system with 1 gal/cfm of storage. The updated rule of thumb for effective demand side control is 3–5 gal/cfm of compressor capacity, but caution should be used in applying this rule. As previously stated, the primary receiver size requirement will depend not only on the demand but also on the type of compressors and compressor control.

Pressure/flow controllers can be used with storage for systems or process applications that require tight pressure bandwidths. Compressed air is stored at a higher pressure in the receiver upstream of the pressure/flow controller. The controller monitors the downstream pressure and opens quickly during a demand event, using the stored-compressed air to maintain the line pressure. Typically, the set point pressure is held within $\pm 1\%$.

Drastic demand swings in the system are often caused by high volume, intermittent users. These applications use large amounts of air over a short period of time. One example of this type of use is a baghouse or dust collector. Dust is collected on the filter media inside the baghouse. Once the dust has accumulated, compressed air is used to expand the bag filter and knock off the dust. This uses a large amount of air in a short amount of time, often causing a drastic drop in the compressed air system pressure. Secondary storage can be used to satisfy these demand events. In this case, the air receiver is placed close to the location of the intermittent user, and is sized to meet this individual demand. The stored air is used to minimize the adverse effect of the high demand on the system.

Reducing System Pressure

High volume intermittent demands and other end uses that cause pressure fluctuations demand higher system pressure to meet the end user requirements. Likewise, high-pressure applications also require a higher system pressure. In many systems, these high-pressure requirements make up only a small percentage of the total compressed air consumption but they determine the operating pressure for the entire system. Higher system pressure also increases "artificial demand." Artificial demand is the term used to describe the effect of unregulated air leaks in the system. The higher the pressure, the greater the flow through an air leak. As the system pressure increases, the effects of air leaks increase, causing additional artificially created demand and increased energy consumption. Stabilizing system pressure and addressing high-pressure demands enable a reduction in system pressure. The effects of reducing pressure follows the same rule of thumb as pressure drops in the system: for every 2 psi decrease in system pressure, there is a 1% reduction in energy consumption. For example, lowering the pressure on a 100 hp compressor, running 24/7, from 110 to 90 psi would save approximately $6500/year at 10 cents/kWh electricity cost. Operating the air system at the lowest possible pressure is well worth the effort.

One of the enablers for lowering the system pressure is to address the high-pressure end users. Sometimes the required pressure is "perception" rather than reality. Production may say "we have problems with the equipment if it drops below this pressure." This perception can be caused by a number of things, including system swings, air leaks on the equipment, worn cylinders or tools, and/or pressure drops going to the equipment. If it is a "perceived" high demand, the cause should be identified and addressed. If the end user truly requires higher pressure, should it be allowed to drive up the entire system pressure? A localized high-pressure requirement can be addressed by either modifying the equipment or isolating the end user.

Since equipment modification is very specific and may include modifying controls, changing cylinders, tooling, end effectors and so on, it cannot be addressed in detail in this chapter, but the key is understanding what in the equipment is driving the high-pressure requirement and making modifications to allow lower pressure, or replacing the compressed air by converting the demand to another energy source (e.g., electric or hydraulic).

If the user cannot be modified or replaced, then there are techniques that can be used to isolate higher pressure loads. Air boosters or intensifiers convert lower pressure, higher volume air to lower volume at higher pressure and are designed for intermittent or noncontinuous loads. Booster compressors or separate smaller compressors can be used for continuous or high-duty cycle loads. If multiple high-pressure loads exist, then the system can be split into a high- pressure system and a low-pressure system. The high-pressure system would be served by one or two compressors, and the remaining compressors would be used to feed the low-pressure system, thus allowing the main system to run at a lower pressure. If one compressor is too big for high-pressure loads, a pressure reducer can be used as a "spillover" into low- pressure systems to load up the compressor.

System Design: First Cost Versus Life Cycle Cost

System design is crucial in the ability to not only reduce pressure but also to continue to maintain the lower pressure. Smaller is often cheaper and in most companies the first cost wins out over life cycle costs. Smaller and less expensive cylinders requiring higher pressure are used on many applications instead of using larger, more costly cylinders that require a lower pressure. One extreme example of this involves two transfer presses of the same size, purchased from the same manufacturer, and for two separate locations. One purchaser let the supplier dictate the operating pressure and the other location specified the operating pressure. The result was that the press with the buyer-specified pressure operated at 60 psi and the one with the supplier-dictated pressure operated at

80 psi. Because of the size of the system, if the plant that purchased the 80 psi press could operate at 60 psi, it would save over $300,000 per year.

Pressure drops are not often considered when purchasing or designing equipment and systems. Pressure drops across dryers and filters, or even piping systems, can have a dramatic effect on energy costs, as indicated by the 2 psi = 1% efficiency rule of thumb. The incremental cost of increasing the size of the equipment to reduce the system pressure drop must be analyzed.

The total cost versus first cost analysis is important for compressors as well. Over a 10-year period, the initial cost of an air compressor is typically 5–15% of its life cycle cost, and the cost of energy is 70–90% or more, depending on the cost of electricity. The total cost and benefits must be weighted because 80–90% of the operational costs are determined by the system design and the type of equipment purchased. Total cost, not first cost, is important to an efficient operation. Selecting the most cost-effective option for both supply side and demand side components, based on life cycle cost, is the right decision.

Eliminate Compressed Air Users

Since the demand drives the supply requirements, the greatest area of compressed air efficiency savings lies in the demand side and in the end users. User requirements may not be easy to change because they potentially affect production, but the potential improvements are great. One of the greatest potential opportunities is to eliminate the use of compressed air in some applications—especially where it is used to provide shaft work. Compressed air is not a very efficient energy source. It takes 7–8 hp of input power to deliver 1 hp of work at a compressed air end user, as illustrated in Figure 22.3. The overall efficiency of a typical-compressed air system can be as low as 10–15%.

The efficiency losses in Figure 22.3 can be summarized as follows:

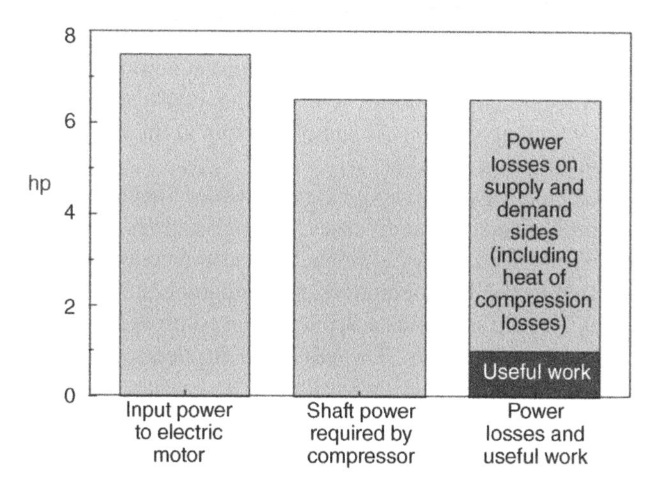

Figure 22.3. Efficiency losses using compressed air to provide shaft work. ([*Source:* [1]. Courtesy The Compressed Air Challenge®.])

- 30 scfm @ 90 psig is required by the 1 hp air motor.
- 6–7 bhp at the compressor shaft is required for 30 scfm.
- 7–8 hp electrical power is required for 6–7 bhp at shaft.

Based on these numbers, the annual energy costs for a 1 hp air motor versus a 1 hp electric motor are **$2330 (compressed air) versus $390 (electric)**, assuming 5 days per week, two shift operations, with electricity at $0.10/kWh.

As seen in Figure 22.3, converting from compressed air to electric power provides opportunities for energy and cost reduction—but where to look and where to start? The first step is to take "inventory" of the applications where compressed air is used. Many times compressed air is used because it is convenient, has a lower first cost, or is simply "the way we have always done it." Take a critical look to see if each use is appropriate and if it can be converted. Blowing, drying, sparging, and cooling are just some examples where compressed air may be used inappropriately. These applications may be by design or just an easy "quick fix" to a production problem. Regardless of the reason, these applications should be analyzed for better alternatives. Low-pressure electric blowers can often be used as a viable option to replace compressed air.

If compressed air is required, then often it can be provided at a lower pressure. Its use can be drastically reduced by either regulating the pressure and/or using the high-efficiency nozzles. These nozzles use atmospheric air along with compressed air to accomplish the task and can reduce compressed air consumption by 75%. Personnel cooling and cabinet cooling are two other examples of incorrect uses of compressed air. Purchasing a fan or a cooling unit for these applications will usually have less than a 1-year payback, and often the payback will be within several months. Vacuum generation, vacuum cups[1], and diaphragm pumps also have a potential to be replaced with electrically driven equipment or to have their air consumption reduced with newer technology. Diaphragm pumps can often be replaced by electric "trash" pumps, or at the very least regulated with speed controls and sump shutoffs. Vacuum pumps can be used for vacuum generation and used in conjunction with vacuum cups. Venturi style vacuum cups can also be replaced with vortex style cups, which operate on the same principle as the engineered nozzles and reduce energy consumption by 75%.

Other electrotechnology conversions to consider include electrically driven mixers, dryers, and blowers. In many cases, DC nut runners are replacing air tools not just because of improved energy efficiency but also increased quality by allowing torque feedback tied to the line operation. Air actuators, stops, and cylinders may also be replaced by electric solenoids and pneumatic controls converted either totally or partially to electronic controls. The advances in electrotechnology have been

[1] Vacuum cups are lift and carry devices used for holding parts during transport. Compressed air is blown over a port in the top of a rubber cup to create a vacuum between the cup and the part, holding it in place until the compressed air is turned off and the part is released. Although vacuum cups can be effective, their energy efficiency is typically poor.

dramatic, and they give us many more efficient options for replacing compressed air applications, so we do not have to do a thing because "that is the way we have always done it."

Maintenance

Maintenance is often considered a "necessary evil," and it is one of the first places that spending is cut. However, a lack of maintenance in a compressed air system costs money. Proper supply-side and demand-side maintenance is critical to efficient operation. Pressure drops across dryers and filters adversely affect system operation. Poor air filter maintenance increases pressure drop, and the rule of thumb of 2 psi equal to 1% efficiency applies. It is critical to change filters and to clean dryers to minimize pressure drops. Inlet air filters are another area on the supply side that is often ignored. Dirty inlet filters act like an inlet valve closing, causing a reduction in air compressor capacity and efficiency. The rule of thumb for inlet air filters is that every 4 in. of water pressure drop across the inlet air filters loses 1% efficiency. But air leaks are by far the biggest loss due to inadequate maintenance.

Leak Repairs

Air leaks are usually the largest cause of wasted energy in a compressed air system, often wasting 20–30% of compressed air production. Even a well-maintained system may have a leak load of 10%. There are two kinds of air leaks: intentional and unintentional. Intentional leaks are designed into the system, added to the system, or used as a matter of convenience or "quick fixes." These include items like cooling, blowing, drying, sparging, and so on, covered earlier in this chapter under the discussion of reducing system pressure. Unintentional leaks are the ones created by "wear and tear" due to normal operation and lack of effective maintenance. The following are some of the most likely areas for air leaks:

- Couplings, hoses, tubes, and fittings.
- Filters, regulators, and lubricators (FRLs).
- Disconnects.
- Open condensate traps or blowdown valves because of fouled condensate traps.
- Point of use devices, like poorly maintained tools.
- Pipe joints and flanges.
- Incorrect and/or improperly applied thread sealants.
- Cylinder rod packing.

The cost of these air leaks is also very significant. For example, as seen in Table 22.1, at $0.10/kWh, losses from a 0.25 in. nozzle @ 100 psi will cost over $18,000/year.

TABLE 22.1. Air Flow Rates from a 100 psi Compressed Air System Through Nozzles of Various Sizes

Leak Rate (SCFM)	Orifice Size (in.)	Annual Cost ($/year) @ $0.05/kWh	Annual Cost ($/year) @ $0.10/kWh	Annual Cost ($/year) @ $0.15/kWh
6.5	1/16	589	1,178	1,766
26	1/8	2,359	4,718	7,077
104	1/4	9,436	18,871	28,307
415	1/2	37,652	75,303	112,955

Note: Costs were calculated assuming 100 psi, motor efficiency of 90%, 8760 operating hours per year, and 100% coefficient of flow. For well-rounded entrance, multiply values by 0.97. For sharp-edged orifices, multiply values by 0.65. All results are approximate.

This highlights the need for a good air leak reduction program. The key elements of a successful air leak program are:

- establish a baseline of both compressed air usage and leak loss,
- calculate the cost of air leaks,
- identify and document the leaks,
- prioritize the leak repairs, fix the leaks, and document the repairs,
- compare baselines, calculate the savings, and "tell the world" or publish the results, and
- repeat the process.

Whether you are using the simple find and fix/seek and repair process or a more elaborate leak tag program, the basics are the same. It is an ongoing process, with the most important part being, **"Fix the Leaks!"**

Leak Detection

To fix the leaks you first have to find them, so another important aspect of the air leak program is leak detection. Did you know that at 100 psi and $0.10/kWh

- a $200/year leak cannot be felt or heard?
- an $800/year leak can be felt, but not heard?
- a $1400/year leak can be felt and heard?

This reflects the importance of the type and effectiveness of the leak detection method used. The most common and effective method used is ultrasonic leak detection, which is described in Figure 22.4.

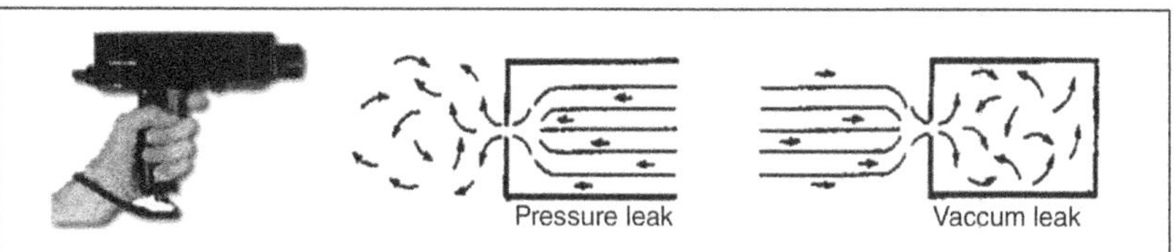

- During a leak, a fluid (liquid or gas) moves from a high pressure to a low pressure.
- As it passes through the leak site, a turbulent flow is generated with strong ultrasonic components, which are heard through headphones and seen as intensity increments on the meter.
- It can be generally noted that the larger the leak, the greater the ultrasound level.
- Ultrasound is a high frequency, short-wave signal with an intensity that drops off rapidly as the sound moves away from its source.
- The leak sound will be loudest at the leak site, which makes locating the source (i.e., the location) of the leak quite simple.

Figure 22.4. How ultrasonic leak detection works. ([*Source:* [1]. Courtesy The Compressed Air Challenge®.])

As seen in Figure 22.4, ultrasonic leak detection uses high-frequency sound waves to locate leaks that the human ear cannot hear and the hand cannot feel. This makes it possible to find and fix those $200/year air leaks. Furthermore, because the larger leaks produce a greater ultrasonic noise level, air leaks can not only be found but also prioritized.

Heat Recovery

Heat recovery is another opportunity for added efficiency in a compressed air system. Eighty to 93% of the electrical energy used by an air compressor is converted into heat. In many cases, 50–90% of this heat can be recovered by properly designing a heat recovery system to take advantage of the available thermal energy and use it to heat air or water. Areas of opportunity for heat recovery include

- space heating (applicable only in cold weather),
- water heating for both domestic and makeup water,
- drying compressed air,
- industrial process heating (typically year-round),
- makeup air heating,
- combustion air preheating, and
- boiler makeup water preheating.

Depending on the type of compressor, heat can be recovered from the intercoolers, aftercoolers, oil coolers, and water jackets (cylinders). Strategies for capturing heat from air-cooled rotary screw compressors include adding ductwork with auxiliary fans to the compressor package to recover the heat for space heating. For water-cooled compressors, heat exchangers are installed to recover the heat and use it for space heating and/or to produce nonpotable (gray) or potable hot water. Lube-free compressors can use "heat of compression dryers," which recover the heat generated by compressing the air and use it to dry the compressed air. Compressors using water-cooled motors offer additional opportunity for heat recovery because the heat produced by the motor can also be recovered. Engine-driven compressors offer even more opportunities for heat recovery, because heat can be recovered from engine jackets and engine exhaust.

A heat recovery project was implemented at the Ford Chicago Assembly plant. The project included

- installing new equipment (four 200,000 cfm direct fired units, four indirect unit heaters),
- utilizing existing equipment (1 direct fired, 24 indirect fired, 18 unit heaters, and 6 infrared),
- installing a new direct contact hot water heater,
- installing three new 5000 cfm water-cooled centrifugal compressors and dryers, and
- installing a global control system.

The system-wide control scheme linked the three water-cooled compressors with a heat recovery system, allowing the heat to be recovered from the motors, intercoolers, aftercoolers, and oil coolers, and from the water-cooled refrigerated air dryers. There are three modes for the three air compressors: (1) full production, 70 °F, three compressors running; (2) exhaust off, 70 °F, two compressors running; and (3) exhaust off, 65 °F, one compressor running. In this way, only the necessary compressors are used, saving the energy of the other one or two when the plant is not in full production mode. This positive pressure heating system replaced a 70-year old steam system and upgraded the compressed air system, saving over $1.8 million in energy, with the air compressors being a large contributor.

Case Study: Benefits of Compressed Air Systems Approach

So, how can these actions translate into savings? When the compressed air systems approach was applied at the Ford Woodhaven Stamping plant, significant savings were achieved. The team implemented an air leak repair program that identified and repaired a significant number of leaks. Counterbalance cylinders were rebuilt and leaking seals on the stamping press die automation valves were replaced. Orifice plates used for measuring flow were replaced with low-loss venturis and averaging pitot tubes to reduce pressure drop. A perceived high-pressure requirement for the robot venturi vacuum cup was removed, eliminating the need for high-pressure satellite compressors and their associated dryers. The header pressure was lowered, reducing the artificial demand. The results were that the compressed air consumption was reduced by approximately 18%. One 800 hp and six small (~30 hp each) satellite compressors with dryers were shut down. The controls on the remaining compressors were adjusted to reduce energy consumption. The aggregate energy savings were 7.9 million kWh, reducing the site's energy costs by $360,000, or more than 3.5%.

Employee Engagement

The final area of opportunity in this chapter relates not only to compressed air systems but also to all energy efficiency, and that is people and employee involvement. People use compressed air and are a big part of the equation to reduce energy. While I have several examples of this during my career at Ford Motor Company, the Monroe Stamping Energy team is by far my favorite and the most effective. What would you say if I told you that you could reduce compressed air usage and energy consumption by buying things like jackets, hats, and key chains? Would you buy it? Well that is what the Monroe Stamping plant did! The team took the following actions:

- The Hourly Energy Team (in their red jackets so they would stand out) implemented an aggressive energy awareness and air leak repair program. The team
  - gave away buttons, key chains, hats, and tee shirts for reporting and getting air leaks fixed as well as for passing the "Red Coats" energy audit,
  - posted "Leak Boards" throughout the plant to track progress,
  - placed equipment shutdown stickers on equipment to show the cost of leaving equipment on, and
  - used the Ford Communication Network to broadcast messages on energy costs throughout the plant.

And they achieved these results:

- Air use declined from 17.4 to 9 million ft^3/day.
- Nonproduction usage was reduced from 5400 to <600 cfm.
- Electricity savings of over $2000/day were achieved.
- Most importantly, the program created a cultural change in the plant for awareness of energy cost, usage, and waste.

CLOSING THOUGHT

In conclusion, taking a systems approach will drive compressed air energy savings, but never forget that it truly is people that make the difference, so get them involved!

REFERENCE

1. Fundamentals of Compressed Air Systems (2013) *Training course, the compressed air challenge®*, Available at: http://www.compressedairchallenge.org/.

23

LIGHTING SYSTEMS

Bruce Bremer

Bremer Energy Consulting Services, Inc., Union, KY, USA

Lighting systems are a costly energy consumer in most facilities, but their energy cost is often overlooked or underestimated. Typical lighting systems are installed for human comfort, safety, and quality, and have a major impact on the operation of the facility as well as on the process equipment. In some types of manufacturing facilities, the annual energy costs for lighting can be as much as 15–20% of the total annual energy spent. Lighting types include incandescent, fluorescent, induction, light-emitting diode (LED), metal halide, sodium, and many more. The intent of this chapter is to focus on typical lighting systems—indoor and outdoor—used in a facility, including the office areas, and to identify some basic energy-saving opportunities.

A facility lighting system directly affects the comfort, mood, productivity, health, and safety of its occupants. Moreover, as the most visible facility system, it also directly affects the aesthetics and image of the facility. Although such effects are difficult to quantify, the effect on employees should be considered as part of every lighting system. Improved lighting enhances visual comfort, reduces eye fatigue, and improves performance on visual tasks, and research efforts are helping to pin down these benefits. Lighting also contributes to the safety of occupants and the security of facilities. Emergency lighting must be available during power outages, and minimum levels of light must be available at night when most of the main lighting is turned off. In addition, safety codes

Energy Management and Efficiency for the Process Industries, First Edition. Edited by Alan P. Rossiter and Beth P. Jones.

require exit signs to highlight escape routes during fires or other emergencies. Outside lighting and indoor lighting can deter crime and permit employees to move safely through the facilities or to their cars. Outdoor light levels and indoor light levels can depend on local ordinances, but both the Illuminating Engineering Society of North America (IESNA) [1] and ENERGY STAR [2] provide general lighting guidelines that should be followed.

Concentrating on the proper operation and maintenance of a typical lighting system will ensure the highest operating efficiency and lowest energy cost. Many functions of lighting systems are important to energy efficiency; the following functions will be analyzed in this chapter:

- Supply-side with demand-side alignment
- Operational control
- Maintenance
- Technologies today and tomorrow

SUPPLY-SIDE WITH DEMAND-SIDE ALIGNMENT

In any production system, the supply requirements are driven by the demand-side requirements, and the lighting system is no exception. Knowing the demand requirements of the manufacturing facility is critical to understanding how the lighting system should function. Individual lights can operate and function independently, but all the lights combined together need to be evaluated and function as an overall system. Understanding the needs of the facility is a critical first step for developing appropriate system control. This understanding sets the stage for the following questions: What are the lighting requirements of the facility? What lights are needed at what time? How can I control the lights in the facility?

Lighting systems must operate in many different modes and conditions. These conditions are critical information to understand and use to maintain the facility lighting levels and overall facility environment while operating the system efficiently. As when considering any type of improvement, understanding the current baseline operation of the system is critical before making any changes to the system.

OPERATION

In general, lighting systems are very similar from facility to facility and these systems always seem to have opportunities for energy efficiency improvements. Two key focus areas for improving lighting system operations are manual control and automatic control.

Manual Control

In many facilities the lights operate 24 × 7, even if the facility does not operate continuously, to ensure lighting levels are maintained in case anyone is working in

TABLE 23.1. Lighting Operating Time Saving Example

Parameters	Mode	Mon–Fri	Sat–Sun	Holidays
Time	Production	On: 6:00–2:00 a.m.		
	Nonproduction	Off: 2:00–6:00 a.m.	Off: 6:00–6:00 a.m.	Off: 6:00–6:00 a.m.

any location on the site. However, individual lights usually serve different areas and have different requirements. With the facility baseline information gathered in the previous step, it should be easy to understand the lighting profiles of all the different locations. Understanding this profile will determine when the lighting system needs to turn on or turn off to meet the requirements of the facility. As an example, if the facility production mode of operation is from 7:00 a.m. to 1:00 a.m. or 18 h, Monday through Friday, when should the lighting system turn on and turn off? Breaking the facility into smaller segments and time frames provide the greatest energy-reduction opportunities (see point "Zoning" discussed later). The key parameters to evaluate are production time, non-production time, and weekend time. Determining operating modes, as shown in the example in Table 23.1, will help quantify the possible energy savings and create a standardized method for controlling lighting operations.

Based on the information shown in Table 23.1, and using this operational methodology, it is very easy to systematically match the lighting need with the facility need. With the implementation of a lighting schedule control concept, it is not uncommon to achieve annual energy savings of 20–30%.

On/Off Switch Control. The basic concept of turning equipment on and off also has much merit and impact on lighting systems. Switches are the old-fashioned control, where everyone has the responsibility to turn off or on the lights in their area of work. The key focus for energy efficiency is to train people to turn on only the lights that are needed in the work area and to turn off the lights when they are not needed. This sounds very basic, but implementation can be a challenge. Key operating components for effective on-off control are zoning, lighting requirements, employee engagement, and plant visualization.

ZONING. Different areas of a facility have different lighting needs. A large production area might need the majority of lights on the entire day or week to meet production needs. But a warehouse area may only need lights on when workers are present. A large open office might need the lights on all day, but could save energy by turning most of the lights off at the end of production, nights, and weekends. Well-planned lighting zones and lighting control strategies should be grouped together and switched together based on how much light is needed, where the light is needed, and when the light is needed. The basic information on facility lighting needs that was developed earlier will aid in the zoning configuration and operational use of the lighting systems.

LIGHTING REQUIREMENTS. Light location and light levels (or foot candle requirements) are also very important when focusing on lighting-reduction opportunities. Optimizing light location often provides a very short payback with minimal investment. A careful analysis of the facility light location and the floor equipment location is the first place to start. Ask yourself some basic questions: Is the light located in an area that supports the lighting needs on the floor? Is the light even needed? If the answer is no, then relocate the light or remove the light. Many plant lighting systems are designed as a grid network with minimal consideration of the light location compared to the equipment layout on the floor. Matching the light location to the lighting needs on the floor is very important for good energy management.

The requirements for adequate lighting levels are also a consideration. Guidelines require different foot candle intensities for different activities in any facility. As an example, more light is needed if an employee is inspecting a product than if an employee is driving a fork truck. Installing lights that have the correct foot candles needed for the job can also provide great energy-saving opportunities. For example, lighting level guidelines can be located at the IESNA [1] or at ENERGY STAR [2].

EMPLOYEE ENGAGEMENT. Manually controlling lights to reduce energy consumption requires employee participation and engagement. Changing employee behavior is not always an easy task because this is very much related to habit and culture. The first step is to educate employees so that they understand the cost associated with lighting the facility or their work area of the facility. This education can be very basic:

- The first step is to communicate the concept that turning a light out for so many hours equates to a certain dollar savings. The dollar savings will have a greater impact than any type of engineering units such as kWh.
- The second step is to ensure that systems or switches are in place to turn the lights off or on. These switches must be in an accessible and convenient place.
- The third step is to establish some type of standardized work procedure that makes it very clear who has the responsibility for the lights and when the lights should be turned off or on.

PLANT VISUALIZATION. It is very important to design the facility layout and show what lighting zone covers what area of the plant and what switches control what lights. A facility lighting zone map as a visual aid will provide the employees an easy way to identify which lights need to be turned on and off. These visual aids can be posted by each light panel and/or switch for easy identification. A typical example is shown in Figure 23.1.

Automatic Control

Manually controlling the lighting system so that employees are able to turn lights on and off represents only one part of the potential for maximizing lighting energy savings. In situations where lighting can be on longer than needed, left on in unoccupied areas, or

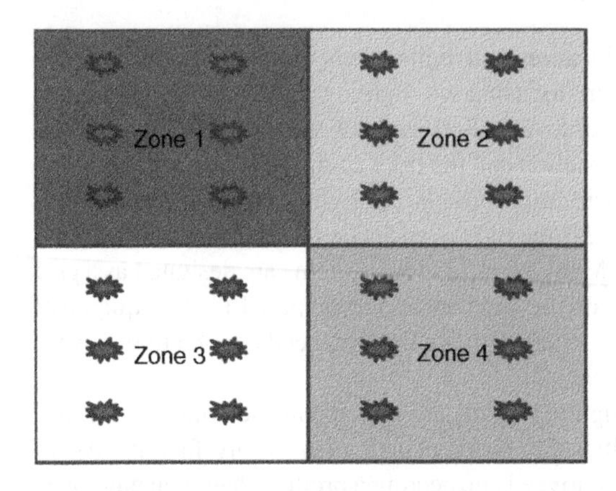

Figure 23.1. Typical facility lighting zone layout.

used when sufficient daylight exists, consider installing automatic controls as a supplement or replacement for manual controls. Automatic control switches such as occupancy sensors, photocells, timed switches, dimming control, addressable ballast, wireless controls, and demand response, or a combination of several, provide additional opportunities. We will analyze each of these seven options in more detail. With the implementation of automatic lighting control, it is not uncommon to achieve additional annual energy savings of 20–30%.

Occupancy Sensors. Occupancy sensors save energy by automatically turning lights off when areas are vacant. Occupancy sensors add convenience by turning lights on when the device senses human presence through motion or by body heat. These devices can range from a simple passive infrared detector built into a light switch to a networked detector that controls lights and also contributes to an integrated facility automation strategy. Sharing occupancy/vacancy information with other components of a facility automation system can also save additional energy. These sensors can be wall mounted to replace an existing light switch, or ceiling mounted or fixture mounted for ease of operation. It is very important to understand the lighting system zoning so that it is easy to determine the area impacted when the lights are turned off. It is also important to review where and how many emergency lights are available in the space being controlled by the occupancy sensor to ensure enough light is available in case of an emergency.

Photocells. Photodetectors operate based on the concept of measuring light levels. This type of control works well with outdoor lighting, either for exterior facility wall lights or exterior parking lot lights. This type of control operates per the following guideline: When the outside natural light diminishes, the photocell can turn on exterior lights to brighten the areas needed, but when the outside natural light increases, the photocell can turn off the exterior lights. Photocells can also be used with indoor lighting, but the control system becomes much more complicated and complex. Photo cells are a type of automatic control system that works well and can provide great energy-saving opportunities.

Timed Switches. Timed switches operate based on either elapsed time after triggering or on programmed schedules using a time clock.

Elapsed time switches, also called timer switches, typically fit into or over a standard wall-switch box and allow occupants to turn lights on for a predetermined period of time that is set either by the occupant or by the installer. Lights go off at the end of that interval unless the cycle has been restarted by the occupant or manually turned off sooner. Time intervals typically range from 10 min to 12 h. Elapsed time switches are much simpler to specify than occupancy sensors, are less prone to faulty user adjustment, and cost less. Elapsed time switches can be mechanical or electronic. Mechanical units are typically set by the user and electronic switches are typically set by the installer. These electronic devices look like conventional toggle switches, so occupants are usually unaware of the presence of the device. Elapsed time switches are also an easy and economical means of complying with energy codes that call for automatic lighting controls.

Clock switches control lights by turning them on and off at prearranged times, regardless of occupancy. They are most useful in locations where occupancy follows a well-defined pattern. These devices cost relatively little to install and can control large loads with a single set of contactors. Equipment can consist of mechanical devices—motors, springs, and relays—or sophisticated electronic systems that handle several schedules simultaneously ([3], p. 26).

Dimming Controls. Dimming can maximize the energy savings of lighting control strategies by lowering light levels to the most effective level. Fluorescent lighting in typical office areas, although efficient, typically does not respond well to dimming. However, the new fluorescent high bay lighting systems in facility production areas or warehouses work well with dimming control. If people are not present and the detectors do not sense movement, the lights can dim to approximately 60% of their normal output. New LED technology offers enhanced dimming that not only saves energy but also improves the comfort of a space by providing lighting levels that best suit the use of the room.

Addressable Ballast. Most dimming is accomplished by controlling banks of dimmable ballasts together. Digitally controlled ballasts with control protocols provide more flexibility. Every ballast is assigned an identifier, or "address," and can be controlled individually or in clusters that can easily be grouped. With some systems, two-way communications are also possible. This capability not only gives users the ability to tailor local lighting conditions to their individual needs but also provides energy managers a tool for tracking and controlling energy use and responding to load-shedding signals. Conventional low-voltage controlled-dimming ballasts can be added to digital lighting control systems via special interfaces ([3], p. 29).

Wireless Controls. One solution to the expense of running wires is wireless lighting controls. A typical wireless lighting control system consists of a set of sensors, actuators, and controllers that communicate via radio waves rather than wires. Although wires are still required for the lighting equipment itself, using radio waves instead of wires to transmit control signals offers a number of potential advantages, in terms of both the ease of installation and maintenance and the flexibility. Newer, more capable wireless systems, some of which are available today, can broaden the wireless lighting controls market considerably if costs can be decreased. One promising technique uses a concept known as a mesh network, which is a decentralized set of wireless nodes that are linked to one another to form a self-organizing, self-healing network. Control is split up among the different nodes so that there are multiple, redundant paths throughout the network. Each device on the network is designed to transmit over short distances, which reduces power requirements and minimizes the potential for interference ([3], p. 30).

Demand Response. To reduce the electrical peak demand and reduce energy cost, many utilities offer load shedding or demand response programs in exchange for incentives or lower energy costs. Networked lighting control systems can take advantage of these programs by temporarily reducing light levels in response to a signal from the utility company. Many companies do not have the ability to implement such technical programs in the response time requested from the utility company, but even manual intervention can reduce energy use if there is a formal program established and executed when energy reductions are required.

MAINTENANCE

There are many opportunities for energy efficiency improvements in lighting system maintenance. Maintenance can encompass cleaning, adjusting, checking components, calibrating components, and many other items. We will consider opportunities in both preventive and predictive maintenance. In preventive maintenance, we will focus on cleaning and adjusting lighting preventively, and on bulb replacement in predictive maintenance.

Preventive Maintenance

As lamps and luminaries age, they accumulate dirt and dust that reduce light output. Lighting systems can accumulate dirt even faster than surrounding surfaces due to the static charge associated with the lamps. This avoidable reduction in light output can prompt calls for lamp or fixture replacement in advance of the actual need for replacement. Cleaning lamps, luminaries, and room surfaces will help retain the original designed light output. This action itself can have minimal energy savings, but it does help the lighting system avoid degraded light output, which can reduce occupant complaints and help defer equipment replacement or upgrade costs. A further benefit can be that fewer occupants will have a need for supplemental lighting such as task lighting, desk lamps, or specialty lights, which will also save energy.

Predictive Maintenance

Reactive maintenance (i.e., replacing lamps when they fail) may not effectively keep illumination at the designed levels, but a proactive maintenance program can be important to the success of any lighting system and its energy efficiency. Some of the tasks associated with predictive maintenance are as follows:

- Regular cleaning of lamps and luminaries
- Scheduled group revamping of luminaries
- Regular inspection and repair of lighting equipment
- Inspection and recalibration of lighting controls
- Reevaluation of lighting system for upgrades

Depending on the type and size of the facility, there can be a cost benefit to initiating a group relamping maintenance schedule in place of an on-demand lamp replacement policy. The major element of savings associated with lamp replacement is in the maintenance labor cost. However, there are also some energy-saving opportunities with this strategy. An important factor in determining the most efficient scheduled relamping plan is the rated life of a lamp provided by the manufacturer. This is the point where 50% of a group of lamps are expected to burn out under normal conditions. The most effective group relamping point is when lamps in an area start burning out on a regular basis. This is commonly at 70–80% of rated lamp life. When 10–15% of the lamps burn out, the optimum relamping period should be adjusted to actual operating conditions and occupant needs. Group relamping will also maintain improved light levels by replacing lamps before their lumen output degrades further ([3], pp. 31–33).

LIGHTING TECHNOLOGIES TODAY AND TOMORROW

Efficient lighting technology is continuing to develop, and it is changing very quickly. There are many types of lighting technologies that can be applied to a variety of different applications. It can be confusing for businesses and facility operators to stay up-to-date on the latest lighting technology options. Right now, the product life cycle for energy-efficient lighting technology is in the range of 6 months, which means that available technology increases potential savings every 6 months. If you cannot find a lighting retrofit option that suits you now, just wait a while and you are likely to find something soon. For this reason, it is critical to keep a focus on what is available in the marketplace.

Lighting technology is quickly becoming more efficient and more affordable. For example, compact fluorescent light (CFL) bulbs cost about $5 just a few years ago, and today they often can be found for as little as $2. Light-emitting diodes are still more expensive than CFLs, but the price is quickly dropping as the market grows and more companies offer them in their product mix. Solid-state lighting is nothing new and has been in use for half a century in applications such as traffic signals. White-light illumination is the kind of light most used in living spaces, work environments, streets, and parking lots, and is responsible for most of the consumed lighting energy. LEDs for white-light applications are quickly getting more efficient and reducing costs. Right now, lighting developers have achieved about half the theoretical limit of lumens per watt (how much light is released for every watt used by the fixture) in LEDs. This means that

LEDs still have huge potential for energy improvements, which will further boost the cost-per-watt ratio and provide savings. LEDs use a minimal amount of energy compared to their counterparts. They use one-fifth the amount of energy as an incandescent light source and about half that of low-energy CFL technology. At this point, the upfront costs of LEDs are still higher than the upfront costs of their fluorescent counterparts, but LEDs actually save the end user money over the life of the bulb. Reviewing the total cost of ownership can overcome the higher purchase price of LEDs. Currently, LEDs last for 25,000–50,000 h, and some can last beyond 100,000 h. This is a huge improvement over incandescent bulbs, which last approximately 2000 h, and CFL bulbs, which last approximately 10,000 h. The long life of the bulb not only reduces the number of bulbs needed in the future but also means the bulbs are more reliable over their lifetime. LEDs require less maintenance and fewer repairs, which can save money. LEDs work well in outdoor lighting applications such as parking lots and parking garages and in commercial spaces such as office areas ([3], pp. 12–22).

Fluorescent bulbs use 20–40% less electricity than the traditional incandescent bulbs and have a 10,000 h life. These types of tube lights are very common in commercial, industrial, and office settings and provide businesses with an easy way to save money and energy. There are many types of fluorescent lights in a variety of sizes, shapes, and colors. The fluorescent bulbs and light fixtures are becoming much more common in facilities for building lighting and are quickly replacing the metal-halide-type light fixture.

Another type of fluorescent light that offers energy savings is induction lighting. Induction lighting is a type of fluorescent light and is more energy efficient than the comparable existing technology such as high-pressure sodium or metal halide. It is comparable in energy efficiency to existing LED products. Induction lighting has a longer life span than many existing technologies, and about the same life as LED technology. Induction lighting today offers a good solution for parking lots and other outdoor lighting applications and can turn on immediately, which is an improvement over other fluorescent lights that have a start-up time before the maximum light yield is reached. They also work well in cold temperatures, but cost five to six times more than a metal-halide light. Induction lighting last 50,000–100,000 h and will consume only half the energy of a metal halide, which makes the system less expensive over its lifetime.

SUMMARY

Lighting is a necessary system for any facility, but lighting systems also have many opportunities for energy savings. The typical philosophy is to let the lights operate 24/7 and 365 days a year, regardless of production time or nonproduction activities. As can be seen from the above examples, there are many opportunities for energy saving related to how the lighting equipment is operated and maintained. The basic concept is to take the lighting equipment already in place and make it more energy efficient and thus obtain the energy savings. Based on this philosophy, it is very cost-effective to focus on operational changes and techniques. Energy efficiency can be obtained through the philosophy of only operating the lighting equipment needed to meet the production needs.

The technology of lighting systems is improving every day and becoming more robust and cost-effective. The continuous activity of monitoring newer lighting technologies is critical, but it is just as important not to randomly replace lighting equipment just because a new technology has come on the market. Many of the new technologies are still not proven and installing these technologies in your facility could have a negative consequence.

The four concepts discussed above—aligning supply side with demand side, operating improvements, maintenance improvements, and new technologies—will provide the framework needed to enhance energy efficiency savings in your lighting systems.

REFERENCES

1. Illuminating Engineering Society (2011) *IES Lighting Handbook*. Available at http://www.loveitlighting.com/ies_candle.html.
2. U.S. Environmental Protection Agency (2010) *Plant Lighting Level Best Practice*. Available at http://www.energystar.gov/index.cfm?c=in_focus.bus_motorveh_manuf_focus.
3. U.S. Environmental Protection Agency (2006) *Energy Star Building Upgrade Manual*, Chapter 6, p. 26. Available at http://www.energystar.gov/buildings/tools-and-resources/energy-star-building-upgrade-manual-chapter-6-lighting.

24

HEATING, VENTILATION, AND AIR-CONDITIONING SYSTEMS

Bruce Bremer

Bremer Energy Consulting Services, Inc., Union, KY, USA

Heating, ventilation, and air-conditioning (HVAC) systems are a costly energy source used in most facilities, but that energy cost is often overlooked or underestimated. An HVAC system is usually installed for human comfort, but can also have a major impact on the operation of the process equipment. In many manufacturing facilities, the HVAC annual energy costs are in the range of 10–15% of the total annual energy spent. The basic energy cost of an HVAC system includes not only the electrical cost of the fan and pump system but also the heating and cooling energy source costs to support the operational needs. There are many varieties and types of HVAC systems, including heating-only units, cooling-only units (see Chapter 21 for a discussion of cooling and chilling), package units, makeup air units, air rotation units, and many more. The intent of this chapter is to focus on a typical HVAC system used in a manufacturing facility, including the office areas, to condition the environment. Typical HVAC equipment components include fans, motors, pumps, heating section, cooling section, dampers, filters, ductwork, piping, and monitoring and controls, which can be seen in Figure 24.1.

Concentrating on the proper operation and maintenance of an HVAC system will ensure the highest operating efficiency and lowest energy cost. Many functions of an

Energy Management and Efficiency for the Process Industries, First Edition. Edited by Alan P. Rossiter and Beth P. Jones.

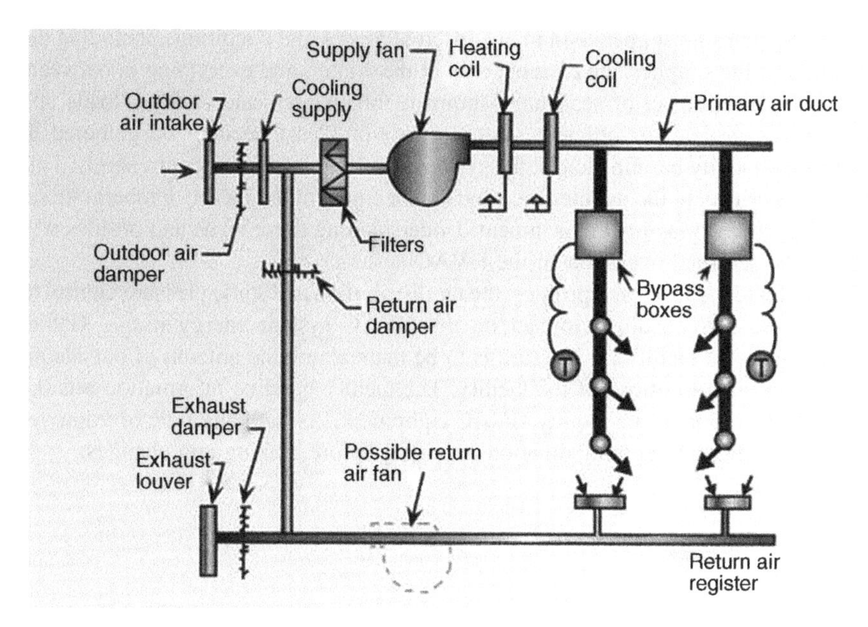

Figure 24.1. Typical HVAC equipment layout. (*Source:* Courtesy of Technology Transfer Services, Inc.)

HVAC are important to energy efficiency, but the following four functions will be analyzed in this chapter:

- Supply side with demand side alignment
- Operation
- Maintenance
- Recommissioning

SUPPLY-SIDE WITH DEMAND-SIDE ALIGNMENT

As in any production system, the supply requirements are driven by the demand-side requirements, and the HVAC system is no exception. Knowing the demand requirements of the manufacturing facility is critical to understanding how the HVAC system should function. Individual HVACs can operate and function independently, but the combination of all HVACs should function and should be evaluated as an overall system. Understanding the needs of the facility is a critical first step for developing appropriate system control. This understanding sets the stage for the following questions: What are the temperature set point requirements of the facility? What are the air distribution requirements of the facility? What are the operating parameters of the facility?

HVAC systems must operate in many different modes and conditions, including the extreme heat of the summer, the extreme cold of the winter, and everything in between. Understanding the impact of seasonal—summer, fall, winter, and spring—loads and requirements is critical for efficient system operation. Data need to be gathered to understand the facility baseline loads, the system profiles, and the air requirements. This information is critical to the maintenance and management of the facility temperature set points and the overall facility environment. Understanding these loads and profiles will also aid in the seasonal operation of the HVAC system.

In addition to the loads and profiles, the air distribution and static pressure control in the facility also have a direct impact on the HVAC system energy usage. Gather information on what facility air pressure is to be maintained, the amount of outside air needed, and in what locations of the facility. The facility baseline information sets the stage for the operation of the facility HVAC equipment. As with any type of improvement, understanding the current situation is critical before making any changes.

OPERATION

In general, facility HVAC systems are very similar from facility to facility, and these systems always seem to have many opportunities for energy efficiency improvements. Two key focus areas for improving operational components of the HVAC system are operating time/set point and operating parameters.

Operating Time/Set Point

At many facilities, all the HVAC units run 24 × 7, even if the facility itself does not operate continuously. Moreover, individual HVACs within the overall facility HVAC system usually serve different areas and have different load requirements. The basic operational information previously gathered should identify the HVAC profiles of all the different locations in the facility. Understanding this profile will determine when the HVACs should turn on or turn off to meet the requirements of the facility.

As an example, if the facility production mode of operation is from 7:00 a.m. to 1:00 a.m. or 18 h, Monday through Friday, and the temperature set point is 65 °F in the winter and 75 °F in the summer, when should the HVAC system turn on and turn off? Breaking the facility into smaller segments and time frames provides the greatest energy-reduction opportunities. The key factors that need to be evaluated are production time, nonproduction time, weekend time for each season, and temperature set points for each season. Dividing the above parameters into modes and seasons, as shown in the example in Table 24.1, helps quantify the possible energy saving and create a standardized method for HVAC operation.

Based on the information shown in Table 24.1, and using this operational methodology, it is very easy to save a large percent of the HVAC energy usage based on a simple systematic approach. With the implementation of an HVAC schedule control concept, it is not uncommon to achieve annual energy savings of 30–40%.

TABLE 24.1. HVAC Operating Time Savings Example

Parameters	Mode	Season			
		Summer	Fall	Winter	Spring
Time	Production	6:00–2:00 a.m.	6:30–1:00 a.m.	6:00–2:00 a.m.	6:30–1:00 a.m.
	Nonproduction	2:00–6:00 a.m.	1:00–6:30 a.m.	2:00–6:00 a.m.	1:00–6:30 a.m.
	Weekend	7:00–7:00 a.m.	7:00–7:00 a.m.	7:00–7:00 a.m.	7:00–7:00 a.m.
Temperature set point in Fahrenheit (°F)	Production	65 Heating	65 Heating	65 Heating	65 Heating
		75 Cooling	75 Cooling	75 Cooling	75 Cooling
	Nonproduction	55 Heating	55 Heating	55 Heating	55 Heating
		85 Cooling	85 Cooling	85 Cooling	85 Cooling
	Weekend	50 Heating	50 Heating	50 Heating	50 Heating
		90 Cooling	90 Cooling	90 Cooling	90 Cooling

Operating Parameters

Operating parameters—typical equipment set points used in the control of an HVAC—are also very important to evaluate and analyze, and can provide additional energy-saving opportunities. The HVAC parameters include mixed air temperature, return air temperature, static pressure, discharge air temperature, damper position, heating and cooling temperatures, and many others. Three key operating parameters are discussed in this chapter: damper position (damper economizer control), static pressure (variable speed drive operation), and discharge air temperature (resetting discharge supply air temperature control).

Damper Economizer Control

Damper economizer control has some great benefits for energy efficiency, but let us first look at some of the operational basics for this type of control. When a typical HVAC system is trying to cool the facility and the outside air temperature is cooler than the inside temperature, the controls should open the outside dampers and use the outside air for free cooling. It is also important to analyze the outside air humidity or enthalpy so that the HVAC is not bringing in saturated air from the outside and creating a humidity problem in the facility. When the outside air temperature gets too warm, the controls should close the outside air dampers to a minimum and use the return air so that the HVAC is not cooling warm outside air.

The other end of the spectrum of damper economizer control is when the HVAC is heating the facility. The outside air dampers should be set to a minimum outside air position so that the HVAC is not trying to heat 100% cold outside air but rather warmer return air and a small percentage of outside air. Minimum outside air requirements for the facility can be determined from the facility specification requirements or the American Society of Heating, Refrigerating and Air-Conditioning Engineers (ASHRAE) standards. With the use of an automatic control system for the damper economizer control, it is not uncommon to achieve annual energy savings of 5–10%.

Variable Frequency Drive Control

Variable frequency drives (VFD) provide great benefit for energy efficiency improvement, but first we need to establish some of the operational basics. A VFD is an electric control module that controls the speed of the HVAC motor, which in turn controls the speed of the fan. Many HVAC systems operate at 100% speed at all times, whether in the summer mode when much air is required for cooling or in the winter mode when much less air is required for heating. The purpose of the VFD is to automatically adjust the HVAC motor speed based on need, which in turn adjusts the air volume of the fan based on a pressure-sensing device in the facility. The control mechanism for the VFD is usually some type of static pressure-sensing device installed in the HVAC ductwork that sends a signal back to the VFD to automatically increase or decrease the motor speed. With the operation of a VFD in a typical HVAC control system, it is not uncommon to achieve annual energy savings of 15–20%. For further discussion of VFDs, see Chapter 15.

Resetting Discharge Supply Temperature Control

Typical HVAC systems use automatic controls to maintain a certain space temperature inside a facility. In the summer, the HVAC supply air will typically be cooled to about 55 °F, depending on the space temperature requirements. In the winter, the supply air will be heated to the range of 70–100 °F. As an energy conservation method, the HVAC supply air temperature can be adjusted or reset based on the actual space temperature in the facility. As an example, if the space temperature inside the facility during the cooling mode is satisfied, then there is no need to continue to cool the supply air to 55 °F. The supply temperature can be automatically raised until the facility temperature set point again requires cooling. This concept of resetting the supply temperature can also be used for the heating mode of an HVAC system. By having this type of control system in place for a typical HVAC system, it is not uncommon to achieve annual energy savings of 2–5%.

MAINTENANCE

There are many opportunities for efficiency improvements in an HVAC system related to maintenance. Maintenance can encompass cleaning, adjusting, greasing, checking components, calibrating components, and many other items. We will consider opportunities in both preventive and predictive maintenance. In preventive maintenance (PM), we will discuss burner calibration, cogged belts, and sensor calibration, and in predictive maintenance, filter replacement.

Preventive Maintenance

Preventive maintenance is critical not only for efficient daily operation of equipment but also for maintaining or improving energy efficiency. Protocols for PM usually consist of a list of manufacturers' recommendations of equipment items to be checked to ensure that the system operates efficiently. The next step is to provide an example of where to focus for energy efficiency, but not actually identify the step-by-step process for a PM.

Three examples of HVAC PMs are being discussed herein: gas burner calibration, belt calibration, and sensor calibration.

Gas Burner Calibration

Many HVAC systems have some type of heating system as part of the overall unit. This heating system can include many types of heat sources such as hot water, natural gas, electricity, and a variety of other energy sources. We will focus on the natural gas system methodology because that is one of the most common types of heating systems. A natural gas heating system typically consists of a natural gas train assembly with a variety of components such as regulators, shutoff valves, igniters and flame supervision control, and safety interlocks. We should calibrate the assembly regularly to verify that the gas burner is operating efficiently to minimize energy use. Air velocity across a natural gas line burner must be kept within an acceptable range because too much or too little airflow can cause the release of raw gas, incomplete combustion of fuel, generation of high levels of waste gases, and overall wasted energy. Implementing a maintenance plan to calibrate the gas train based on a predetermined schedule often achieves annual energy savings of 5–8%.

Belt Calibration

Belt-driven components are common in an HVAC system. Belt drives are simple and allow manual speed control of the motor and fan of an HVAC system. Control is accomplished through the adjustment of different-size pulleys. While belt-drive systems are generally considered to be efficient, certain belts are more efficient than others. Standard belt drives typically use V-belts that have a trapezoidal cross section and operate by wedging themselves into the pulley. These V-belts have initial efficiencies on the order of 95%, which can degrade as much as 5% over the life of the system if the belts are not periodically retensioned. Fans with standard V-belts can be retrofitted by replacing the V-belt with an energy-efficient cogged V-belt.

Installing an energy-efficient cogged V-belt in a typical HVAC control system often achieves annual energy savings of 1–2%.

Sensor Calibration

The calibration of sensors is often overlooked, but the philosophy of "if it doesn't 'look' broken—don't fix it" is not the best energy efficiency methodology. Unfortunately, sensors out of calibration can lead to enormous energy losses and can go undetected for years without a proactive maintenance program. To minimize such energy losses, a standard sensor calibration program is critical. The following sensors should have an annual calibration program [1]:

- Outside air temperature sensor
- Mixed air temperature sensor

- Discharge or supply air temperature sensor
- Chilled water supply temperature sensor
- Heating water supply temperature sensor
- Wet bulb temperature or relative humidity sensor
- Space temperature sensor
- Return air temperature sensor
- Economizer and related dampers
- Cooling and heating coil valve
- Static pressure transmitter
- Air and water flow rates transmitter

HVAC sensors have defined operating limits, and the accuracy of a given sensor is primarily a function of the sensor type. As with any sensor device, accuracy can degrade over time. Accordingly, sensor assessment and calibration should be a routine maintenance function. Refer to manufacturer's data for assessment and calibration recommendations. During calibration, equipment can be activated, such as running the dampers to the fully open and then fully closed position, to verify the sensor function. During this process, field staff should also verify that all moving parts are properly lubricated and seals are in good shape. Because economizers are dampers that interact with outside air, facilities where economizers are installed should receive special attention. In addition to the normal PM procedures, economizer dampers should be checked at a higher frequency to ensure proper modulation, sealing, and sensor calibration. The temperature and/or humidity (i.e., enthalpy) sensors used to control the economizers should also be included in a routine calibration schedule. Proper calibration of the HVAC system sensors usually achieves annual energy savings of 1–2%.

Predictive Maintenance

Air filters play a critical role in maintaining indoor air quality and protecting the downstream components from dirt and dust that affect equipment efficiency. In the worst case, dirty filters can result in supply air bypassing the filter and depositing dirt on the heating/cooling coils rather than on the filter. This results in dirty coils, poor heat transfer, and in general inefficiency in performance. As a rule, sites should routinely change filters based on the pressure drop across the filter, calendar scheduling, or visual inspection. Schedule intervals should be developed based on the dirt loading from indoor and outdoor air. Measuring the pressure drop across the filter is the most reliable way to assess filter condition and can be used to establish guidelines for changing the filters based on a predictive maintenance philosophy. This type of process can reduce the energy inefficiencies associated with manually predicting the time to change filters. This concept will reduce the loading on the fan and improve energy efficiency so that energy saving can be obtained. The use of predictive maintenance for the filters often achieves annual energy savings of 1–2%.

RECOMMISSIONING

Commissioning is the process of ensuring that systems are designed, installed, functionally tested, and capable of being operated and maintained according to the owner's operational needs. *Recommissioning*, or ongoing commissioning, is the term for applying the commissioning process to a facility that has been previously commissioned (either during construction or as an existing facility). Recommissioning is normally done every 3–5 years to maintain top levels of facility performance and/or after other stages of the upgrade process to identify new opportunities for improvement. Four key areas to focus on during the recommissioning process are understanding the current situation, identifying opportunities, applying countermeasures for improvement, and planning next steps [2].

Understanding the Current Situation

The first step in recommissioning is to understand how the equipment is to function based on the design and to analyze the current situation against those criteria. A variety of possible problems can be identified by reviewing facility documentation during the recommissioning process. Facility documentation includes operating requirements, original design documents, equipment lists, drawing of the facility's main energy-using systems, control documentation, operation and maintenance manuals, testing, adjusting, and balancing reports. Evaluating this information can lead to possible energy improvement opportunities.

Opportunity Identification

Identifying possible problems in an HVAC system can be a tedious and complex task. It is important to divide the HVAC system into components to better understand the problems. As an example, problem areas of an HVAC system can relate to monitoring, control, air systems, heating systems, or cooling systems.

Once the current situation is understood, the next step in recommissioning is to look at possible field indicator opportunities, which are commonly found during a walkthrough. The indicators are as follows ([2], Sections 5.2 and 5.3, pp.5–11):

- Systems that simultaneously heat and cool, such as constant and variable air volume reheat
- Equipment that has been bypassed and does not operate automatically
- Inaccurate temperature, pressure, or humidity sensor readings
- Improper facility pressurization (either negative or positive), that is, doors that stand open or are difficult to get open
- Equipment or piping that is hot or cold when it should not be
- Short cycling of equipment
- Variable frequency drives that operate at unnecessarily high speeds

- Variable frequency drives that operate at a constant speed even though the load being served should vary
- Economizers in need of repair or adjustment. Potential problems include frozen dampers, broken or disconnected linkages, malfunctioning actuators and sensors, and improper control settings.

Two other important activities can aid understanding of the existing system and help in the identification of improvements:

- Diagnostic monitoring of energy systems can help determine where particular problems lie. Data are typically gathered using the facility's existing energy management system (EMS) along with portable data loggers to obtain any data not available through the EMS. Variables typically monitored include whole-facility energy consumption (including electricity, gas, steam, and chilled water), end-use energy consumption, operating parameters (e.g., temperatures, flow rates, and pressures), weather data, equipment status and run times, actuator positions, and set points.
- Functional testing takes a system or piece of equipment through its paces while personnel observe, measure, and record its performance in all key operating modes. Functional testing can also be used to help verify whether a particular improvement is really needed and will be effective. For example, an observation that the throttling valve on a pump is not fully open can indicate that energy savings could be achieved by trimming the impeller so that the valve can be fully open. A functional pump test will determine the value of this possible improvement.

Countermeasures for Improvement

Both the controls and the components of the heating and cooling systems present saving opportunities during the recommissioning process. The EMS and controls within a facility play a crucial role in providing a comfortable facility environment. Over time, temperature sensors or thermostats can drift out of tune. Wall thermostats are frequently adjusted by occupants, throwing off controls and causing unintended energy consumption within a facility. Poorly calibrated sensors can increase heating and cooling loads and lead to occupant discomfort. To tune the heating and cooling controls, take the following steps ([1], Section 9.6.8, pp. 9.72–9.76):

- Calibrate the indoor and outdoor facility sensors, including room thermostats, duct thermostats, humidistats, and pressure and temperature sensors, in accordance with the original design specifications.
- Inspect dampers and valve controls to ensure that they are functioning properly. Check pneumatically controlled dampers for leaks in the compressed-air hoses. Also examine dampers to ensure that they open and close properly. Stiff dampers can cause improper modulation of the amount of outside air being used in the supply airstream. Dampers are sometimes actually found wired in a single position or even disconnected, violating minimum outside air requirements.

- Review facility-operating schedules. HVAC controls must be adjusted to heat and cool the facility properly during occupied hours. Occupancy schedules can change frequently over the life of a facility, and control schedules should be adjusted accordingly. Operating schedules should also be adjusted to reflect daylight saving time. When the facility is unoccupied, set the temperature back to save some heating or cooling energy, but keep in mind that some minimum heating and cooling can be required when the facility is unoccupied. In cold climates, for example, heating can be needed to keep water pipes from freezing.

Next Steps

The goal of the recommissioning process is to ensure that the facility operates as intended and meets current operational needs. Recommissioning can be very cost-effective with field experience showing typical energy savings of about 10–15%. A well-planned and executed recommissioning project generally consists of planning and execution phases as well as plans to ensure that benefits persist and even increase through such measures as training, operations and preventive maintenance, and performance tracking. Plans should also be made for periodic recommissioning of the facility at least once every 3–5 years.

SUMMARY

HVACs are a necessary system for any facility, but an HVAC system also has many opportunities for energy savings. The typical philosophy is to let the HVAC equipment operate all the time to maintain facility comfort, regardless of production time or non-production activities. As can be seen from the above examples, there are many opportunities for energy savings by improving the operation and maintenance, and energy efficiency of existing equipment is operated and maintained. Energy efficiency can also be improved through the philosophy of only operating the equipment to meet production needs. The four concepts discussed above, namely aligning supply side with demand side, implementing operating and maintenance improvements, and recommissioning the HVAC system, provide the framework needed to enhance energy efficiency savings.

REFERENCES

1. U.S. Department of Energy (2010) *Operations & Maintenance Best Practices: A Guide to Achieving Operational Efficiency*. Release 3.0, Section 9.6.6, pp. 9.71–9.72. Available at http://energy.gov/eere/femp/downloads/operations-maintenance-best-practices-guide-chapter-9 (accessed February 8, 2015).
2. U.S. Environmental Protection Agency (2007) *Energy Star Building Upgrade Manual*. Chapter 5, Section 5.1, pp. 2–4. Available at http://www.energystar.gov/sites/default/files/buildings/tools/EPA_BUM_CH5_RetroComm.pdf (accessed February 8, 2015).

PROCESS

25

IDENTIFYING PROCESS IMPROVEMENTS FOR ENERGY EFFICIENCY

Alan P. Rossiter[1] and Joe L. Davis[2]

[1]*Rossiter & Associates, Bellaire, TX, USA*
[2]*PSC Industrial Outsourcing, LP, Houston, TX, USA*

Somewhat ironically, the most neglected area for energy efficiency evaluations in the process industries is within the process itself. In part, this reflects a reluctance to tamper with facilities that directly impact product quality and production rates. While this can be a legitimate concern, it is unfortunate, because many viable energy-saving opportunities are left on the table. A second issue is that each process is to some degree unique, so it is often not possible simply to replicate ideas. Instead, each plant needs to be evaluated and understood in order to define and develop improvement options, and this typically requires specialized expertise. The tendency, instead, is to focus on utility systems (e.g., steam) and specific items of equipment (such as furnaces) where there is a well-known set of parameters to assess and a recognized set of options for improving energy efficiency.

In this chapter, we introduce a general approach for identifying energy efficiency improvements within the process area, and we illustrate this with a few simple examples. We also discuss how to evaluate opportunities once they have been identified. The focus is on improving the operation of existing process facilities and identifying and evaluating retrofit projects, but many of the principles are also applicable to evaluating and improving new plant designs.

Energy Management and Efficiency for the Process Industries, First Edition. Edited by Alan P. Rossiter and Beth P. Jones.

PROCESS FLOW DIAGRAM REVIEWS

The most effective method for identifying process unit energy efficiency improvements is to carry out a process flow diagram (PFD) review [1]. This can be considered a "structured brainstorming" activity, similar to hierarchical process reviews [2]. A PFD should be developed for each process unit (e.g., hydrotreater and crude unit) to show the major equipment items and their interconnections, together with the basic heat and material balance across the unit. All fired heaters, columns, and heat exchangers should be shown, with temperatures, flow rates, and pressures labeled on the inlets and outlets of each piece of equipment. Ideally, the duty for each fired heater, heat exchanger, or block of exchangers should be shown. In addition, any place where steam is used for heating or stripping, or where steam is generated, should be indicated on the PFD with flow rates labeled.

The procedure for a PFD review is similar to that customarily used for "HAZOP" studies. With the marked-up PFD, plant operations and technical support personnel, with assistance from energy efficiency specialists, review each of the main streams, equipment items, and systems to identify inefficiencies and areas of opportunity. The plant operations and technical staff bring their knowledge of day-to-day plant issues to the table. The specialists bring their knowledge of similar processes at different locations and the types of energy efficiency opportunities that have worked elsewhere. Together they brainstorm ideas for the process unit under consideration.

Typically, a large number of ideas will be identified during a PFD review. These can range from adjusting set points and operating targets, through new control schemes, minor piping changes, and equipment modifications, to completely new processes and novel technologies. All of the opportunities should be documented during the PFD review and then later reviewed to quantify the potential savings, estimate the implementation costs, identify technical risks, and determine applicability.

Variations on the PFD Review Approach

There are numerous variations on the PFD review approach that various organizations have used successfully. For example, setting up face-to-face meetings with plant operations, technical support, and energy efficiency personnel can be logistically difficult, and the meetings themselves are time-consuming. An alternative approach is to provide the PFD to an energy efficiency specialist who generates a list of questions and ideas that is then distributed to the other personnel for review and comment. Where possible, the energy efficiency specialist(s) should meet face to face with the other personnel when they have had a chance to review the initial findings—or, failing this, they should at least have a conference call or web meeting. This approach obviously lacks the benefit of face-to-face discussion during the idea generation phase, but it has proved effective in many situations.

Another related approach is kaizen "treasure hunts" [3]. Kaizen is a Japanese word meaning "continuous improvement." This technique was pioneered in the United States by the Japanese carmaker Toyota, and it has been used primarily in the discrete manufacturing sector. However, it has also been applied successfully in other sectors, including the process industries—although typically on smaller facilities. Cross-functional teams (site employees and external experts) meet on-site to investigate a facility's energy use. Over a fairly short period—typically 3 days—the team observes the way equipment is operated, collects data, compares against experience from other facilities, identifies potential improvements, and completes "detail sheets" that describe each opportunity together with estimated costs and savings. This approach provides a good way of engaging personnel at all levels across a facility in energy efficiency activities and has yielded some excellent results [4].

It is important to understand that none of these approaches is an isolated activity. Even though a series of PFD reviews or a kaizen treasure hunt typically takes just a few days, a great deal of preparation is needed to ensure that both the right people are included and trained where necessary and the necessary data are available. Moreover, after the PFD review or treasure hunt is over, it is essential that the ideas that have been generated are progressed and implemented where appropriate.

Process Heat Integration

One key area that is often explored during PFD reviews is heat integration. We noted earlier that it is often not possible simply to replicate ideas between processes. One area where replication is often possible is heat integration, as there are several well-known "standard" heat integration configurations. For example, Figure 25.1 shows a reactor with a feed–effluent heat exchanger (FEHE). In this case, heat from a hot reactor effluent stream is used to preheat a cold feed stream. This reduces both the heat load on the feed heater (H) and the cooling duty on the effluent cooler (C) by an amount that is equivalent to the heat transferred in the FEHE.

Figure 25.2 shows a distillation column with a feed–bottoms heat exchanger (FBHE). The cold feed stream is heated by the hot stream from the bottom of the column, and this reduces the amount of heat needed in the reboiler. This arrangement is very common, and it can save a lot of energy in columns. However, care must be taken, because in this case there is not necessarily a one-to-one correlation between heat recovered in the feed and the reduction of heat load on the reboiler. Moreover, heating the feed can result in additional vapor traffic above the feed tray, which can overload the condenser. These effects depend on both the nature of the material being processed and

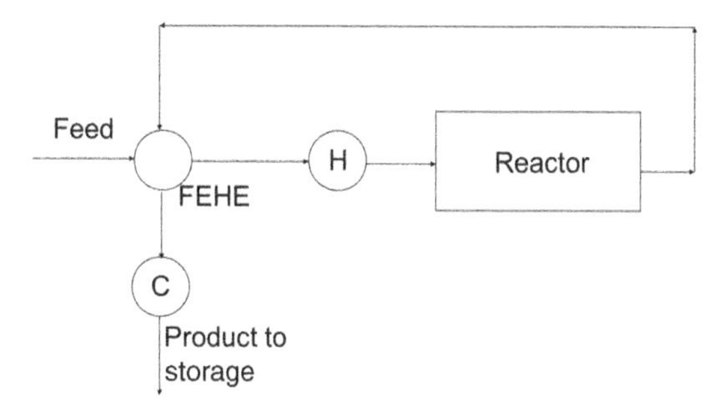

Figure 25.1. Reactor with feed–effluent heat exchanger.

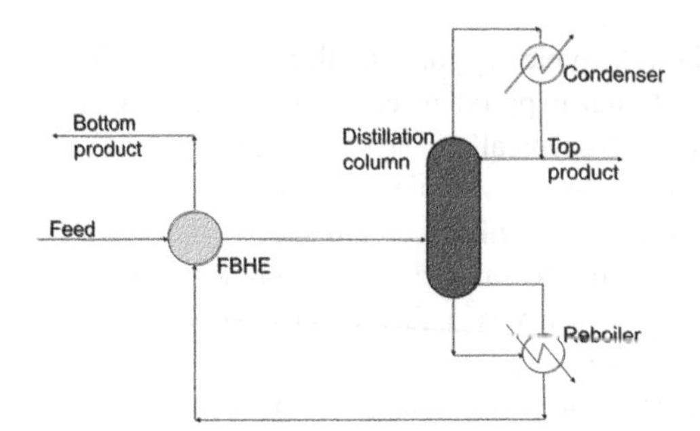

Figure 25.2. Distillation column with feed–bottoms heat exchanger.

the design of the distillation column, and simulations and/or plant trials are needed to quantify them.

There are numerous variants on these standard configurations; for example, in some cases, it is possible to recover heat from a distillation column overhead as well as heat from the bottoms stream, and some columns have pumparound streams that also provide heat recovery opportunities (see Chapter 26).

In some cases, the standard configurations are inadequate for the task of heat integrating a process. This typically occurs where there are many large streams being heated or cooled, for example, in oil refinery crude units or fluid catalytic cracking units (FCCUs). In these cases, more complex heat exchanger networks are needed. A number of good tools are available for these situations, and pinch analysis is the most widely used of these (see Chapter 26).

Typical Results from PFD Reviews

As noted earlier, a wide range of different types of energy efficiency opportunities can be identified in PFD reviews. Typical examples, mostly drawn from distillation applications and experience in oil refineries and chemical plants, are given below, together with comments on how they are handled.

Operating Targets. In virtually all processes, there are a number of key process variables—often termed "key performance indicators" (KPIs)—that can be adjusted to optimize energy use. These typically come to light during PFD review discussions and are documented for further evaluation.

One of the biggest challenges in identifying and stewarding KPIs is setting targets for them—an activity that would be carried out after the PFD review. Without a thorough understanding of the processes, it would be easy to set targets that reduce energy consumption, but jeopardize yields. Pumparound rates, for example, must be set at just the right point to simultaneously optimize both yield and energy.

Another important consideration in setting targets is the recognition that they must be flexible. As feed slates change, or as operating modes shift, energy requirements will be impacted. Heavier crudes, for example, will generally require more energy to process than light crudes. Operating modes are often very different in summer and winter, and these differences can significantly impact energy consumption. For example, a refinery may send its vacuum tower gas oil straight to distillate blending in the wintertime, whereas in the summertime the gas oil will be sent to the FCCU or hydrocracker for further upgrading to gasoline. Therefore, in the summertime, gas oil cut points will not be as important, and thus the vacuum column pumparounds should be set to optimize energy recovery, rather than cut point. For some KPIs, then, it will be appropriate to assign several targets depending on the mode of operation.

Finally, where advanced process control (APC) or real-time optimization (RTO) is applied, it may be more important to set operating strategies rather than hard targets. The PFD review process can be a very effective process for verifying that APC applications are correctly set to optimize both yield and energy.

There are a number of ways to set targets for KPIs. The simplest way is through experience. Energy experts who have seen multiple units worldwide may be able to apply experience-based targets for some of the simpler energy metrics such as stripping steam rates by comparing current operation against industry standards and best practices. For more complex metrics, such as reboiler rates, process simulation is required in order to set the target. Test runs can then be used to confirm the target. Alternatively, a 12-month trend of the process variable could be run to identify the "best achieved" and set a target around that value. This can help take out the variation that comes with different operators running the unit over changing shifts. This topic is considered further in Chapter 27, where tools for making KPIs accessible to plant operators are discussed.

Inappropriate Operating Practices. Targets for KPIs are needed to minimize energy use within a normal operating envelope. However, PFD reviews often identify situations where current plant operations are either wholly inappropriate or at least significantly suboptimal in terms of energy use. Operators often place equipment in service inappropriately, and once these inappropriate operating norms have been established, they sometimes stay in place for years. PFD reviews often identify these situations and they can go a long way toward minimizing this type of equipment misuse, as illustrated by the following two examples.

Example 25.1. A common inefficiency in process plants is the cooling of streams that should not be cooled. A petrochemical facility installed an air cooler on the feed line of a distillation column (Figure 25.3) to prevent the condenser from being overloaded during certain abnormal operating conditions [5]. Although the air cooler was designed for use only during abnormal conditions, it became an accepted practice to run it continuously during normal operations. Removing heat from the feed during normal conditions required the reboiler to work harder, thereby increasing the steam load.

Changing operating procedures to reflect the original intent of the air cooler—shutting off the air cooler fan during normal operations—reduced the reboiler duty by more than 30% and saved the plant more than $1 million/year with no reduction in throughput or loss of product quality, and no investment required.

Turning the air cooler off during normal operations was a good first step. However, a significant amount of heat was still being wasted due to convection in the air cooler. This loss was eliminated by installing a bypass around the air cooler (Figure 25.3, after modifications)—a small project that saved an additional $200,000/year.

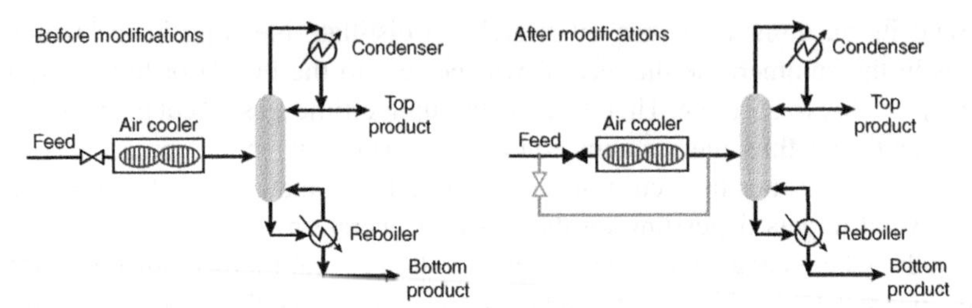

Figure 25.3. A distillation column was designed with an air cooler on the feed line. Although the cooler was designed for use only during abnormal conditions, it was run continuously, resulting in an increased reboiler heat load. Shutting off the air cooler during normal operations did not completely solve the problem, as a significant amount of heat was wasted as a result of convection in the cooler. A simple fix: install a bypass line around the air cooler. (Reprinted with permission from *Chemical Engineering Progress* (CEP), December 2012 [6]. Copyright 2012, American Institute of Chemical Engineers (AIChE).)

The overall solution in this case therefore includes both

- a "no cost" operating change (shut off the fan) and
- a small project or "low investment cost" component (add the bypass).

Example 25.2. This second example, from KBC Advanced Technologies (http://www.kbcat.com/), was presented at the 27th Industrial Energy Technology Conference, New Orleans, LA, May 10–13, 2005, and it is adapted with permission. It also illustrates how a PFD review can identify inappropriate use of equipment associated with a distillation column, but in this case it is rather more subtle [1]. The example comes from a European refinery that was already a very good energy performer. A PFD review was carried out on the highest energy consuming units.

The isomerization unit, as shown in Figure 25.4, has two naphtha splitters, T-1 and T-2. T-1 has two reboilers, E-1 and E-2. E-1 uses low-pressure (LP) steam as the heating medium. E-2 uses the hot naphtha feed to T-2 as the reboiling medium. T-2, on the other hand, has only one reboiler, which uses medium-pressure (MP) steam as the heating medium.

In the base case, heat to T-1 was supplied predominantly from LP steam in E-1. The logic for this was simple: maximizing LP steam use in T-1 allows for a reduced requirement of MP steam in T-2.

During the PFD review, however, this logic was challenged. First, it was pointed out that increasing the feed preheat to T-2 does not save E-3 reboiler duty in direct proportion. However, using naphtha in E-2 does save LP steam in direct proportion. Furthermore, the 150 °C feed temperature to the naphtha splitter appeared unnecessarily high. The assumed pricing of MP and LP steam were also challenged.

Outside of the PFD review, the steam system was modeled, as discussed in Chapter 18. This showed that MP steam was only slightly more valuable than LP

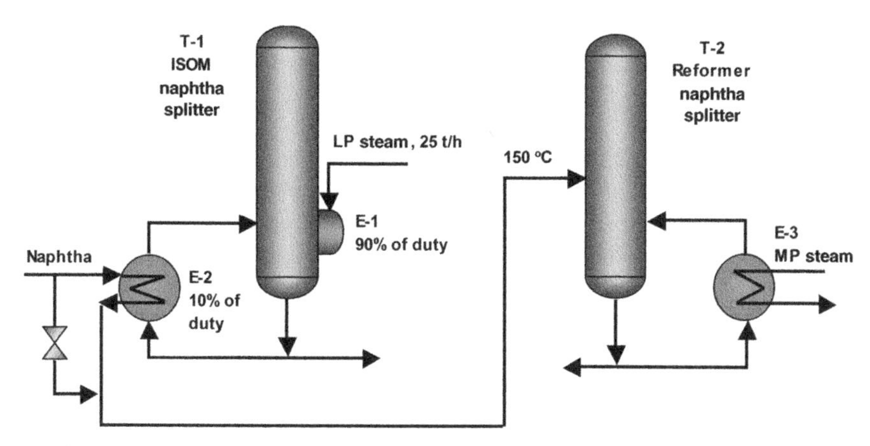

Figure 25.4. Isomerization unit splitters: base case operation. (*Source*: Ref. [1].)

steam. In addition, a simulation of the naphtha splitter showed that the feed preheat recovery factor was only 0.5–0.6. Therefore, the better solution, as shown in Figure 25.5, was to increase the duty of E-2 and reduce the T-2 feed temperature to 120 °C. Even though this resulted in a 4 t/h increase of MP steam to E-3, it reduced the amount of LP steam to E-1 by 7 t/h. The annual savings were $215,000.

In summary, the following steps were carried out to identify this energy opportunity:

1. Understand and question the existing heat integration pattern via PFD review.
2. Identify marginal mechanisms in the steam and power system.
3. Calculate marginal steam costs.
4. Model T-2 to find E-3 reboiler duty with a reduced feed temperature.

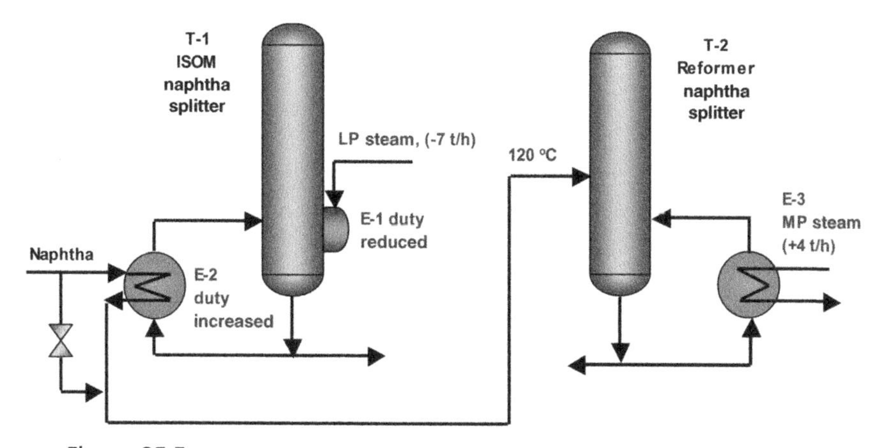

Figure 25.5. Isomerization unit splitters: after optimization. (*Source*: Ref. [1].)

As a result of this work, the feed preheat temperature to T-2, together with the steam rates to E-1 and E-3, became stewardable energy metrics for the isomerization unit.

Equipment Modifications and Additions. As Example 25.1 shows, it may be possible to achieve some of the potential energy savings in a particular situation simply by changing operating practices, and then achieve the remaining savings with some investment in new or modified facilities. There are obvious advantages in the former scenario (operating changes only): They require no investment, and they can usually be implemented very quickly. However, during a PFD review it is often not possible to determine how much can be achieved using only existing facilities, and how much of the potential savings will require investment in new equipment. Ideas are documented during the review, and evaluated afterward.

When the process under review includes reactors or distillation columns, the PFD review will invariably check to see whether FEHEs (Figure 25.1) and FBHEs (Figure 25.2) are included. If not, these are likely to be documented as ideas for evaluation. Even if they are present in the existing design, they are often documented to evaluate whether there is an economic incentive to install additional heat exchangers, or to modify the existing heat exchangers to increase heat recovery (see Chapter 10).

All of the examples discussed thus far have focused on small portions of a process—single distillation columns and their peripherals, reactor systems, or distillation systems. Sometimes PFD reviews identify opportunities that span much larger process areas, even the entire process [2]. This is illustrated in Figure 25.6, which shows a combined reaction/separation recycle structure.

In many processes, the conversion in the reaction step is low, and unconverted feed material has to be recovered and recycled. This often entails significant energy costs, as energy is consumed in the separation processes that recover the unconverted material

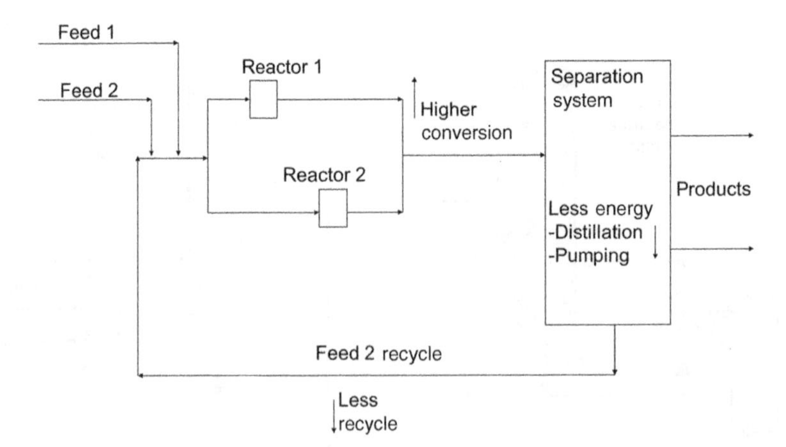

Figure 25.6. The reaction/separation recycle structure of the process, and the parameters within it, should be challenged as part of a PFD review.

(e.g., distillation, crystallization, or extraction). In addition, energy is consumed in recycling the recovered material to the reactor (pumping, compressing, or conveying). It follows that, as a generalization, low conversion/high recycle processes are high energy consumers, so there is an incentive to increase conversion and reduce recycle rates to save energy.

However, energy is not the only issue. Driving the reaction to higher conversions often results in poorer selectivity, which can lead to overwhelming costs associated with additional feedstock requirements and disposal of unwanted by-products. For example, high severity operation of FCCUs in oil refineries results in increased low-value fuel gas production, with can render additional energy-saving projects uneconomical (see Chapter 20). Furthermore, the performance of an existing separation system may deteriorate as recycle flows go down (e.g., distillation column trays may start to weep as vapor traffic is reduced). This is therefore a complex optimization, and during a PFD it is likely to be identified and documented for subsequent evaluation. The results of these evaluations can vary a great deal, and can sometimes be counterintuitive, as the following two cases illustrate.

Case 25.1. In a study at a chemical plant, the value of the feed material was very high and dominated all other costs. The optimum solution was to *reduce* conversion and *increase* the recycle rate, in order to minimize losses of the feed material. This action increased energy consumption, but saved $1,000,000/year in overall operating costs due to feedstock savings. No new facilities were required.

The optimum value for the recycle rate depends on the costs of energy and feedstock, and also on several other parameters. The recycle rate should therefore be treated as a KPI and monitored routinely.

Case 25.2. In an oil refinery study, a comparable evaluation led to a reconfiguration of the reaction system to increase conversion. The internals of a distillation column also had to be changed to handle much reduced flow rates. This project saved more than $1,000,000/year in energy, and it also debottlenecked the process. However, it required a significant capital investment.

PFD Reviews as a Motivational Tool

PFD reviews often provide an opportunity for site personnel to showcase their ideas. In a recent instance at a chemical plant, the plant's control engineer explained a new control algorithm that he had written to optimize the operation of a large compressor. The new application had been ready for several months, but it had not been turned on because the Operations Department had concerns about how it might impact the operability of the plant.

The operations supervisor was also in the meeting, and initially he very vocally expressed his objections to any changes to the existing control scheme. However, the control engineer demonstrated that the energy savings with the new operating mode were far greater than the operations supervisor had realized. Furthermore, the visiting energy management specialist in the PFD review meeting was able to endorse the new control scheme based on experience at other facilities. There followed a very animated discussion on strategies for testing the new algorithm and steps that that should be taken to safeguard plant operations. By the end of the meeting, the operations supervisor was committed not only to testing the new control scheme, but also to making it work.

This incident is by no means an isolated case, and it highlights once again that successful energy management is not purely about good technological solutions. It is also about human behavior: engaging people in the process of energy efficiency and motivating them to succeed at it.

EVALUATING OPPORTUNITIES AFTER A PFD REVIEW

In most cases, the ideas that come out of a PFD review require a considerable amount of scrutiny before being accepted as viable projects. The exact form that this takes can vary considerably, but at a minimum the energy savings need to be quantified and, if new facilities are needed, costs must be estimated.

In many energy assessment programs, time is set aside to evaluate the list of PFD review ideas immediately after the review meetings are completed. This is desirable, as the thinking behind the ideas is then fresh and accessible. The goal at this stage is not to develop definitive designs and extremely accurate cost estimates. Rather, the purpose is to screen the ideas with sufficient accuracy to weed out those that cannot realistically result in a viable return on investment. The end result is a shortlist of the most attractive projects, ranked by estimated payback. Viable options that do not require new facilities can then be implemented by the plant operations group, and projects that require investment can be incorporated into the work plan for the corporate project engineering group.

Estimating Energy Savings

Three main issues need to be considered when considering energy savings:

1. How much energy is saved?
2. What type of energy is saved?
3. What is the energy worth?

Sometimes the answer to the first question (How much energy is saved?) is so simple that it requires no calculation. For example, if the only energy impact of a particular idea is to shut off a pump, the direct energy saving is most likely the elimination of the electric power used in the pump motor. Care is always needed, however, because closer scrutiny sometimes shows that there are hidden energy savings or penalties:

- Shutting off one pump may increase the power requirement on another pump.
- Reductions in flow or discharge pressure from compressors do not necessarily result in realizable savings. If the compressor is operating near its surge limit, any

attempt to reduce its power consumption can lead to instability (see Appendix 15.1).

- In other cases, it is necessary to simulate a section of the process, or at least to set up a simple spreadsheet model, to produce a credible savings estimate. This is generally true, for example, when adding a feed–bottoms heat exchanger or adding heat exchangers to a preheat train.

The second question is: What type of energy is being saved? The two main classes are thermal energy (heat, which may be provided either directly from a fired heater or indirectly in steam or some other heat transfer medium) and electric power. Many sites have steam at several different pressure levels, so it is important to identify which level is impacted. In some cases, the steam demand is impacted at more than one pressure level, as illustrated in the isomerization splitter example earlier in this chapter.

The third question (What is energy worth?) can also be more difficult to answer than it might at first appear. In the case of electric power, savings usually translate into a reduction of power import from the grid, so we need to know how much the power bill changes as the electric load goes down. Conceptually, this seems easy: We just need to know the cost of imported power in $/kWh. However, electricity contracts can be complex. They may include demand charges and time-of-use components, as well as various other factors, in addition to simple energy charges. It is therefore important to understand the power contract before estimating the value of electrical savings.

Thermal energy savings can also be complicated. We have already noted that the marginal cost of steam on many sites depends on its pressure level (see Chapter 17). The credit for saving steam also depends on whether or not the condensate is recovered and recycled, as both energy and water are lost if it is not.

When thermal energy is supplied directly to a process stream with a fired heater, the most important consideration is the cost of the fuel. However, the efficiency of the furnace is also important; for example, if the furnace has an 80% thermal efficiency, then the amount of fuel that is needed is 1.25 times the process heat load. In some cases, an energy efficiency project can affect the efficiency of a furnace. For example, increasing feed preheat ahead of a charge furnace may increase the stack gas temperature, which in turn reduces the furnace efficiency. This change in efficiency should be included in the evaluation of the energy savings.

Cost Estimating

It is not necessary to develop extremely accurate cost estimates during the project screening activity, but the estimates must at least be realistic. A common error is simply to obtain the cost for the main equipment item in a small project and assume that the final cost of a project will be roughly the same amount. In reality, there are many additional elements in the cost of any project, including foundations, piping, and control, as well as engineering and various labor disciplines and overheads. Furthermore, most energy efficiency projects are revamps, and these tend to be more costly than new installations because they require working in and around existing facilities. If the work has to be carried out during turnarounds, then the costs are generally even higher. In addition, costs can vary considerably from location to location, due to labor rates and other factors.

While it is not necessary to quantify each of these components at the scoping stage, it is important to provide an overall cost estimate that reflects the total cost of implementation. In some cases, corporate cost estimating groups are able to provide total installed cost data or simple correlations for key equipment items that facilitate the development of acceptable cost estimates. Alternatively, recent data from similar projects at the same site can sometimes be used as the basis for a "ballpark" estimate for a new project idea from a PFD review. There are also various software tools and literature sources that can be used to estimate costs.

Technical Credibility

Many of the ideas that come out of PFD reviews challenge existing operating philosophies and operating boundaries. Often these challenges are justified, but it is always important to verify that the existing equipment and systems can tolerate the proposed operating conditions. If not, then the cost of equipment upgrades may have to be included in the project.

Some PFD review ideas involve "transformational technologies"—concepts or applications that are not yet proven. While this should not disqualify an idea from consideration, it does mean that it is likely to take a significant amount of time and money to bring the idea to fruition, and there is a significant risk that it will never be implemented. The company sponsoring the PFD review needs to decide whether to make this type of investment in research and development.

A small number of the ideas documented in PFD reviews are manifestly infeasible, and the person carrying out the screening needs to be alert to this possibility. Screening out infeasible projects is an important part of the evaluation process.

PFD REVIEWS IN CONTEXT

PFD reviews can be used as a stand-alone technique for identifying and organizing opportunities for improving energy efficiency on virtually any type of process plant. However, most often they form part of a larger energy efficiency initiative, such as an overall site energy assessment, or a step in the development of an energy management system. They are also often used in conjunction with a pinch analysis to explore a wide range of energy efficiency options for a process or production site.

While there are many considerations to an effective PFD review, and hidden traps that must be avoided, the investment of time and resources in conducting a thorough review such as the one described in this chapter invariably pays big dividends, as there are invariably numerous economically attractive opportunities for improvement in today's complex refining, petrochemical, and other process plants.

REFERENCES

1. Davis, J.L., Jr. and Knight, N. (2005) Integrating process unit energy metrics into plant energy management systems. *27th Industrial Energy Technology Conference*, New Orleans, LA, May 10–13, 2005.
2. Rossiter, A.P. (ed.) (1995) *Waste Minimization Through Process Design*, McGraw-Hill, New York, pp. 149–163.
3. Bremer, B. (2005) *Toyota Treasure Hunt System Turns Up Savings and Uses the Expertise of Process Engineers*. Available at http://www.energystar.gov/ia/business/industry/Bremmer_Toyota.pdf (accessed January 11, 2014).
4. Environmental Defense Fund (2014) *Manufacturing Energy Productivity Pilot: Promising Results*. Available at http://www.edf.org/sites/default/files/Cobasys_Pilot_Case_Study_8_30_11.pdf (accessed January 11, 2014).
5. Rossiter, A.P. (2007) Back to the basics. *Hydrocarbon Engineering*, 12 (9), 69–73.
6. Rossiter, A.P. and Venkatesan, V. (2012) Easy ways to improve energy efficiency. *Chemical Engineering Progress*, 108(12), 16–20.

26

PINCH ANALYSIS AND PROCESS HEAT INTEGRATION*

Alan P. Rossiter

Rossiter & Associates, Bellaire, TX, USA

Process heat integration using pinch analysis is a respected tool for achieving energy efficiency. It is particularly effective in situations that require complex heat exchanger networks, especially where the thermal loads are moderate to large (>1 MBtu/h) on each of the heat exchangers. Preheat trains in oil refinery crude units provide a good example of the type of process where pinch analysis is most applicable.

Other approaches are also applied to complex heat integration problems. Most notable among these are various numerical optimization methods [2,3]. However, thus far these have not received the same level of success or acceptance as pinch analysis.

FUNDAMENTALS OF PINCH ANALYSIS

Pinch analysis is a systematic technique for analyzing heat flow through an industrial process based on fundamental thermodynamics. The second law of thermodynamics

* This chapter has been adapted with permission from *Chemical Engineering Progress* (CEP), December 2010 [1].

Energy Management and Efficiency for the Process Industries, First Edition. Edited by Alan P. Rossiter and Beth P. Jones.

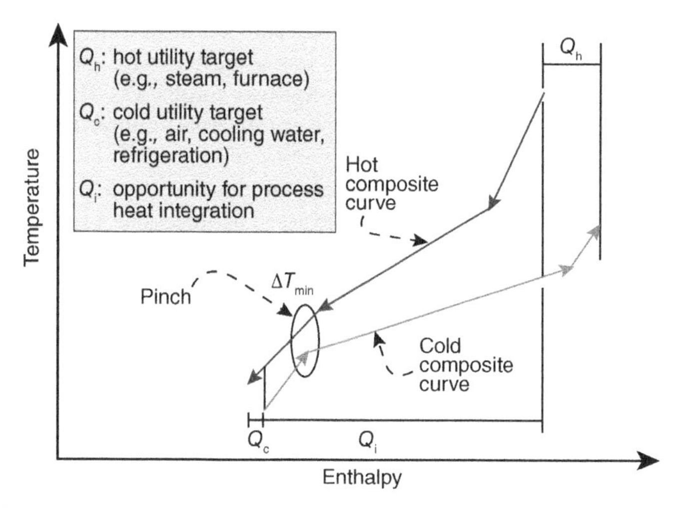

Figure 26.1. Composite curves show the combined heat sources and combined heat sinks in any process as temperature versus enthalpy plots. These can be used to show how much heat can be recovered by heat integration, and how much has to be supplied or removed by external utilities.

requires that heat flow naturally from hot to cold objects. This key concept is illustrated in the hot and cold composite curves (Figure 26.1), which represent the overall heat release and heat demand of a process as a function of temperature (see Appendix 26A).

The hot composite curve represents the sum of all the heat sources (hot streams) within the process in terms of heat load and temperature level. Similarly, the cold composite curve represents the sum of all the heat sinks (cold streams) within the process. When these curves are placed together on a single temperature–enthalpy plot (as in Figure 26.1), it is apparent that heat can be recovered within the process wherever there is a portion of the hot composite curve above a portion of the cold composite curve; that is, heat can flow from a higher temperature part of the process to a lower temperature part (see Appendix 26A). To keep the size of the heat recovery equipment reasonable, the temperature difference (approach) must be larger than a defined minimum allowable temperature approach, ΔT_{min}.

Most processes display a pinch—a region where the vertical separation of the curves approaches and eventually reaches the ΔT_{min} value. The pinch divides the process into two distinct regions:

- Above the pinch (i.e., in the higher temperature range), some heat integration is possible (where the hot composite curve sits above the cold composite curve), but there is a net heat deficit and an external utility heat source (Q_h) is required.
- Below the pinch (i.e., in the lower temperature range), some heat integration is possible (where the hot composite curve sits above the cold composite curve), but there is a net heat surplus and an external utility heat sink (Q_c) is required.

The distinction between net heat source and net heat sink regions is a key characteristic of the pinch approach, and it forms the basis for the pinch principle: *Do not transfer heat across the pinch.* This has two corollaries—do not use external (utility) cooling above the pinch, and do not use external (utility) heating below the pinch.

Composite curves and the pinch principle are the most widely recognized pinch tools. Many other pinch-based techniques also assist in process heat integration, such as heat exchanger grid diagrams ([4], pp. 32–35), grand composite curves ([4], pp. 52–61), and the CP rules ([4], p. 36). (CP, in this context, is the product of flow rate and specific heat capacity. The CP rules dictate when it is feasible to match a given hot stream with a given cold stream, based on their respective CP values.)

Related tools and techniques include algorithms to define the trade-off between energy consumption and capital investment, as well as pressure drop trade-offs in heat recovery, distillation column optimization, and total site analysis [5]. Many of the more recent developments in pinch analysis have focused on the management of material resources, such as water and wastewater [6,7] and hydrogen [8]. The primary purpose of this chapter, however, is to show how simple pinch techniques can be applied to real-world problems to improve energy efficiency, as illustrated in the following example.

RETROFIT PINCH PROCEDURE

Pinch analysis was initially developed for new plant designs. For retrofit work, the techniques need to be modified. The key difference is that in retrofit situations, the revamp must take into account existing equipment and plot space, whereas the designer of a new plant has greater flexibility to add or delete equipment at will.

Many different approaches to retrofit pinch analysis are possible. The example discussed here uses one of the simplest:

1. Obtain data.
2. Generate energy targets and utility targets.
3. Identify major inefficiencies in the existing heat exchanger network.
4. Define options for reducing or eliminating the largest inefficiencies.
5. Evaluate options.
6. Select the best option or combination of options.

RETROFITTING AN OIL REFINERY'S CRUDE DISTILLATION UNIT

This example describes a retrofit pinch analysis of a 90,000 bbl/day crude distillation unit (CDU), including both atmospheric and vacuum towers. The process flow diagram in Figure 26.2 shows the main process streams, equipment items, and heaters and coolers in relation to the heat recovery network, and Figure 26.3 gives details of the crude preheat train.

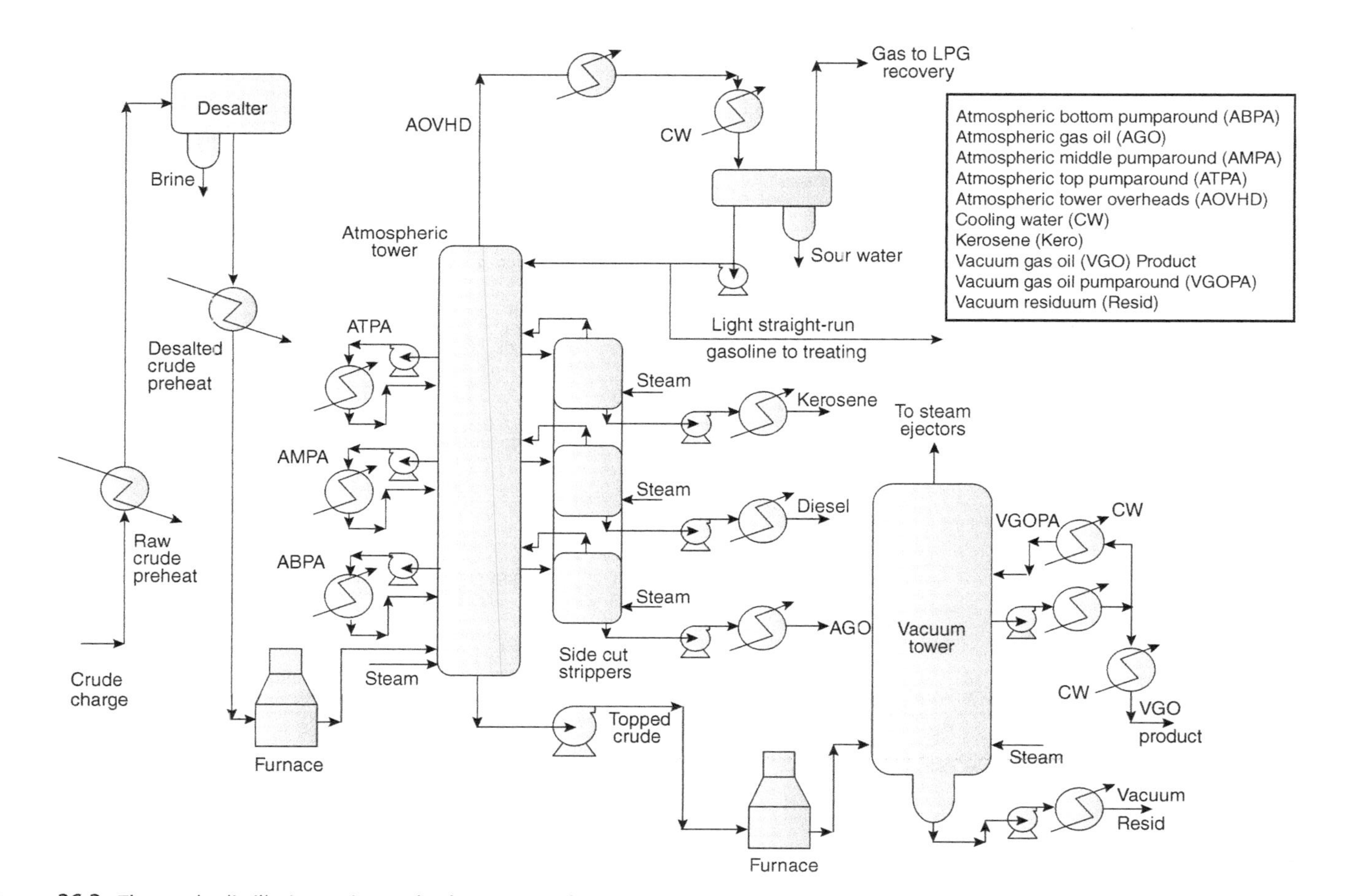

Figure 26.2. The crude distillation unit uses both an atmospheric tower and a vacuum tower to fractionate crude oil. Pumparounds are used to control cut points and remove heat at the highest practical temperature levels.

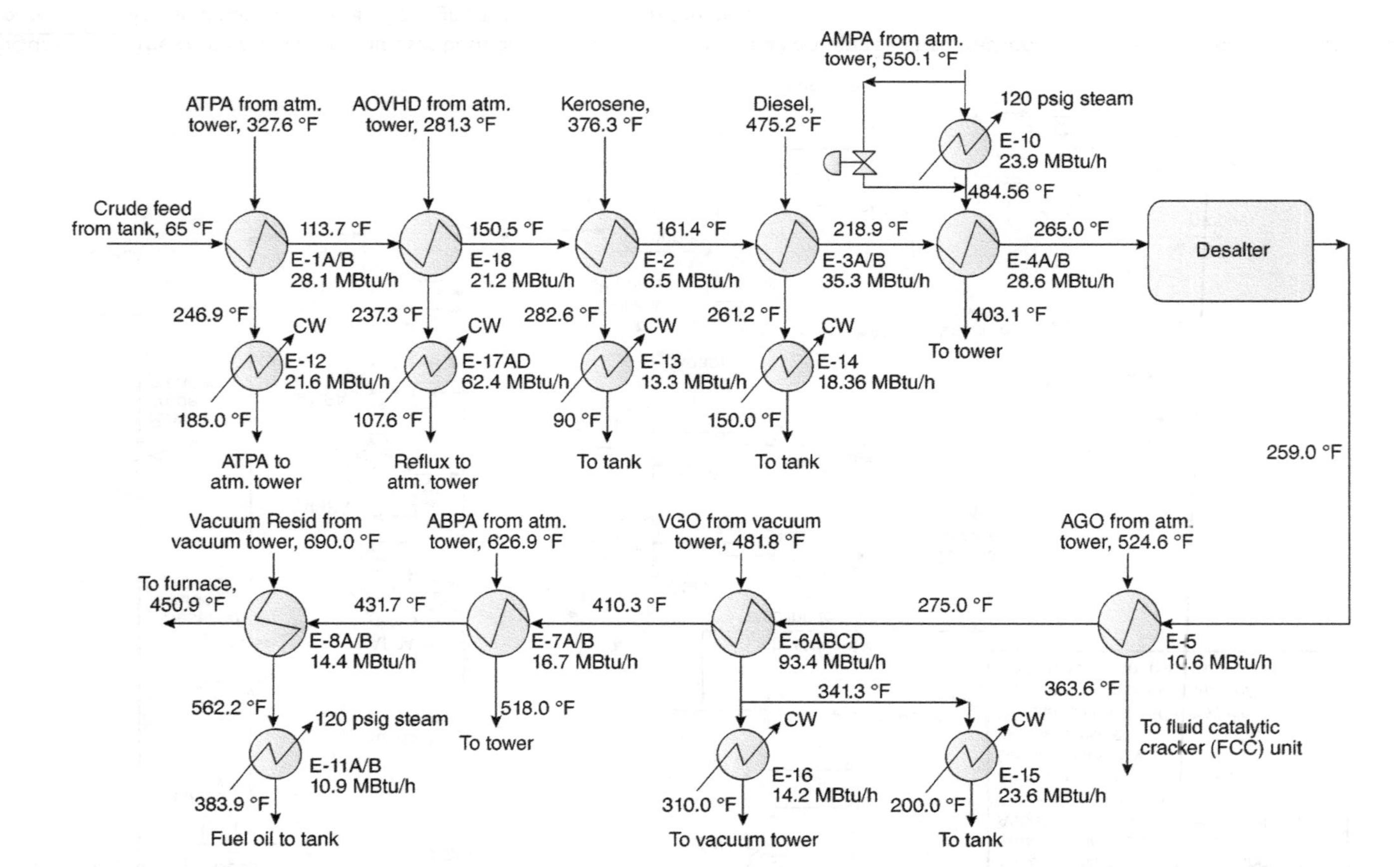

Figure 26.3. Heat from the pumparounds, product rundowns, and atmospheric tower overheads is recovered in the existing crude preheat train. Excess heat is removed in steam generation and cooling water.

The first major processing units in a refinery [9], CDUs separate crude oil by distillation into fractions based on boiling range. The separation is generally performed in two steps (Figure 26.2). First, the raw crude oil is fractionated at close to atmospheric pressure in an atmospheric tower. Then, the high-boiling bottom fraction (topped crude or atmospheric reduced crude) from the atmospheric tower goes to a second fractionator (vacuum tower) operating at high vacuum, which is generally provided by steam ejectors. A vacuum is used because the high temperatures required to vaporize topped crude at atmospheric pressure would cause thermal cracking.

The products from the CDU include the overheads, bottoms, and side streams from the two towers. Some of the fractions, especially those drawn as side cuts from the atmospheric tower, contain excessive amounts of low-boiling components that must be removed by steam stripping in order to meet product specifications.

Crude fractionation requires a large amount of energy. The highest temperature heat input is provided by feed furnaces. Lower temperature heat invariably comes from a network of exchangers that recover heat to preheat the crude oil.

The heat sources to the crude preheat train in this example include the atmospheric tower overheads (AOVHD) and product rundown streams: kerosene (Kero), diesel, atmospheric gas oil (AGO), vacuum gas oil (VGO) product, and vacuum residuum (Resid).

In addition, the atmospheric and vacuum towers have pumparound circuits, specifically the atmospheric top pumparound (ATPA), atmospheric middle pumparound (AMPA), vacuum gas oil pumparound (VGOPA), and atmospheric bottom pumparound (ABPA). These pumparounds provide a mechanism for removing heat from the towers at intermediate temperature levels, rather than taking all of the distillation heat out in the overheads. The amount of heat that can be removed in each pumparound is governed by the cut points required for the various product fractions. Most of the pumparound heat is used to preheat the crude feed, and excess heat from the pumparounds is either used to generate 120 psig steam or rejected to cooling water.

The VGO product and the VGO pumparound streams are combined through the preheat train, and are only split into separate streams downstream of exchanger E-6ABCD.

Another important component of most CDUs is the desalter. Raw crude oil contains inorganic salts, metals, and various organic compounds that can cause fouling, corrosion, and catalyst deactivation in downstream equipment. These undesirable materials are removed to acceptable levels by adding water (typically 3–10% by volume) and separating the aqueous and oily phases in the desalter vessel. This also removes suspended solids from the oil. The desalter typically operates at about 270 °F, so it is placed partway through the preheat train.

Step 1: Obtain Data

The most important data for a pinch study are the heat loads and temperatures for all of the process streams and utilities. In most cases, this information is obtained from a combination of test data, measured plant data, and simulations, often supported by original design data. Heat exchangers in crude preheat trains, especially the higher temperature exchangers, are subject to significant fouling, and most companies perform scheduled cleaning to maintain heat transfer rates and minimize blockages. It is important that the data for the pinch study represent realistic, sustainable heat exchanger conditions.

Once the data required for the analysis have been collected, they must be organized in the proper format for the pinch study—a process that is referred to as data extraction. The requirements vary somewhat depending on which software package is being used, but in general, the extracted data provide a simplified representation of the heat duties and inlet and outlet temperatures associated with all of the heaters, coolers, and process-to-process heat exchangers in Figure 26.3. Any data that are not potentially useful for heat integration purposes are omitted. Table 26.1 presents the resulting data set for input to pinch software.

Heating and cooling utilities must also be specified. Fired heaters are typically represented simply as heat sources at a single temperature (in practice, most software require a small temperature range) that is hot enough to satisfy any anticipated heat load in the unit. Ambient cooling (water or air) can also usually be represented as a heat sink at a single temperature. Steam generation is more complex and is generally represented as a segmented utility. The colder segment (230–350 °F in this case) represents boiler feed water (BFW) preheat, and the hotter segment (at a constant 350 °F) represents the latent heat. The utility data for the pinch study are summarized in Table 26.2.

The furnace efficiency and on-stream factor are derived from historical plant data. The fuel and steam cost data for evaluating projects are typically specified by the company's economics group. The ambient cooling uses cooling water, which is comparatively inexpensive (relative to furnace firing or steam), and it is ignored in utility cost calculations.

Simple equipment cost correlations are used for initial screening. Ideally, these should be agreed on with cost estimators at the site where the study is performed, since site-specific factors are often significant. In the absence of site-specific data, literature values can be used.

For this example, the installed cost of the heat exchangers (including foundations, local piping, valves, and instrumentation) is $\$200/ft^2$. Additional allowances are needed for piping costs if significant pipe runs are required for any of the options.

Companies generally specify investment criteria for their projects (e.g., hurdle rates). In this example, the economic cutoff for investments is a 4-year simple payback.

Step 2: Generate Energy and Utility Targets

1. *Set the value of* ΔT_{min}. Targets for minimum energy consumption are calculated based on the value chosen for ΔT_{min}. This parameter reflects the trade-off between capital investment (which usually increases as ΔT_{min} gets smaller) and energy cost (which decreases as ΔT_{min} gets smaller). It is possible to explore this trade-off quantitatively using pinch targeting methods, but in practice that is rarely done. Rather, rule-of-thumb values for ΔT_{min} that optimize the trade-off

TABLE 26.1. The Data Extracted for the Pinch Analysis Represents the Heating and Cooling of the Process Streams and the Utility Requirements of the Existing Process

			Hot Side			Cold Side		
	Heat Exchanger	Duty (MBtu/h)	Stream Name	T_s (°F)	T_t (°F)	Stream Name	T_s (°F)	T_t (°F)
E-1A-D	ATPA versus raw crude	28.1	ATPA	327.6	246.9	Raw crude	65.0	113.7
E-18	AOVHD versus raw crude	21.2	AOVHD	281.3	237.3	Raw crude	113.7	150.5
E-2	Kero versus raw crude	6.5	Kero	376.3	282.6	Raw crude	150.5	161.4
E-3A/B	Diesel versus raw crude	35.3	Diesel	475.2	261.2	Raw crude	161.4	218.9
E-4A/B	AMPA versus raw crude	28.6	AMPA	484.5	403.1	Raw crude	218.9	265.0
E-5	AGO versus desalted crude	10.6	AGO	524.6	363.6	Desalted crude	259.0	275.0
E-6A-F	VGO versus desalted crude	93.4	VGO	481.8	341.5	Desalted crude	275.0	410.3
E-7A/B	ABPA versus desalted crude	16.7	ABPA	626.9	518.0	Desalted crude	410.3	431.7
E-8A/B	Vacuum Resid versus desalted crude	14.4	Vacuum Resid	693.0	562.2	Desalted crude	431.7	450.9
E-10	AMPA steam generator	23.9	AMPA	550.1	484.5	120 psig steam generator		
E-11A/B	Vacuum Resid steam generator	10.9	Vacuum Resid	562.2	383.9	120 psig steam generator		
F-1	Feed furnace	200.0	Furnace			Desalted crude	450.9	665.0
E-12	ATPA cooler	21.6	ATPA	246.9	185.0	Cooling water		
E-13	Kero cooler	13.3	Kero	282.6	90.0	Cooling water		
E-14	Diesel cooler	18.4	Diesel	261.2	150.0	Cooling water		
E-15	VGO product cooler	23.6	VGO product	341.5	200.0	Cooling water		
E-16	VGO PA cooler	14.2	VGO PA	341.5	310.0	Cooling water		
E-17A-D	AOVHD condenser	62.4	ATM overheads	237.3	107.6	Cooling water		

The shaded area represents utility data and the unshaded area represents process data. T_s: supply temperature; T_t: target temperature.

TABLE 26.2. Heating and Cooling Utilities Are Defined in Terms of Temperature, Specific Enthalpy, and Unit Cost

	Temperature			
Utility	T_s (°F)	T_t (°F)	Δh (Btu/lb)	Cost ($/MBtu/h per year)
Furnace	750	749	n/a	49,400
120 psig steam generation	230	350	124	−36,500
	350	351	871	
Cooling water	60	61	n/a	n/a

Steam generation has a negative cost as it reduces net energy costs. Furnace efficiency = 85%; on-stream factor = 96% or 8400 h/year; fuel cost = $5.00/MBtu; 120 psig steam cost = $4.50/MBtu.

for different classes of processes, and between process streams and utilities, can be applied, in most instances, with a high level of confidence. Table 26.3 shows the rule-of-thumb values for CDUs and the actual values that were selected for this example. Similar values are also appropriate for many other refinery processes, such as fluid catalytic cracking (FCC) units, coker units, hydrotreaters, and reformers.

2. *Determine targets*. The next step, energy targeting, involves (conceptually) placing the hot and cold composite curves on a set of x–y axes and moving them horizontally until the smallest vertical distance between the curves is equal to the ΔT_{min} value (see Appendix 26A). In practice, however, the energy targets can be calculated more directly using what is known as the problem table algorithm ([4], pp. 25–30). Variants of the problem table algorithm are encoded in the commercially available pinch analysis software tools. The targets for the current example are shown in the form of composite curves (Figure 26.4) and a summary table (Table 26.4).

The composite curves show overall minimum hot and cold utility targets. Comparing these with the existing utility consumption yields the overall scope for energy saving. Most commercial software tools also quantify targets for each individual utility, as shown in Table 26.4.

TABLE 26.3. Rule-of-Thumb ΔT_{min} Values Have Been Developed from Experience in Many Pinch Studies, and They Have Been Used to Guide the Selection of Conservative ΔT_{min} Values in the Current Example

Type of Heat Transfer	Rule-of-Thumb ΔT_{min} Values (°F)	Selected ΔT_{min} Values (°F)
Process streams against process streams	50–70	70
Process streams against steam	15–35	35
Process streams against cooling water	10–35	30

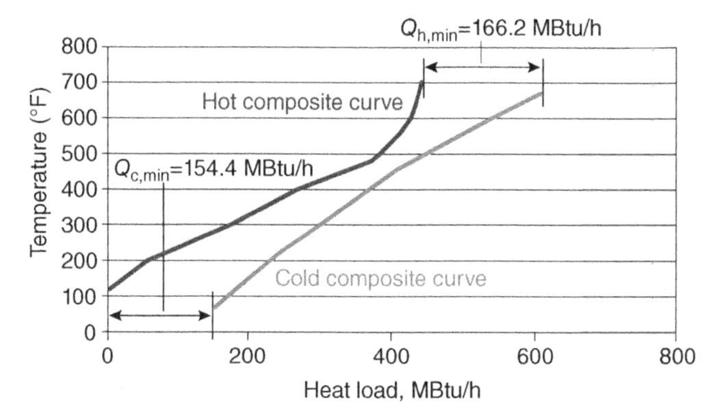

Figure 26.4. The composite curves for the crude unit example show the combined heat sources and combined heat sinks as temperature versus enthalpy plots. These curves are used to show how much heat can be recovered by heat integration, and how much has to be supplied or removed by external utilities.

The heat integration opportunities in the CDU are best understood from the summary information in Table 26.4. The first two columns show the existing heat loads for each utility and the corresponding target loads. In the case of 120 psig steam, more than 12% of each of these duties is attributable to the sensible heat used to raise the BFW from its supply temperature to the saturation temperature. The third column shows the scope for reducing each utility (existing load – target load). The 120 psig steam is exported, so a negative scope implies added value.

The following broad conclusions can be drawn from Table 26.4:

TABLE 26.4. Pinch Analysis Yields Overall Targets for Energy Use and for Individual Heating and Cooling Utilities

	Existing (MBtu/h)	Target (MBtu/h)	Scope (MBtu/h)	Saving (k$/year)
Total hot demand	200.0	166.2	33.8	
Total cold demand	188.2	154.4	33.8	
Hot utilities				
Fired heater	200.0	166.2	33.8	1670
Cold utilities				
120 psig steam generation	34.8	59.7	–24.9	909
Cooling water	153.4	94.7	58.7	0
Total				2579

These are used to calculate the scope for reducing utility consumption and the potential for energy cost reductions. Steam generation has a negative cost as steam generation reduces net energy costs.

- The furnace duty (absorbed heat) can be reduced by 33.8 MBtu/h through additional preheating of the crude oil. The savings are seen in reduced firing in the atmospheric feed furnace, F-1. The credit is $1,670,000/year.
- The 120 psig steam generation can be increased by up to 24.9 MBtu/h. This is worth $909,000/year.
- If these savings in furnace duty and steam generation are achieved, the cooling water duty is reduced by 58.7 MBtu/h.

Step 3: Identify Major Inefficiencies in the Heat Exchanger Network

This step incorporates design considerations. Most commercial pinch programs have tools to identify major inefficiencies and determine where heat crosses each pinch in a heat exchanger network (HEN). Two different pinches must be considered: the process pinch and a utility pinch. The process pinch (the type of pinch described earlier in the chapter) divides the process into a net heat sink region (above the process pinch temperature) and a net heat source region (below the process pinch temperature).

In this example, the process pinch temperature for hot streams is 481.8 °F. The process pinch temperature for cold streams, by definition, must be less than this by ΔT_{min} (i.e., 411.8 °F).

However, it is more convenient to quote a single pinch "interval temperature," which is determined as part of the problem table analysis. This is usually the average of the hot and cold stream pinch temperatures, and in this case it is 446.8 °F. If heat crosses this pinch, the furnace duty increases and additional heat goes to one of the two cold utilities (120 psig steam or cooling water).

The utility pinch occurs at 350.0 °F (interval temperature). It arises because there are two cold utilities. Any excess heat should be removed from the process in 120 psig steam between the process pinch and the utility pinch. Below the utility pinch temperature, excess heat has to be removed in cooling water. If heat crosses the utility pinch, less of the valuable 120 psig steam is produced, and more heat is rejected in cooling water.

The results from studying the pinches may be presented either as a grid diagram (Figure 26.5) ([4], pp. 32–35), or as a cross-pinch summary table (Table 26.5). Both provide substantially the same information, but in different formats.

In the grid diagram (Figure 26.5), the pinches appear as broken vertical lines. The hot and cold stream pinch temperatures are shown at the top and the bottom of the diagram, respectively. The process streams are shown as horizontal lines, with the hot streams running from left to right and the cold streams from right to left; that is, high temperatures are generally on the left of the diagram and low temperatures are on the right. The temperature scale does not need to be precise. The key is that the initial and final temperatures of each stream are appropriately related to the pinches. It is then apparent which streams have segments above the process pinch temperature, between the two pinches, and below the utility pinch.

Heat exchangers can now be added to the diagram. Process-to-process heat exchangers are shown as dumbbells linking a hot stream to a cold stream. Utility heat exchangers are shown as circles with a label identifying the type of utility used. If

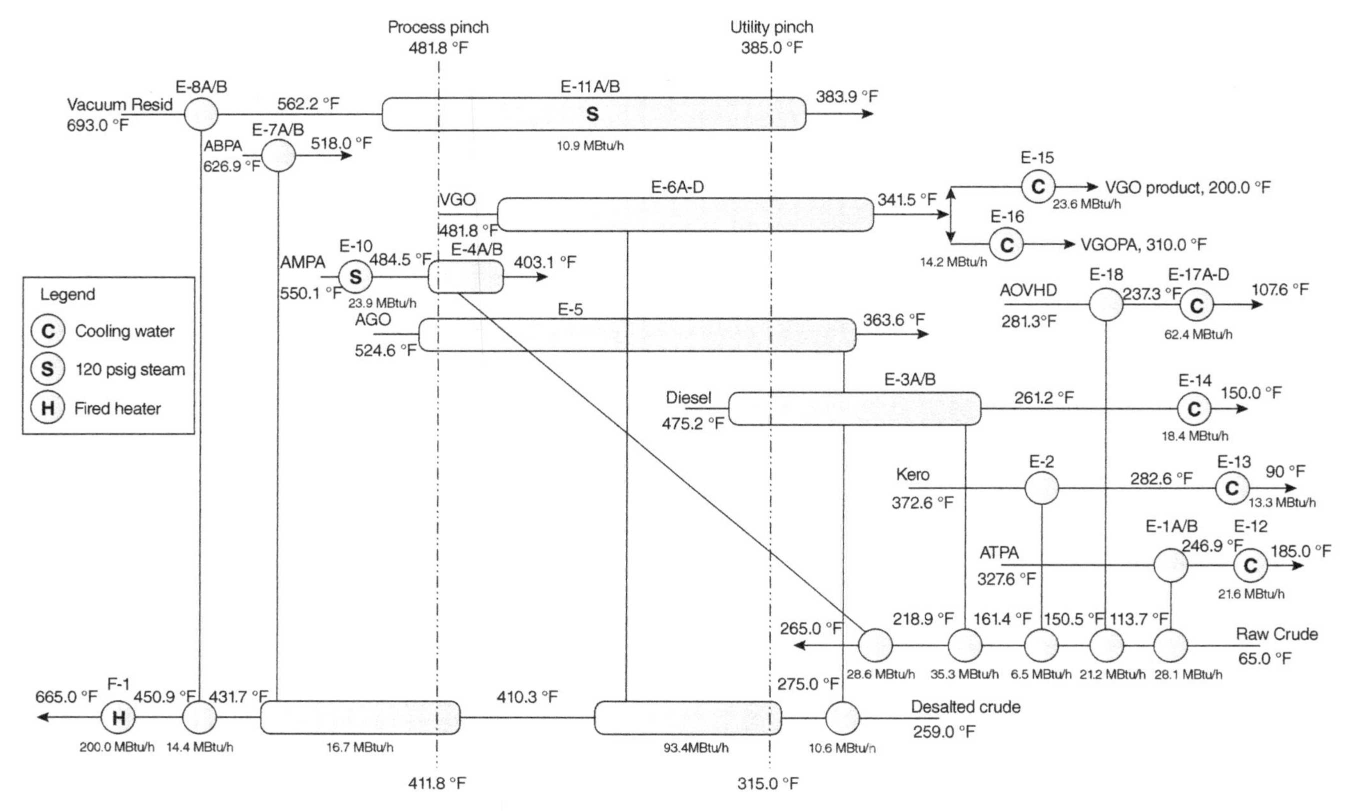

Figure 26.5. Representing the existing preheat train as a grid diagram highlights inefficient cross-pinch heat exchangers.

TABLE 26.5. Tabulating the Amount of Cross-Pinch Heat Transfer in Each Heat Exchanger Focuses Attention on the Largest Inefficiencies in the Preheat Train

Heat Exchanger		Hot Stream	Cold Stream	Cross-Pinch Duties (MBtu/h)	
				Process (446.8 °F)	120 psig Steam Generation (350.0 °F)
E-3A/B	Diesel versus raw crude	Diesel	Raw crude		14.9
E-4A/B	AMPA versus raw crude	AMPA	Raw crude	0.9	28.6
E-5	AGO versus desalted crude	AGO	Desalted crude	2.8	9.2
E-6A-F	VGO versus desalted crude	VGO	Desalted crude		−1.3[a]
E-7A/B	ABPA versus desalted crude	ABPA	Desalted crude	1.2	
E-10	AMPA steam generator	AMPA	120 psig steam generator	23.9	3.0
E-11A/B	Vacuum Resid steam generator	Vacuum Resid	120 psig steam generator	4.9	1.3
Total				33.8	55.6

[a]A negative cross-pinch duty in a heat exchanger means that the minimum temperature difference between the hot and cold streams is less than the specified ΔT_{min}.

either the hot or cold portion of a heat exchanger extends across one or more pinch boundaries, the appropriate circle is elongated to show the temperature range relative to the pinches. Wherever possible, the bar of a dumbbell is drawn vertically. However, this is not possible when the entire duty crosses a pinch. This is the case for E-4A/B, which consists of two heat exchanger shells, E-4A and E-4B, placed in series.

Table 26.5 reveals that the largest inefficiencies are in the two AMPA heat exchanger services, E-4A/B (AMPA versus raw crude, with 28.6 MBtu/h crossing the 120 psig steam pinch) and E-10 (AMPA steam generator, with 23.9 MBtu/h crossing the process pinch). In Figure 26.5, this inefficiency in E-4A/B is indicated by the diagonal cross-pinch line. However, the problem with E-10 is less obvious. The user must recognize that 120 psig steam generation should only occur at temperatures below the utility pinch.

These inefficiencies occur for two reasons. The highest temperature heat in the AMPA is used to generate 120 psig steam; it would be more beneficial to use it to preheat desalted crude at the hot end of the preheat train. The rest of the AMPA heat is used to preheat much colder raw crude, although it is hot enough to generate 120 psig steam. Therefore, the preheat train redesign must focus on redistributing the AMPA heat.

The next largest inefficiency (14.9 MBtu/h crossing the steam generation pinch) is in E-3A/B, the diesel versus raw crude exchanger. This indicates an opportunity to generate 120 psig steam using part of the diesel rundown heat, and using lower temperature heat sources to replace heat from the diesel stream to preheat the raw crude.

All of the remaining cross-pinch duties are significantly smaller (<10 MBtu/h). Although there are sometimes viable projects for savings of this magnitude, it is best to focus, at least initially, on the larger opportunities, which in this case account for more than 70% of the heat crossing the process pinch and more than 80% of the heat crossing the utility pinch.

Step 4: Define Options for Reducing or Eliminating the Largest Inefficiencies

Three types of opportunities should generally be considered in retrofit projects: rearranging existing heat exchangers to increase feed preheat and/or steam generation; adding heat transfer area to existing matches, for example, by adding new heat exchanger shells; or adding new heat exchangers to match streams that are not currently matched.

The inefficiencies in Table 26.5 and the stream data in Table 26.1 provide the information required to generate specific ideas. The most promising opportunities are as follows:

- Rearrange the existing heat exchangers in the AMPA circuit (E-4A/B and E-10).
- Add process-to-process heat exchangers to increase the feed preheat. The VGO (product plus pumparound) is the best heat source since it rejects a large amount of heat at high temperatures to cooling water in E-15 and E-16.
- Add 120 psig steam generators. Step 3 showed that the largest opportunity of this type is on the diesel stream, ahead of E-3A/B.
- Add a BFW preheater to increase production of 120 psig steam (the existing design has no BFW preheating). If the BFW can be preheated with waste heat, steam production can be increased by up to 12% in the existing steam generators.

Step 5: Evaluate the Options

A technical and economic comparison of the various options that have been identified is now needed to see which ones meet the investment criteria and which of those are the most attractive.

In any heat exchanger network, each change in any particular heat exchanger is likely to have knock-on effects on other heat exchangers. This is particularly true in CDU preheat trains, which are generally the most complex HENs in a refinery.

Some commercially available pinch analysis software packages incorporate tools for evaluating these effects, although many practitioners prefer to use spreadsheets or other simulation tools to assess the interactions. Regardless of which tools are used, some type of model is needed to assess the performance of the HEN and to quantify the utility savings attributable to each option and combination of options evaluated.

It is also important to consider any constraints that could affect the viability of a new heat integration scheme. For example, hydraulic constraints may limit the number of heat exchangers that can be added. Minimum flow rates must be maintained in all heat exchangers to keep fouling rates to acceptable levels. Better heat integration often results in closer temperature approaches that may reduce the heat loads in existing pumparound heat exchangers, but the pumparounds must still be able to remove sufficient heat to control fractionation cut points. The desalter temperature must be kept within an acceptable range. There must not be excessive vaporization of the crude before it reaches the furnace. It is often necessary to make significant compromises to accommodate all of these factors, so the final design is often very different from an ideal pinch design.

Using the model and allowing for all known constraints, screening-quality economic evaluations are carried out for the opportunities identified in step 4. These calculations accomplish the following:

- Quantify the utility savings attributable to each option and combination of options. The utility savings are converted to monetary savings using the utility costs data in Table 26.2.
- Estimate the cost of implementing each option. Generally, this requires estimating the sizes of new heat exchangers and any other new equipment needed, and the lengths of new pipe runs, and then using cost correlations to develop cost estimates.
- Assess economic viability. The estimated costs and savings for an option are used to calculate the simple payback (cost/annual savings), as well as other measures of value, such as return on investment (ROI) or net present value (NPV), to quantify the attractiveness of each option.

Step 6: Select the Best Option or Combination of Options

As noted in steps 3 and 4, the largest opportunity involves realigning the AMPA stream so that its hottest portion is matched against the desalted crude above the process pinch and its lower temperature portion is used to generate 120 psig steam. This can be achieved fairly easily. The existing AMPA versus raw crude heat exchangers, E-4A/B, have a high enough temperature rating to be reused in this hotter service, and the sequence of E-4A/B and the existing AMPA steam generator, E-10, can be reversed with a minor piping change, as shown in Figure 26.6.

However, moving E-4A/B from its existing location between E-3A/B and the desalter reduces heat input to the raw crude, and this heat must be replaced. The easiest replacement is by adding a new service, E-X1A/B, to recover heat from VGO and apply it to the raw crude immediately ahead of the desalter, where E-4A/B is currently located. E-X1A/B consists of two large shells (approximately 7000 ft^2 each). These changes (rearrange AMPA and add E-X1A/B) constitute the core modifications for this revamp project.

The core modifications increase crude preheat by 14.0 MBtu/h, which raises the coil inlet temperature by 17.1 °F and translates into equivalent feed furnace savings. In

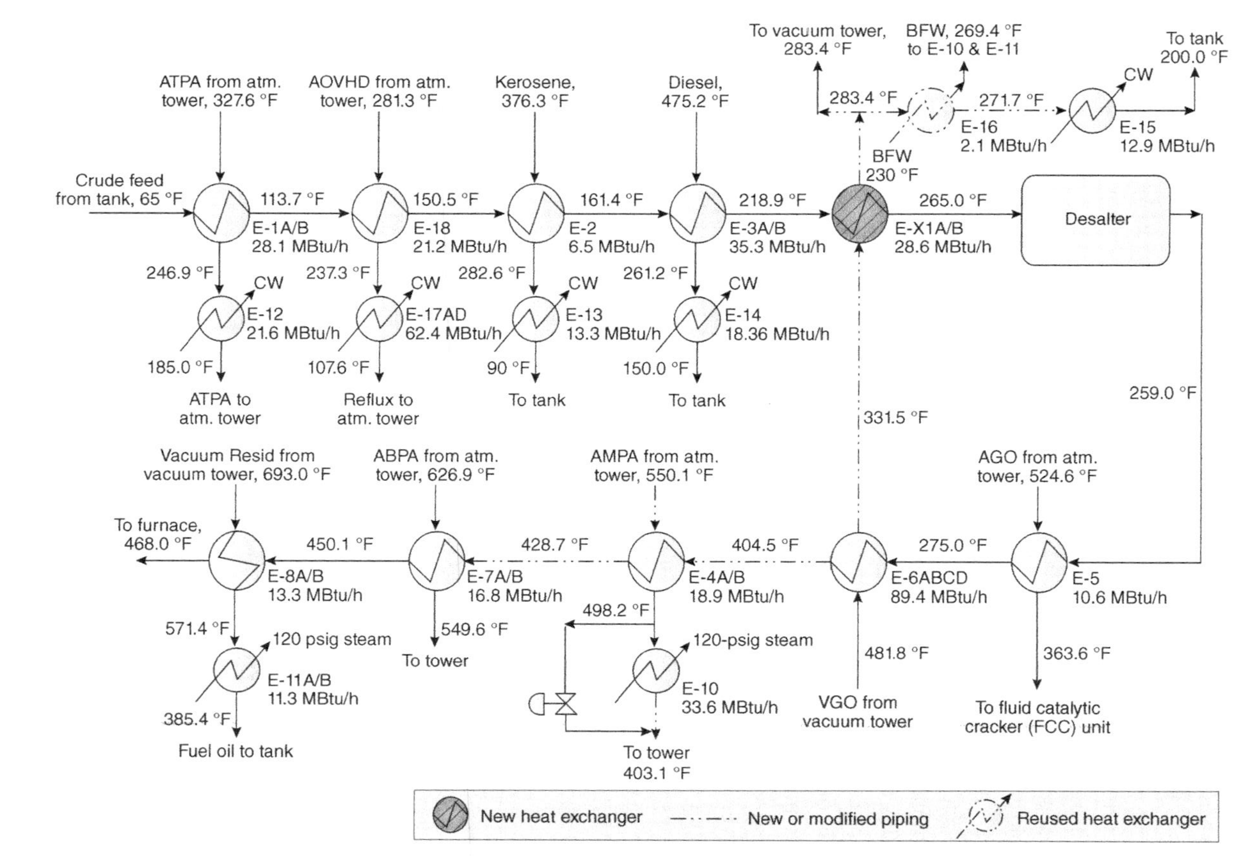

Figure 26.6. The final design requires just one new pair of heat exchanger shells, plus repiping of some existing equipment.

addition, these changes also increase heat to 120 psig steam generation (in E-10 and E-11A/B) by a total of 10.1 MBtu/h. These combined savings are worth $1,060,000/year. The cost for the modifications is $3,335,000, most of which is associated with the addition of E-X1A/B. The simple payback is 3.1 years.

When E-1XA/B is added to recover additional heat from the VGO (pumparound and product), the VGO pumparound cooler, E-16, is no longer needed. The E-16 shell is therefore available for an alternative use, such as preheating BFW using the VGO product rundown ahead of cooler E-15. This arrangement is also shown in Figure 26.6. This change recovers 2.1 MBtu/h of rundown heat to BFW, which results in an increase in steam generation worth $76,000/year. The cost of the modification is $250,000, giving a simple payback of 3.3 years. This covers the cost of changing the tube bundle in E-16 (which is necessary because the existing bundle is not rated for the pressure of boiler feed water) and some piping modifications.

Ordinarily, it would not be economical to add a heat recovery exchanger in a CDU for a duty as low as 2.1 MBtu/h. However, in this case, the option of reusing an existing piece of equipment greatly improves the economics.

The other options described in step 4 were also evaluated. However, they either were uneconomical or had technical problems and were impractical to implement. Table 26.6 summarizes the two economical project options.

Overall, these changes save 14.0 MBtu/h in crude preheat and recover 12.2 MBtu/h for additional 120 psig steam generation (26.2 MBtu/h total), worth $1,136,000/year. These results compare with an "ideal" target crude preheat saving of 33.8 MBtu/h and a target increase of 24.9 MBtu/h in heat recovery for 120 psig steam generation (58.7 MBtu/h total), with a net monetary target saving of $2,579,000/year (Table 26.4). Therefore, the selected design achieves about 45% of the energy and economic target savings.

The economically achievable saving is a relatively low percentage of the target because of the interactions in the preheat train. Adding heat recovery in a preheat train reduces temperature differences in all downstream heat exchangers, which causes a reduction of heat load in these heat exchangers. Additional heat transfer area is needed to recover this lost heat and increases the project cost, making it less attractive.

It is usually simpler and much less expensive to incorporate energy efficiency measures in the initial design of a CDU than it is to improve its energy efficiency in a revamp. However, as this example shows, some significant revamp opportunities do exist, and pinch analysis is an effective tool for identifying them.

TABLE 26.6. Evaluation of the Available Options Showed that Two Projects Satisfy the Economic Requirement for a Payback Not Exceeding 4 Years

ID	Project Description	Duty (MBtu/h)	Credit ($/year)	Investment ($)	Payback (Years)
1	Core modifications	24.1	1,060,000	3,335,000	3.1
2	BFW preheat	2.1	76,000	250,000	3.3
Total		26.2	1,136,000	3,585,000	3.2

OBSERVATIONS FROM CRUDE UNIT EXAMPLE

Pinch analysis is a very powerful technique for identifying minimum energy consumption targets for heating and cooling, and for identifying projects to achieve significant energy savings. Properly calculated pinch targets are always thermodynamically achievable, and ΔT_{min} values are selected with the intention of generating economically realistic targets. However, achieving savings requires not just targets, but actual projects. In many cases, practical process constraints and interactions limit what can be achieved economically to something significantly less than the pinch targets. This is an important fact that needs to be kept in mind when using pinch analysis.

COMMERCIAL USE OF PINCH ANALYSIS

A wide range of pinch software is now available, ranging from shareware and simple spreadsheets to highly sophisticated commercial packages. Because of this availability, engineers in a wide range of organizations can carry out pinch analysis, at least at a basic level. However, most advanced pinch analysis work is carried out by specialized consultants, typically in small companies. Some of the larger engineering companies also maintain a capability in the technology, and several academic institutions are engaged in both research and industrial applications. Whoever performs the pinch work, it is important that pinch activities should not be carried out in isolation. Rather, they should be integrated into a larger engineering organization to ensure the proper flow of information, evaluation of results, and ultimate implementation of projects.

APPENDIX 26A. COMPOSITE CURVES

The signature output from pinch targeting studies is a diagram showing "composite curves." This appendix describes how these curves are constructed and briefly discusses their significance.

Heat Transfer in a Single Heat Exchanger

Heat transfer in a simple two-stream heat exchanger can be represented as a two-dimensional temperature versus heat load (*T–H*) diagram, as shown in Figure 26A.1. Temperature (usually in units of °F or °C) is customarily plotted vertically, and heat load (commonly in units of MBtu/h or MW) is plotted horizontally. The hot stream (i.e., the stream that is releasing heat) in the heat exchanger runs diagonally from top right to lower left from its supply temperature (T_{hs}) to its target temperature (T_{ht}), and the cold stream (i.e., the stream that is absorbing heat) runs from bottom left to upper right from its supply temperature (T_{cs}) to its target temperature (T_{ct}). The heat load, Q_{hx}, has the same absolute magnitude for both streams, although the hot stream is giving up heat while the cold stream is absorbing heat. The slopes of the two lines are equal to the reciprocal of the "heat capacity flow rate (CP)" of the corresponding

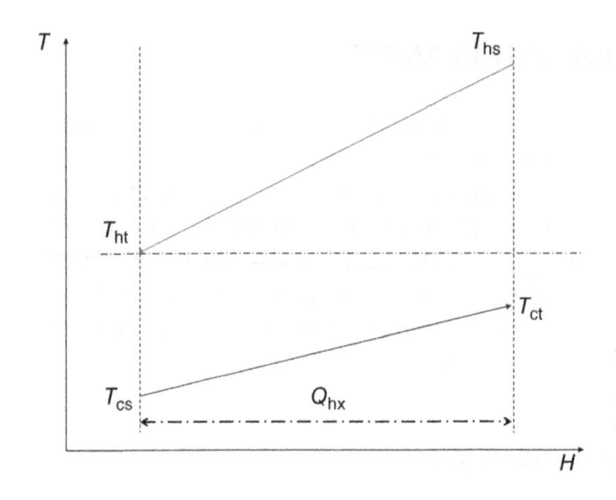

Figure 26A.1. A temperature versus heat load diagram is a convenient way to represent heat transfer in a simple two-stream heat exchanger.

stream, and in most cases the CP values—and hence the slopes—are different for the hot and cold streams.

Data Required for Composite Curves

Pinch analysis allows us to examine the heat transfer in multiple heat exchangers simultaneously. The simplest way to understand this is with composite curves.

Table 26A.1 shows data for a process that has four streams, two hot (H1 and H2) and two cold (C3 and C4). The data for each stream includes the supply and target temperatures (T_s and T_t) and the heat load, and also the "heat capacity flow rate (CP)," which is the magnitude of (heat load)/($T_s - T_t$).

We assume that the minimum temperature approach for heat integration (ΔT_{min}) is 20 °F. We also assume that any heating or cooling requirements that cannot be met by heat integration must be satisfied by using external utilities. The problems that we will now address are as follows:

1. How much of the heat that has to be removed from the two hot streams can be recovered in the cold streams?
2. How much heat has to be supplied from an external hot utility (e.g., furnace firing or steam)?
3. How much heat has to be rejected to a cold utility (e.g., cooling water or air)?

TABLE 26A.1. Data for Four-Stream Problem

Stream	T_s (°F)	T_t (°F)	Heat Load (MBtu/h)	CP (MBtu/h/°F)
H1	388.0	178.0	42.00	0.200
H2	222.0	122.0	40.00	0.400
C3	200.0	260.0	48.00	0.800
C4	104.0	148.0	13.20	0.300

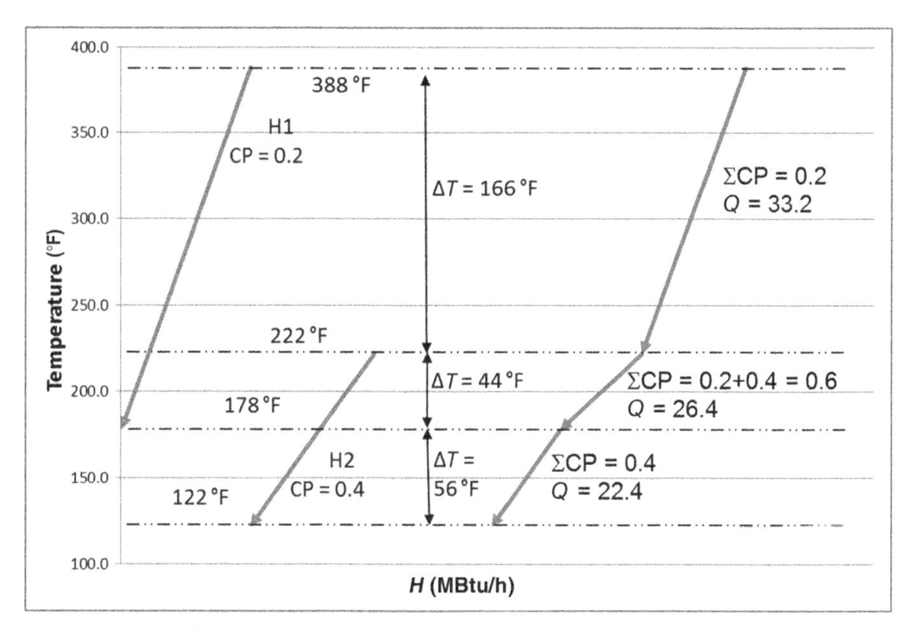

Figure 26A.2. Construction of the hot composite curve.

Constructing the Curves

We can approach this problem by constructing composite curves, and we start with the hot composite curve. The procedure is shown in Figure 26A.2, and the steps are as follows:

- Select the hot streams only (H1 and H2).
- Plot the two hot streams together on a T–H diagram.
- Draw in "boundary temperatures" corresponding to the supply and target temperatures. In this case, the boundary temperatures are 388, 222, 178, and 122 °F.
- Calculate the temperature difference (ΔT) between adjacent boundaries. In this example, the ΔT values, starting from the hottest pair of boundaries, are 166, 44, and 56 °F, as shown in Figure 26A.2.
- Identify the streams between adjacent boundary temperatures. In this example, only stream H1 exists between the hottest two boundaries, H1 and H2 coexist between the next pair, and only H2 exists between the final pair.
- For each pair of boundaries, calculate the combined CP value (ΣCP) for all of the streams between the boundary temperatures. In this example, the ΣCP values are 0.2, $0.2 + 0.4 = 0.6$, and 0.4 MBtu/h/°F.
- Compute the total heat load contribution between each pair of boundary temperatures ($Q = \Sigma CP \cdot \Delta T$). In this example, the Q values are 33.2, 26.4, and 22.4 MBtu/h.

- Each pair of boundary temperatures with its associated heat load can be considered as a "stream" with its own supply and target temperatures and its own heat load, which is represented by a diagonal line on the *T–H* diagram.
- We can now piece together the hot composite curve by placing the segments between each pair of boundaries end to end, as shown in Figure 26A.2.

The same procedure is used to construct the cold composite curve, except that in this case we use the cold streams rather than the hot streams. This is shown in Figure 26A.3. Note that in this particular example only stream C3 exists between the hottest pair of boundary temperatures and only stream C4 exists between the lowest pair of boundary temperatures, but *neither* of the streams exists between the middle pair of boundary temperatures. Consequently, ΣCP and Q are both zero in the middle region.

We can now plot the hot and cold composite curves on the same *T–H* diagram such that the smallest vertical separation between the curves is equal to ΔT_{min}, which in our example is 20 °F. This is shown in Figure 26A.4. The circled region near the center of Figure 26A.4 is the "pinch," that is, the region where the temperature difference is equal to ΔT_{min}.

The hot composite curve represents the combined heat source within our process, in terms of both temperature level and quantity of heat. Similarly, the cold composite curve represents its combined heat sink. When they are plotted together as in Figure 26A.4, wherever a portion of the hot composite sits vertically above a portion of the cold composite curve heat can flow from a heat source to a heat sink, and the available

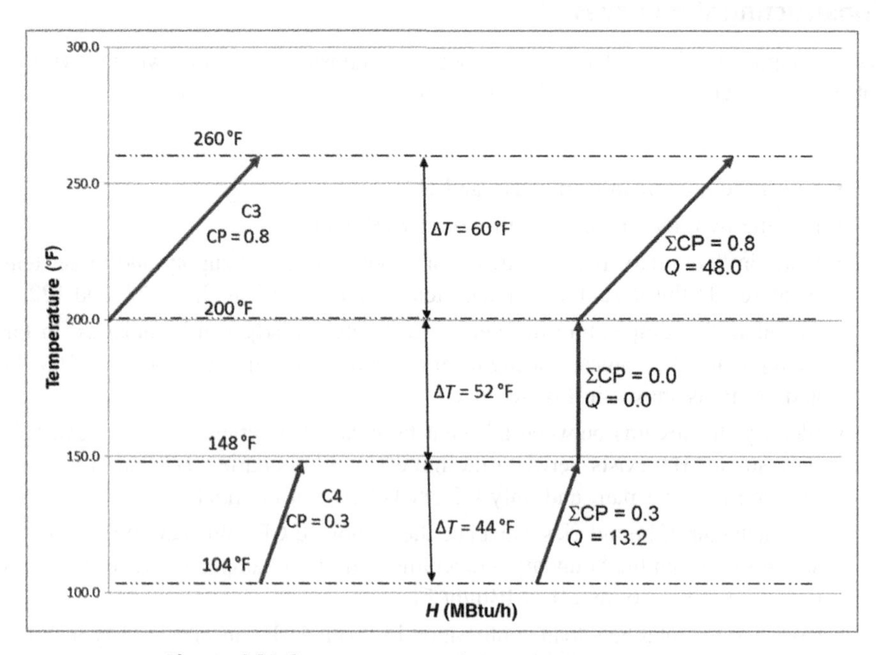

Figure 26A.3. Construction of the cold composite curve.

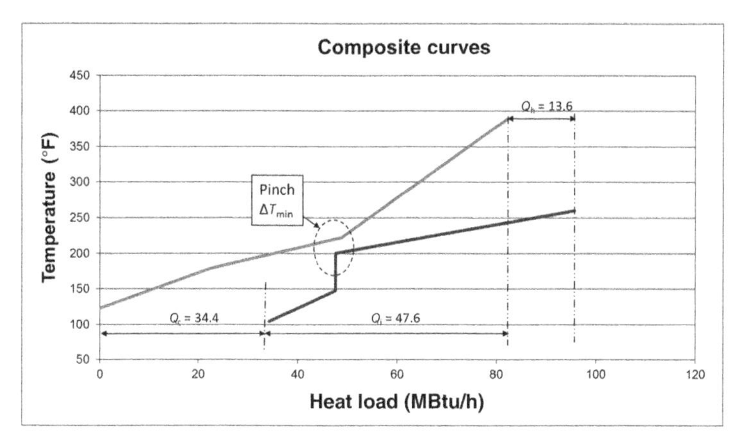

Figure 26A.4. Hot and cold composite curve combined on a single *T–H* diagram.

temperature driving force is guaranteed to be at least ΔT_{min}. Thus, the horizontal projection (Q_i) of the overlap region in Figure 26A.4 represents the maximum scope for heat integration within the process, subject to the prevailing value of ΔT_{min}.

There is a portion of the hot composite curve to the left of the overlap region. This represents heat sources that cannot be recovered within the process, and therefore must be rejected to a cold utility such as air or cooling water. The horizontal projection (Q_c) is the amount of heat that must be rejected. Similarly, there is a portion of the cold composite to the right that represents heat sinks that cannot be satisfied by heat recovered within the process, and that must therefore be supplied by heat Q_h from an external utility heat source. Q_i, Q_h, and Q_c are therefore the answers to the questions we posed earlier, and in this example their values are 47.6, 13.6, and 34.4 MBtu/h, respectively. If we change the value of ΔT_{min}, the values of Q_i, Q_h, and Q_c will all change, reflecting the impact of allowable temperature approach on potential heat recovery.

REFERENCES

1. Rossiter, A. (2010) Improve energy efficiency via heat integration. *Chemical Engineering Progress*, 106(12), 33–42.
2. Gunderson, T. and Naess, L. (1987) The synthesis of cost optimal heat exchanger networks. *XVIII European Federation of Chemical Engineering Congress*, Giardini Naxos, Italy, April 26–30, 1987.
3. Gunderson, T. and Grossmann, I.E. (1990) Improved optimization strategies for automated heat exchanger networks through physical insights. *Computers & Chemical Engineering*, 14(9), 925–944.
4. Linnhoff, B., et al. (1994) *A User Guide on Process Integration for the Efficient Use of Energy*, revised 1st edition, IChemE, Rugby, UK.
5. Linnhoff, B. (1994) Pinch analysis: building on a decade of progress. *Chemical Engineering Progress*, 90(8), 32–57.

6. Wang, Y. P. and Smith, R. (1994) Wastewater minimization. *Chemical Engineering Science*, 49(7), 981–1006.
7. Manan, Z.A., et al. (2004) Targeting the minimum water flow rate using water cascade analysis technique. *AIChE Journal*, 50(12), 3169–3183.
8. Alves, J.J. and Towler, G.P. (2002) Analysis of refinery hydrogen distribution systems. *Industrial & Engineering Chemistry and Research*, 41(23), 5759–5769.
9. Gary, J.H. and Handwerk, G.E. (1994) *Petroleum Refining Technology and Economics*, 3rd edition, Marcel Dekker, New York, pp. 39–69.

27

ENERGY MANAGEMENT KEY PERFORMANCE INDICATORS (EnPIs) AND ENERGY DASHBOARDS

Jon S. Towslee

EFT Energy, New York, NY, USA

WHAT ARE EnPIs?

KPIs or key performance indicators have been around for a long time. Depending on the industry, there are numerous types and uses of KPIs. Traditionally, KPIs have been used for financially based measurements and comparisons. This chapter focuses on the creation and use of KPIs for energy management in manufacturing (energy management performance indicators or EnPIs). At the highest level, businesses are constantly challenged to measure the performance of peer plants and track corporate improvements over time to satisfy stakeholders within the organization. In this chapter, we will discuss the use of EnPIs to measure, track, improve, and compare the energy efficiency of manufacturing processes.

Industries such as commercial real estate and retailing have standard benchmark EnPIs that can be used to measure performance of a given facility with other similar facilities on a normalized basis. For example, the Department of Energy website offers many benchmarks for commercial office buildings and shopping malls. However, similar metrics and benchmarks for manufacturing processes are hard to find. Manufacturing processes and plants are often too varied to have a common set of EnPIs or metrics that can be universally applied across a diversified business, let alone

Energy Management and Efficiency for the Process Industries, First Edition. Edited by Alan P. Rossiter and Beth P. Jones.

a whole industry. It is typically much easier to establish common EnPIs on specific systems and machinery, as these are often similar across manufacturing operations. Meaningful EnPIs are often selected SBU (strategic business unit) by SBU, plant by plant, or process by process.

When setting out to establish energy and production EnPIs, it is important to think through how and by whom these metrics will be used. The best place to start in selecting EnPIs is to determine which business/technical questions you are trying to answer. The following are some common questions addressed in establishing EnPIs using production and energy data:

- Which facilities produce product in the most energy- and cost-efficient manner (peer to peer)?
- Which processes are the most energy and cost efficient (peer to peer)?
- Which machine/process operating modes are most effective (sharing best practices)?
- Are processing lines/machinery performance improving or deteriorating over time (operational tuning, employee training, maintenance, and replacement)?

It is nearly impossible to answer any of these questions in absolute terms, since it is very rare that two plants are built and operated in exactly the same way. This is where the need for EnPIs comes into play. Finance will often define the high-level metrics for a facility based on monthly, weekly, and daily production volumes, but these high-level metrics rarely help answer the business questions associated with improving energy efficiency. Such metrics simply do not provide meaningful and actionable information at a granular enough level to help process and facility engineering personnel to identify specific actions to improve those metrics. Consequently, specific EnPIs must be created to help obtain enough pertinent information to not only answer the relevant business questions, but also improve and maintain the operation and energy efficiency of plants, processes, and machines.

In order to make meaningful comparisons from site to site, process area to process area, line to line, and so on, we must normalize data. Each process area, line, unit, and machine is designed for a different capacity and loading point. Therefore, it is common and necessary to compare how much work or production is obtained from an area, as a function of how much work or energy goes into that area. Whatever normalization calculations are developed should be applied in a consistent fashion across the enterprise.

EnPIs can run the gamut from very simple to very complex. Simple EnPIs often start in the form of total plant production divided by total plant energy consumption, usually on a monthly basis. This is probably the most basic and simplistic EnPI. This simple EnPI, however, does not provide enough insight into what is happening in the plant to allow operators to drive this number up or down. To identify actionable data, you need to obtain more granular measurements within the plant and to develop more complex EnPIs that are dependent on several different variables operating simultaneously. When developing these more complex EnPIs, you may need to employ more sophisticated software platforms that support multivariable regression modeling. The more complex EnPIs will provide more meaningful and actionable information that will help you to drive measurable improvements in energy and operational efficiencies throughout your plant and enterprise.

HOW TO CHOOSE EnPIs

Before determining the EnPIs needed to adequately monitor and measure energy and operational performance, you first need to determine the scope of each metric. Are you attempting to track and manage performance at the plant, area, line, or machine level? You need to determine what business questions you are trying to answer and determine exactly what data points are needed to formulate the EnPIs to answer those questions.

The more comprehensive your plan to track energy within your facilities, the more attention you will need to pay to systems, energy, and heat flows. This is especially true if the products (outputs) of one process become feedstock or energy supplies (inputs) to other processes. The simple way to think about this is to take an "outside-in approach." First, "draw a circle" around your plant and determine which EnPIs and underlying data will be needed to track energy performance at the plant-wide level, whether comparing this site to its peers or simply trying to show time series improvement in the plant. This methodology should be repeated at successively deeper levels within the plant until you have reached the desired level of granularity (e.g., area, unit, line, or equipment levels) (see Figure 27.1).

There are two approaches to planning and selecting the set of EnPIs and underlying measurements that will satisfy your business needs while balancing the practical need to

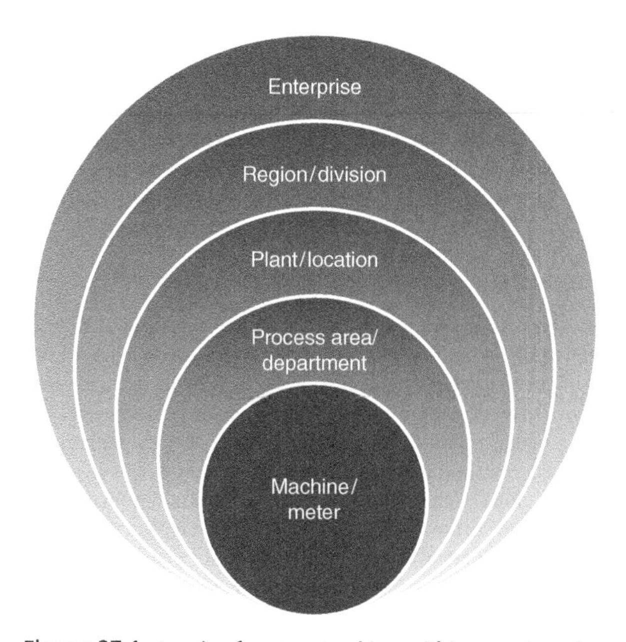

Figure 27.1. Levels of energy tracking within an enterprise.

obtain reliable data in a cost-effective manner. The first approach, which is more of a top-down approach, is to determine what business questions you are trying to answer and then determining exactly which EnPI data points are needed to provide those answers at each level within the site. This approach gives you the theoretically perfect set of data and EnPIs. The second approach, which is more of a bottom-up approach, is to start with data availability. Which meters and process measurements are readily available on a network or data historian? Determine how closely these measurements match the theoretical set of data established in the first approach. This more pragmatic approach will provide a road map for the investment needed to close critical measurement gaps and help you to formulate a list of missing instruments and meters. The cost/benefit analysis associated with this approach should stimulate some spirited discussion with plant stakeholders to arrive at the appropriate balance of data availability and infrastructure cost to obtain the data needed to satisfy EnPI tracking requirements. It is often more useful to start with EnPIs that are "directionally correct" rather than "theoretically perfect."

To measure the energy performance associated with a specific functional area (e.g., machine, line, or process area), it is necessary to account for all of the energy flows in order to create a meaningful EnPI. As mentioned earlier, the methodology involves drawing a boundary around the functional area and making sure that all energy inflows and outflows are metered/measured and quantified. This will provide a complete picture of energy and operational performance when comparing this functional area with similar functional areas either on one site or across your enterprise. Examples include measuring electrical energy or volumes of natural gas or steam flowing into a system while measuring the outflows of steam, compressed air, process gases, and so on from the system.

Ultimately, you are trying to measure and account for the complete "energy balance" at each machine, line unit, system, or process area. While simple EnPIs, which do not include all of the energy flows from a given area, can be created and add some analytical value, the lack of a complete energy balance can restrict their use when trying to compare functional areas and share best practices across different locations.

Another issue to consider in creating EnPIs is whether to calculate simple metrics based on "energy in" divided by "production out" (or the inverse "production out" divided by "energy in") or to create efficiency metrics based on "total energy or work in" divided by "total energy or work out." Consider the following example to illustrate this point:

> Consider an air compressor with a kWh meter on the input and airflow totalizer on the output. The simple EnPI would be total cubic feet of air produced per kWh of electricity consumed. This might be effective if the compressor is fully loaded and runs in a consistent mode of operation. However, a more complex measurement of "work out" divided by "work in" may provide a better, more useful metric to ensure that the compressor is optimally loaded and sequenced. For this later metric, more instrumentation is required. For example, to determine the work done by the compressor, we need to know both the inlet and the outlet pressures, in addition to airflow. This could be further enhanced by considering inlet temperature and humidity content as well. The point is not to unnecessarily complicate the EnPI metric, but rather to "right size" it to address the business/technical issue at hand in the most cost-effective way possible.

DATA COLLECTION

Now that you have determined the set of EnPIs that strikes the right balance between your business objectives and your budget, the next step is to create a measurement road map and to begin connecting existing and new measuring devices to your energy management system to track, report, and forecast EnPIs. It is best to start at the devices themselves. The devices are often transmitters sending either physical 4–20 mA analog signals or register values via a connected network. It is important to keep in mind that these transmitters are often in place to measure instantaneous flows used in control and SCADA systems. When trying to measure energy or work, however, it is more appropriate to capture totalized flow rather than instantaneous flow measurements. Many devices will provide both instantaneous and totalized values, but often at a price premium. Totalizing meters (e.g., utility gas, water, and electric meters) are the most accurate way to aggregate energy usage/consumption. Such meters typically provide pulse outputs that increment with the delivery or consumption of a fixed amount of energy. The next best option is to totalize energy and production flows in programmable logic controllers (PLCs) and distributed control systems (DCSs). However, the farther up the data chain from the source meter, the larger the error that can be introduced into the energy integration calculation.

Once you determine the "what" and the "how" of data collection, you need to consider the frequency of data collection. As mentioned at the outset, monthly data is not granular enough to provide actionable information about what is happening in your operations each week, day, or hour. Many plants find that hourly or 15 min data collection provides the best balance between data storage cost, system performance, and granularity of measurements to really understand how the EnPIs are tracking through each week, day, or hour. Keep in mind that in order to properly control a process, many control systems will scan and update real-time measurement as frequently as in millisecond intervals. For EnPIs, it is not necessary to capture totalized flow at this frequency. EnPIs provide good results at lower data capture rates. We find that most plants choose the 15 min data capture interval as it aligns with the 15 min demand interval period that most electric utilities use to calculate demand charges.

Once the meters and measurement instruments are connected to networks, control systems, data loggers, and/or data historians, many energy management systems (EMSs) can capture and reuse these data to record, track, report, and forecast EnPIs. It is also important to make sure that the energy management system you select and use to track EnPIs provides a means to capture manual data. Fuel sources such as oil and coal are typically delivered in bulk batches and rarely have automated metering systems. In addition, you may still have a few production processes that are tracked on clipboards. Do not ignore these valuable data just because they exist in a manual format. Embrace them and use them to the extent you can to increase the completeness of energy measurement and balance calculations.

Finally, a word about the importance of creating "virtual meters." Meters and instruments at the "top" or supply side of a given utility will often capture and measure total energy supplied to the plant or system. These high-level meters include energy used in loads (both metered and unmetered), energy lost in distribution systems, and energy lost in system and machine inefficiencies. Rarely will plants have every energy system metered to every source and load. Therefore, "bottom-up" EnPIs and "top-down" EnPIs will almost never reconcile perfectly. It is a good practice to have your energy management system capture the losses and unmetered loads in a "virtual meter." Virtual meters are created in an EMS by mathematically combining physical meters. Virtual meters provide useful insights into whether to invest in more meters and/or address efficiency losses in systems. Virtual meters also act as good proxies to identify failed instruments.

DASHBOARDS

When people think about energy management systems, they often equate them with dashboards. Dashboards in and of themselves are oversold and of limited value. Dashboard screens full of gauges and trend graphs are too often designed to "look impressive" rather than to provide actionable data. The key to an effective dashboard involves the customization of the display to provide the right information to the right people at the right time to drive energy and operational efficiency. An effective dashboard is customized to meet the specific needs of the end user. The real question that must be answered is: When displaying EnPIs, measurements, and metrics on a screen, what is the desired outcome, action, or result? In other words, how will a dashboard drive results and what information needs to be provided to evoke those results and move users to action?

The first thing to consider in the customization of a dashboard is the user's specific goals and responsibilities. Site energy leaders may need to see information that helps them drive energy efficiency across the plant. Line operators may only need to see very simple displays of the operating variables that are in their immediate control to indicate whether they are operating a machine or line in the most economical way possible (see Figure 27.2).

A good energy management system will provide both standard and customizable displays that can be configured as needed to meet the specific needs of each user. Energy management systems log data captured from many meters and data historians across a plant to trend both raw measurements and EnPIs for each reporting area of the plant. An energy management system should also provide real-time display screens for dashboards as well as the ability to export data, EnPIs, and reports to external stakeholders, such as finance and plant management personnel. Again, each of these elements must be customizable to meet the specific needs of the end user, whether their focus is operational, financial, or energy engineering related.

REGRESSION MODELING

This section introduces the importance of multivariable regression analysis (MVR) for tracking those EnPIs and processes that rely on the simultaneous interaction of many different factors. As mentioned earlier, simple EnPIs can be calculated based on a single

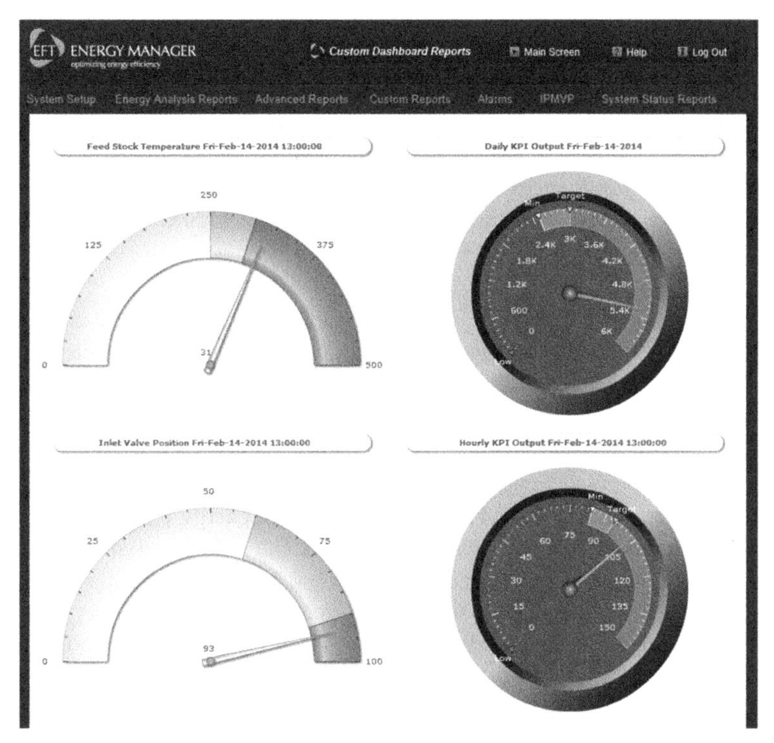

Figure 27.2. A simple dashboard created specifically for a line operator. It shows both EnPIs and the status of variables under his/her control.

set of variables around a process, line, or machine. These EnPIs are useful for tracking performance in "one isolated dimension." However, in complex process industries, there are usually several interrelated influences at play at any given time. For example, if EnPIs include multiple heat flows into and out of a process area, or differing energy densities of different products, or influences on the quality or condition of raw materials, the best way to look at how each of these influencing factors affects each EnPI is to use MVR to study the correlation between these factors. This provides a "multidimensional" view of each EnPI, and this can be used to predict or forecast the likely or expected performance of an area, unit, or machine.

How do single and multivariable regression analyses compare with each other? Don't they both provide a mathematical model that can be used to predict the expected result of an EnPI? The short answer is yes. The real question is how well does a single variable explain the operation of a complex process? Single-variable regression is a very simple and straightforward method of comparing a dependent variable with how it corresponds to one independent variable. For example, if one is looking at the energy consumption of a chiller used for human comfort or space conditioning, it would be meaningful to look at "cooling degree days" as that one single independent variable. Manufacturing processes, however, are often much more complex and finding one single independent variable that explains the variation in the dependent variable is nearly impossible. Multivariable regression applies many of the same mathematical techniques as single-variable regression, but, as the name implies, looks at several independent variables simultaneously to help site engineers understand which variables interact in which ways to drive variation in the process energy consumption and EnPIs.

MVR models are very useful in process industries to calculate complex EnPIs and understand the many influences on energy efficiency. The MVR models can be used in many different ways to provide value and insight for plant engineers. The following are some of the most popular uses of MVR modeling:

- *Complex EnPI calculations:* When simple EnPIs, as described above, do not adequately explain what is happening within a process, MVR models can be used to gain insights into the correlation of several independent influences and thus better understand the interaction of different loads and subprocesses within the plant.
- *Measurement and verification:* When energy conservation measures or capital projects are proposed and justified, they are usually based on some underlying assumptions about expected savings and gains in energy efficiency. With MVR analysis, process conditions can be modeled both before and after the project is completed, allowing plant personnel to view energy savings on a normalized basis, taking into account changes in production volume, runtime, or any other variable included in the model to see how well the project has performed.
- *Forecasting:* Once an MVR analysis has been completed, the resultant model can be used to make predictions, projections, or forecasts on how both energy consumption and EnPIs will be impacted over hours, days, weeks, months, or years. Actual conditions can be tracked in real time and compared with expected results to alarm plant operators on potential conditions leading to energy waste, poor EnPI performance, or demand spikes. As a subset of forecasting, MVR models can be used for budgeting and targeting (see Figure 27.3):

    - **Budgeting:** This application employs the MVR analysis and models over longer periods of time to either forecast site, area, unit, line, or machine energy consumption and EnPIs over weeks, months, or years or establish budgets from both financial and engineering units' perspectives. In the example shown, this site is planning for a partial shutdown in the month of February, thus the lower energy projection for that month. Forecasted production volumes can be fed into energy models to establish future budgets.
    - **Targeting:** Once budgets have been established, the MVR analysis and models can be used to create targets to challenge plant operations personnel to drive energy consumption and EnPIs to new levels and monitor progress against those targets.

- *ISO 50001 and superior energy performance program reporting:* These newer, more holistic energy management programs are quickly gaining traction across the globe and have rigorous reporting requirements to demonstrate that stated goals are indeed being achieved. MVR analysis and modeling are a recognized requirement to meet the reporting objectives of these programs.

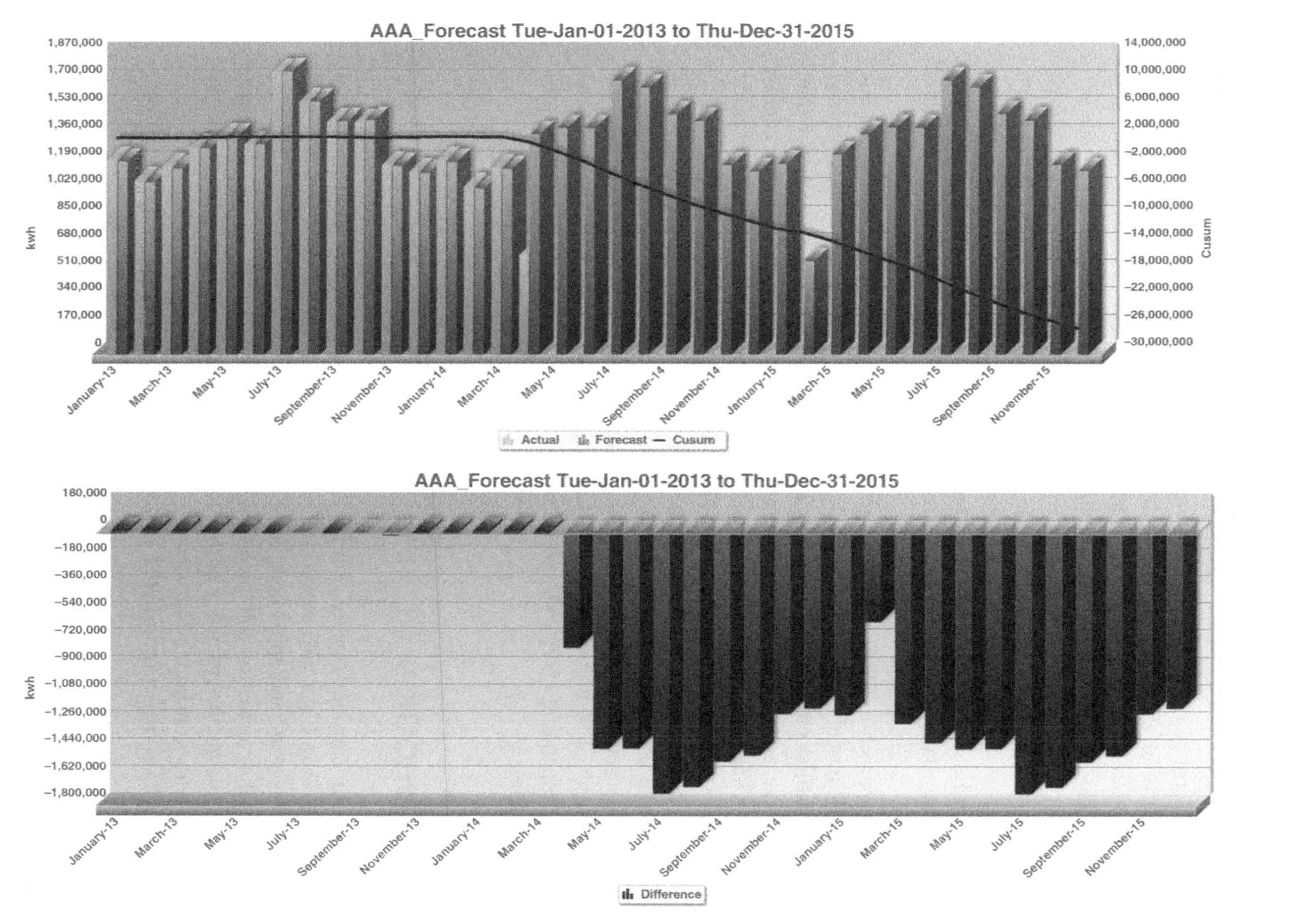

Figure 27.3. An excerpt from a forecast report showing energy projection over the course of 3 years, including a planned shutdown. The top graph shows actual energy for past months in the lighter bars to the left versus forecasted energy for both past and future months in the darker bars to the right. The lower graph shows monthly differences between the forecasted and actual values.

There are many choices when it comes to selecting energy management tools that can assist with MVR analysis in your facilities. At the most basic level, you can use spreadsheet programs or statistical analysis packages often associated with Lean or Six Sigma programs. While these tools can be effective, they are manually intensive and provide only snapshot views of how EnPIs are trending. In addition, many users have found that the level of effort required to obtain data from multiple sources, to align that data in appropriate time intervals, and, finally, to upload and model that data in a statistical analysis package makes it prohibitive to update models more than once per quarter. This approach certainly does not provide the granularity and timeliness required to track EnPIs in near real time, especially on those complex processes that are perhaps among the biggest energy users in your plant. Programs such as ISO 50001 and the DOE's Superior Energy Performance Program have monthly reporting requirements, so a spreadsheet approach may or may not work depending on the complexity of the data sets required.

Another approach to MVR analysis is to select and use an EMS platform that has embedded MVR modeling capabilities. Energy management platforms with embedded modeling capabilities provide many advantages over the manual tools mentioned above. First, they utilize automated data collection. As mentioned earlier, a good energy management package will "sit on top of" existing systems and obtain data from existing meters, control systems, and process data historians. This automated data collection ensures that models are always fed with the most up-to-date data. A second advantage is that MVR analysis can occur in real time to facilitate the automatic tracking and alarming of current energy consumption and EnPI data versus expected or projected values. This approach provides a dynamic view, allowing site operators and engineers to be forewarned of impending energy efficiency issues and proactive in making the appropriate adjustments.

PROCESS APPLICATION

To conclude our consideration of EnPIs, let us look at a process application that demonstrates how an EMS with embedded MVR capability can be used in a cascaded energy system to create and apply complex EnPIs. This example involves the aggregation of data and the calculation and tracking of EnPIs on the efficiency of a set of boilers used to produce steam, the efficiency of the steam distribution system, and the efficiency of the process units consuming that steam.

This system has metering installed at several points (see Figure 27.4), as follows:

- Each boiler has an input natural gas meter (GM) and an input makeup water meter (WM).
- Each boiler has an output steam meter (SM).
- The steam distribution system has a steam meter on each steam-consuming load and on the pressure letdown station.
- Each process area is metering output production (PM) through the DCS.

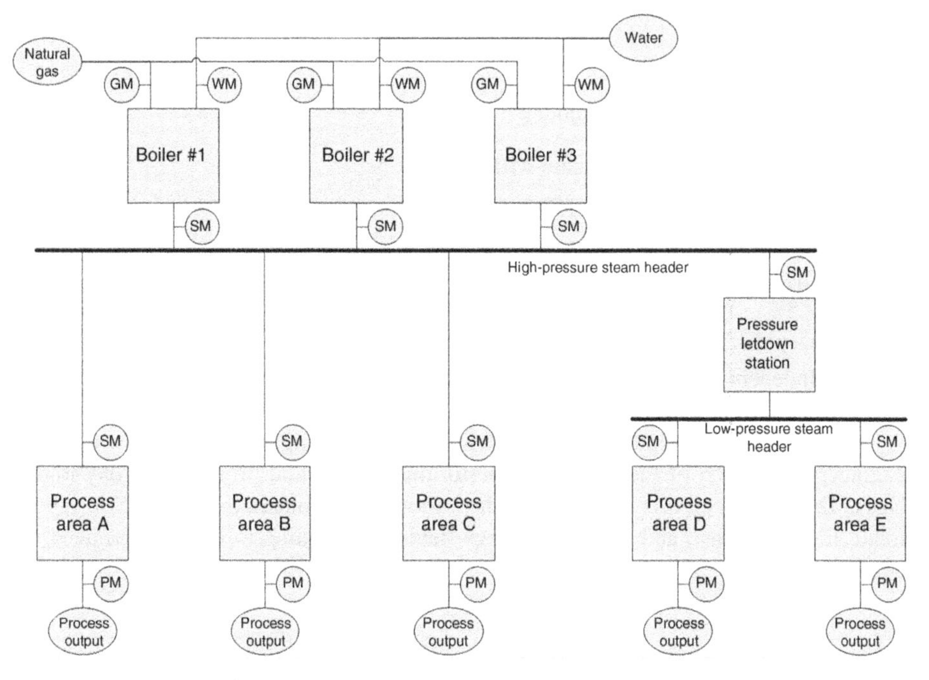

Figure 27.4. Application example: steam system.

Several virtual meters are created by the EMS to aggregate and totalize the energy associated with the boilers and steam distribution system and to calculate real-time EnPIs. These values are logged every 15 min for basic trending, reporting, and alarming:

- Total makeup water.
- Total natural gas.
- Total high-pressure steam produced.
- Total low-pressure steam consumed.
- Energy balance on high-pressure steam header.
- Energy balance on low-pressure steam header.
- Process area "*X*" EnPI = process output/steam input.
- Boiler "*X*" EnPI = MBtu steam out/MBtu gas in.
- Steam letdown station EnPI = MBtu steam out/MBtu steam in.

With the data available from the meters and virtual meters, several MVR models can be created to identify and project energy usage for each process area. Models can be aggregated to understand and project total steam requirements. Models can also be created for each of the boilers to understand and project total natural gas and makeup water supply needs:

- Process area "*X*" MVR model:
  - Dependent variable: steam MBtu.
  - Independent variables: production output, humidity content of feedstock, and so on.
- Boiler "*X*" MVR model:
  - Dependent variable: natural gas MCF.
  - Independent variables: steam MBtu to individual process areas.

Together these and other MVR models and the resulting EnPIs will show the plant engineers how each of the systems is performing on an hour-to-hour and day-to-day basis. This information is presented to the various stakeholders by the EMS to report, trend, track, forecast, and alarm on energy deviations leading to loss, waste, need for adjustment, repair, and so on.

CLOSING THOUGHTS

To obtain the best results from EnPIs and energy dashboards, it is always important to start with end goals in mind. First, it is important to understand how EnPIs will be used to answer specific business and technical questions, and second, when creating dashboards, it is important to consider what actions or outcomes are desired. Knowing this at the outset will allow us to "work backward" to find the right data, based on availability and cost, to establish the most meaningful simple and complex EnPIs, and to set up MVR models that provide insight into the interaction of processes within the monitored facilities. EnPIs will provide "normalized views" of energy consumption, providing the ability to track efficiency gains over time and to compare similar process areas and machinery across peer locations. As a final thought, budget is always a key driver in setting up EnPIs. Budget constraints often lead to compromises between data availability and the need for additional instrumentation. Remember that EnPIs, which may not be theoretically perfect, may still provide significant value as long as they are "directionally correct."

ACKNOWLEDGMENTS

The author wishes to thank Craig Ennis and Bill Boardman of EFT Energy for providing input and reviewing material.

INDEX

Energy Management and Efficiency for the Process Industries, First Edition. Edited by Alan P. Rossiter and Beth P. Jones.